Molecular Biology

made simple and fun

Molecular Biology

made simple and fun

David P. Clark, Ph.D.

and

Lonnie D. Russell, Ph.D.

Cache
River
Press

© **Cache River Press, 1997**

Library of Congress Catalogue Card Number 96–092824
ISBN 0-9627422-9-5

1st Edition

Authors:
David P. Clark, Ph.D.
Lonnie D. Russell, Ph.D.

Publisher:
Cache River Press
2850 Oak Grove Road
Vienna, IL 62995
USA

1-888-862-2243

Printed in the United States of America

Acknowledgements

The authors would like to thank the following individuals whose contributions to the process of developing and producing this book were invaluable: Laurie Achenbach, Neil Billington, Dave Bunick, John Caster, Terri Epstein, Steve Hagan, Carey Krajewski, John Martinko, Dan Nickrent, Marjorie Russell, Benton Tong and Ned Trovillion. The help of Karen Fiorino is gratefully acknowledged. It was Karen who could intuitively interpret the authors' rough sketches and produce the excellent figures contained herein. Special thanks go to Kelby the chimp and Lauren Russell who make excellent cover subjects.

Dedicated to:

a devoted mother, *Jean May Clark*

and

a wonderful child, *Lauren Kristin Russell*

Contents

Introduction

We are in the midst of a molecular biology revolution! Science has advanced to the point where the very genes that control the makeup of all living creatures and what actions they perform, are open for examination, and yes, even for deliberate change! No doubt this will have an enormous impact on the way we and our children live our lives. The impact of this new technology on society and environment is limitless.

The term molecular biology is somewhat of a misnomer, but it is one so commonly used in science that we must learn to live with it. What is really meant by molecular biology is the biology of those molecules related to genes and gene products and heredity, in other words, molecular genetics. As a field of study, genetics has been around for some time. Since the 1800s we have known about classical genetics when the curious stumbled onto and figured out the inheritance patterns of such things as hair or eye color. Techniques revealing how the inherited characteristics that we observe daily are linked to the underlying processes that cause them have only been developed since World War II. These techniques have allowed scientists to determine the molecular basis for inheritance. Thus the term molecular genetics and its closely related cousin, molecular biology.

The rate at which science is examining the molecules of inheritance and how they are controlled and expressed has been increasing throughout the 1980s and 1990s. Some day, what's taking place now may be called the dawn of the age of molecular genetics. It's a revolution in the same sense as the industrial revolution, and the consequences of today's scientific findings will rapidly change our lives. Even if we are not scientific "players" in the development of molecular biology, we all need a basic literacy in the topic to see what is happening and to understand how our lives may change. Society may choose to control scientific developments that might be dangerous if not used properly. More likely is that fear bred of ignorance will result in the banning or over-regulation of useful advances. This book is designed to help you become literate in molecular biology.

The impact of the molecular biology revolution is universal. A major impact appears to be in the realm of human health. Currently the United States government is supporting the Human Genome Project that has as

molecular biology the study of life using techniques that reveal its molecular make-up

molecular genetics genetics as it is viewed at the level of molecules; in particular involving the structure and sequences of the nucleic acids which carry genetic information

Human Genome Project government supported project with the goal of determining the complete sequence of the genetic information in humans and other species

1

two of its goals first, the mapping of all human genes and that of other selected species, and second, determining the sequence of their DNA – all by the year 2005! Some scientists believe that when the sequence of DNA is known and all of the 50,000 to 100,000 human genes have been mapped, the genetic components for virtually every disease will be revealed. Actually, inherited diseases are due to defective versions of certain genes and to understand why they cause problems, we first need to investigate the normal roles of these genes. In 1993 and 1994 more genes responsible for human disease were identified than had been discovered in all previous years. Since almost all disease has a genetic component, the trend is to define our mental and physical health in terms of genetics. Our ability to predict and affect the outcome of disease will greatly increase. Clinical medicine will also change rapidly to reinterpret disease from a genetic standpoint.

The potential resulting from these changes is interesting and even alarming to some. Health insurers could potentially determine the risk to individuals of inherited diseases and set insurance premium rates accordingly. All of us could be required to carry an identification card that represents our genetic profile, a profile even more unique than our signature or fingerprints. Our privacy as humans may be invaded by those trying to determine our genetic characteristics. Someone could surreptitiously scrape a few skin cells from another person and, given a little time, and a lot of technology, know more about them than the victim might desire. Some may marry and plan children based upon the genetic profiles of their partners. The fate of unborn babies may depend upon genetic analysis of samples mailed to commercial laboratories. The potential is present to increase human and animal life spans by preventing disease and slowing the aging process. Those alarmed by these possibilities do not take into consideration that society will work in parallel with these advances to make them a positive and not a negative influence on our lives.

Not only the health profession, but virtually all other professions too, will be affected by the molecular biology revolution. Ethicists will work overtime to determine the ethical implications of our increased genetic knowledge and its use in genetic manipulation. Politicians and lawyers will be concerned with arising legal conflicts and the new laws that must be made to accompany the genetic revolution. Agriculturalists will have the opportunity to expand their products with new varieties of **genetically engineered** plants and animals. Animals and plants used as human food sources will be engineered to adapt them to less favorable conditions so they can be grown in regions presently uncultivatable. Farm animals that are resistant to disease and crop plants that are resistant to pests will be developed in order to increase yields and reduce costs. Ecologists will examine the impact of genetically modified organisms on other species and on the environment. We could go on indefinitely but space limitations prevent us from examining all the ramifications of the genetic revolution. Suffice it to say, no matter who you are, you have a stake in the molecular revolution.

As authors we've taken a common sense approach to molecular biology, a topic that is often considered too technical for presentation to the student or layman. First, we tried to make it simple. Beginning with information that most people already know, we progress in a step-by-step fashion to reach a level that would allow the reader to attend and understand a scientific seminar in molecular biology. Even though we aim for a high technical level, we try to avoid being too technical. Not everyone needs to know everything in this book. Like James Bond, use it on a "need to know" basis. Terminology is defined as we go along. Lots of simple diagrams are used to illustrate one point at a time. There's even corny humor (well, it made *us* laugh!) and tidbits of irrelevant information to make reading enjoyable. Individual chapters are self-contained to some extent, through the repetition of definitions and reminders of earlier material. Although this duplication made the book a bit fatter, it does mean there won't be a need to flip back to earlier chapters as often. In writing this book, we tried to avoid traditional textbook style that causes instant fright in most students. Most of us dislike reading dry and lifeless material so we tried to write informally, and conversationally. Actually, the secret behind our approach is that one of the authors was an absolute dummy when it came to molecular genetics. He bullied the other author, the whiz-kid molecular geneticist, into explaining things in a clear and understandable way, as depicted on the back cover.

So relax while you're reading; after all, you could have chosen to study Chinese with its thousands of characters or even subatomic thermoelectromagnetohydrodynamics. Enjoy the book and be amazed by the intricacies of the biological processes that make you and your natural environment what they are. You will soon be an informed citizen of the molecular age.

WHICH WILL IT BE? CHINESE OR MOLECULAR BIOLOGY

I've decided to take Chinese this semester. It has over 30,000 characters! Let's take it together!

No way! I'm taking molecular biology. It has only five basic characters!

Roses are red violets are blue. Molecular biology is easy for you!

Background and Additional Readings:

Biology by Campbell NA. 4th edition, 1996. The Benjamin Cummings Publishing Company Inc., Menlo Park, California.

Introduction to Cell and Molecular Biology by Wolfe SL. 1995. Wadsworth Publishing Company, Belmont, California.

Molecular Biology of the Cell by Alberts B, Bray D, Lewis J, Raff M, Roberts K, & Watson JD. 3rd edition, 1994. Garland Publishing, Inc., New York & London.

Molecular Biology of the Gene by Watson JD, Hopkins NH, Roberts JW Steitz JA, & Weiner AM. 4th edition, 1987. Benjamin Cummings, Menlo Park, California.

Bacteria: The Molecular Biologist's Guinea Pigs

2

Most of the early experiments providing the basis for modern day molecular biology were done using bacteria. Simple, primitive creatures like bacteria consist of only a single cell, whereas large and complex living organisms like animals and plants are divided into many cells. Many types of bacteria can be grown under controlled conditions in the laboratory on nutrient medium consisting of a few simple ingredients. Better still, for some experiments, bacteria can be grown on a mixture of half a dozen known and purified chemical components. Because bacteria are not complex, they are often referred to as "lower organisms" (Fig. 2.1).

Higher organisms consist of several organs (such as noses, livers, wings, branches, leaves, etc.) and each of these contains cells of many different types. To complicate things even more, cells of higher organisms may have up to 100,000 genes that may be switched on or off depending on the tissue of which the cell is a part. In contrast, a bacterial culture contains identical, single cells with only a couple of thousand genes each.

Consequently, bacteria, especially one known as *Escherichia coli*, (or simply *E. coli*) have been used to clarify the principles of molecular biology. The knowledge derived by examining *E. coli* has been used to untangle the genetic operation of higher organisms. Because of this, it is important to know about these simple, so-called "lower" organisms from which so much essential information has been obtained.

"What applies to E. coli applies to E. lephant"

Jacques Monod

bacteria primitive, relatively simple, single-celled organisms often used by molecular biologists

2.1 REVENGE OF A "LOWER" ORGANISM

Imagine the nerve of these authors to call me "lower." A few of my frends will take care of their "lower" bowels.

Did you know that many researchers now use lawyers as laboratory guinea pigs instead of rats? There are four reasons:

1) There are more lawyers than rats.

2) Even rats have feelings.

3) Biologists sometimes get emotionally attached to their rats.

4) There are some things even a rat won't do!

How are Bacterial Cells Classified?

Bacteria are microscopic organisms existing as single cells. Individual bacterial cells grow and divide but they do not associate to form complex tissues and organs as in higher organisms. A few drops of bacterial culture consists of millions of identical cells floating around, suspended in nutrient soup. In contrast, an organ from a higher organism, say a liver, contains many different cell types. Knowing that bacteria are very much alike is important for research, since all the cells in a bacterial culture respond in a reasonably similar way, whereas those from the higher organism will give a variety of responses making analysis much more difficult.

Bacteria (singular, bacterium) are the simplest living cells and are classified as **prokaryotes**. Bacterial cells (Fig. 2.3) are surrounded by a cell wall and a membrane and, like all cells, contain all the essential chemical and structural components necessary for life. In particular, each bacterial cell has a single chromosome carrying a full set of genes providing it with the genetic information necessary to operate as a living organism.

By definition, prokaryotes are those organisms whose cells are not subdivided by membranes into a separate nucleus and cytoplasm. All prokaryote cell components are together in the same compartment. In contrast, the larger and more complicated cells of higher organisms (animals and plants) are subdivided into separate compartments and are called **eukaryotic** cells. Figure 2.4 shows the different compartments of the prokaryotic and eukaryotic cells.

prokaryotes lower organisms like bacteria with a primitive type of cell containing a single chromosome and having no nuclear membrane

eukaryotes higher organisms with advanced cells which have more than one chromosome within a compartment called the nucleus

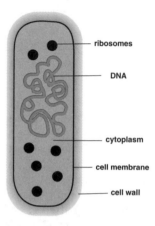

2.3 TYPICAL BACTERIUM

ribosomes

DNA

cytoplasm

cell membrane

cell wall

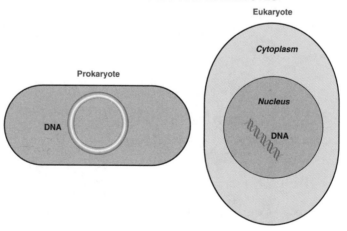

2.4 PROKARYOTES AND EUKARYOTES

Eukaryote

Cytoplasm

Prokaryote

Nucleus

DNA

DNA

How Big are Bacterial Cells?

A typical bacterial cell is about a micrometer long by half a micrometer wide. A micrometer, also known as a micron, is a millionth of a meter and a meter is slightly more than a yard in length.

A million bacteria, end to end, would stretch about a yard in length, or a thousand would just cover the tip of a typical ballpoint pen.

Sizes, in Decimal Units, of Some Typical Objects
(each step is 1,000 times greater than the one below)

1 meter	= 5-year-old child	= 1,000 millimeters
1 millimeter	= fruit fly	= 1,000 micrometers
1 micrometer	= bacterial cell	= 1,000 nanometers
1 nanometer	= sugar molecule	= 1,000 picometers

Bacterial Cells May be Cultured in the Laboratory

Bacteria are often said to be grown in the test tube. In practice they are grown as a suspension in liquid inside flasks or bottles. Many bacteria grow well if they are provided with just a solution of essential minerals plus a single carbon compound, such as a sugar, to provide a source of energy and organic matter. Bacteria can also be grown as colonies (visible clusters of cells) on the surface of an agar layer in flat dishes, known as Petri dishes (Fig. 2.5). Agar is extracted from seaweed and "sets" like gelatin, but unlike gelatin, agar is not normally served for dessert.

2.5 PETRI DISH WITH BACTERIA

Top View **Side View**

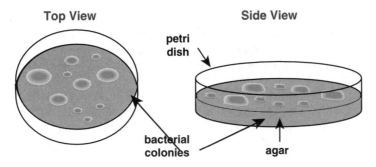

petri dish

bacterial colonies

agar

A Typical Bacterial Culture Medium - M9 Medium:

Chemical component	Formula	Grams per Liter
Sodium hydrogen phosphate	Na_2HPO_4	6.0
Potassium hydrogen phosphate	KH_2PO_4	3.0
Sodium chloride	$NaCl$	0.5
Ammonium chloride	NH_4Cl	1.0
Magnesium sulfate	$MgSO_4$	0.12
Glucose, a simple sugar	$C_6H_{12}O_6$	4.0

Which Kinds of Bacteria are Used for Experiments?

Although many different types of bacteria are used in laboratory investigations, the bacterium used most often in molecular biology research is *Escherichia coli* whose natural home is in your intestines. Bacteria have two names, both printed in italics. The first refers to the **genus,** a group of closely related species. The genus name is often abbreviated to a single letter, as in "*E. coli.*" Next comes the **species,** or individual, name.

genus a group of closely related species

species a group of closely related organisms with a relatively recent common ancestor

They're Everywhere! They're Everywhere!

Bacteria are found almost everywhere. Some bacteria live in the sea, others live in freshwater, and yet others are found growing happily in sewage. Some bacteria live in the soil, some are found living in the roots of plants and some live inside animals - including you. In addition to these normal habitats, bacteria are also found in hot sulfur springs, in the Dead Sea and the Great Salt Lake, and in thermal vents in the deep sea where temperatures approach boiling point. Most other organisms would not want to live under such conditions, not that they could survive anyway!

Do All Bacteria Cause Disease?

No, on the contrary, most bacteria are beneficial and help maintain the environment by degrading waste materials. For example, soil bacteria help degrade the remains of dead plants and animals (including humans!) and take part in the breakdown of animal waste. If "good" bacteria did not maintain the environment, we would not be here. Thus, bacteria started the first recycling programs. Bacteria also help degrade many man-made chemicals and pollutants. Only a few kinds of bacteria cause diseases, just as only a small percentage of humans are responsible for crime.

The First Modern Conflict - The Russo-Japanese War

The Russo-Japanese War marks a turning point in human history. It took place in 1904 and 1905 and was the first truly modern war. In all previous wars, the number of soldiers who died during military campaigns as a result of infectious diseases spread by lack of hygiene was greater than the number killed by enemy action. Only in our own century have we become so civilized that the number of humans killed by other humans has outnumbered those killed by bacteria.

Perhaps the most fascinating historical example was the Iberian War of Liberation, waged in 1350. At that time, the Iberian peninsula, what is now Spain plus Portugal, was still partly controlled by the Islamic empire. The Spanish and Portuguese Christians were fighting to rid themselves of Arab domination when the Black Death passed through. Roughly 50 percent of the population of Europe died. Half of both the Moslem and Christian armies, including Alfonso XI, King of Castile, dropped dead within a couple of months. Despite this, fighting resumed the next year, rather like a time out for injury in a football game! (The outcome, in case you care: the Moslems were driven out of Europe, back to North Africa.)

Which Diseases Do Bacteria Cause?

Cholera, tuberculosis, bubonic plague ("black death"), anthrax, syphilis, gonorrhea, whooping cough, diphtheria and a variety of other diseases are caused by bacteria. These diseases were widespread before modern hygiene and technology largely eliminated them from the advanced nations. Their abolition was mostly due to clean water, sewers, flush toilets and soap rather than any actual medical advances.

Treatment of Bacterial Infections

Typically patients are given antibiotics for the treatment of bacterial infections. These are chemical substances capable of killing most bacteria but which are relatively harmless to people. Most antibiotics are synthesized by a kind of fungus known as mold (see Fig. 2.8).

antibiotic chemical substances that kill bacteria selectively, that is, without killing the patient too

Penicillin, the most famous antibiotic, is made by a mold called *Penicillium* which grows on bread producing a blue layer of fungus.

2.8 BREAD MOLD SUPRESSES GROWTH OF BACTERIA

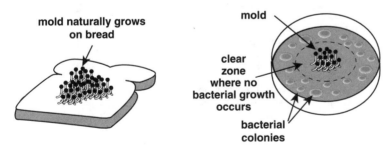

mold naturally grows on bread

mold

clear zone where no bacterial growth occurs

bacterial colonies

Interestingly, a few antibiotics are actually made by bacteria. These are made in nature by one kind of bacteria in order to kill other types of bacteria.

Some antibiotics, like *chloramphenicol*, were originally made by molds but nowadays can be chemically synthesized. Finally, some antibiotics, such as *sulfonamides* are entirely artificial and are synthesized by chemical corporations.

Do Bacteria Themselves Suffer from Disease?

Yes, we all have our problems. Even bacteria can get sick, usually as a result of infection by a **virus**. Bacterial viruses are sometimes referred to as "**bacteriophages.**" Phage comes from a Greek word meaning "to eat." When bacteria catch a virus, they do not merely get a mild infection, like a cold, as we usually do. They are doomed. The bacteriophage takes over the bacterial cell and fills it up by manufacturing more bacteriophage (see Fig. 2.9). Then the cell bursts and liberates the new bacteriophage crop to infect more bacteria. This takes only about an hour or so. In a matter of hours, a bacteriophage epidemic could wipe out a culture of bacteria with a population several times that of the earth's human population.

virus extremely small entity that grows only by infecting another cell

bacteriophage virus that preys on bacteria

2.9 LIFE CYCLE OF A BACTERIOPHAGE

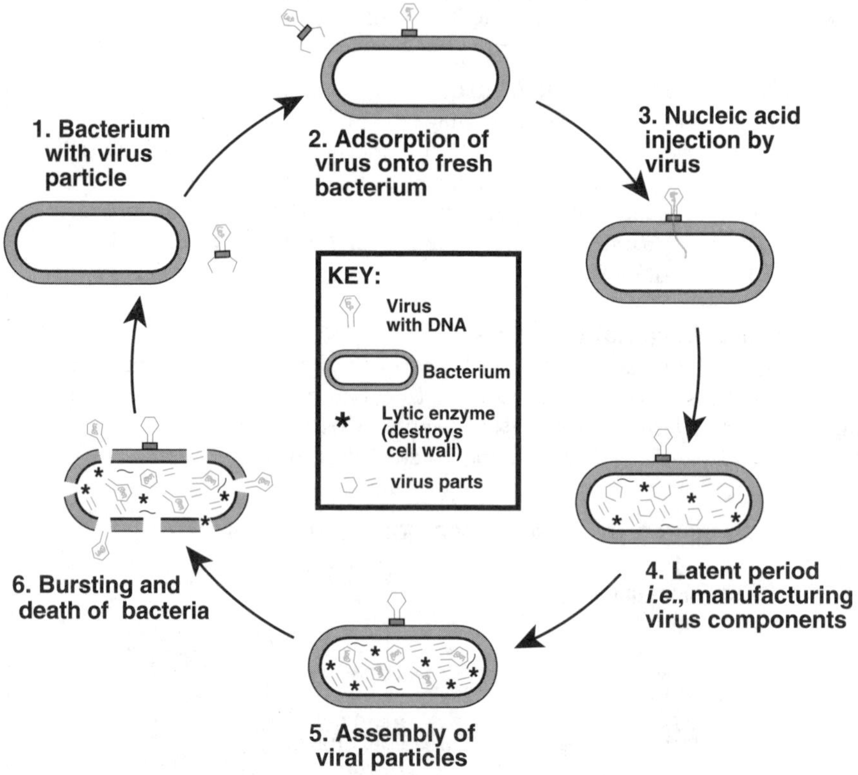

1. Bacterium with virus particle
2. Adsorption of virus onto fresh bacterium
3. Nucleic acid injection by virus

KEY:
- Virus with DNA
- Bacterium
- * Lytic enzyme (destroys cell wall)
- virus parts

4. Latent period *i.e.*, manufacturing virus components
5. Assembly of viral particles
6. Bursting and death of bacteria

Do Bacteria Ever Kill Each Other?

Yes, of course; which civilized species doesn't kill their own kind? Bacteria kill each other in two ways. Very occasionally bacteria which are even tinier than usual infect other, larger bacteria. This results in a bacterial disease of bacteria! Much more often, though, bacteria kill each other by some form of chemical warfare. Some bacterial strains secrete toxic chemicals in order to kill off other types of bacteria which are competing to live in the same habitat. It's a jungle out there!

Great fleas have little fleas upon their backs to bite 'em,
And little fleas have lesser fleas, and so ad infinitum.

from De Morgan:
A Budget of Paradoxes

10

2.10 BACTERIA KILLING OTHER BACTERIA

Can a Bacterial Virus Infect People?

No way. At least this is one thing you don't have to worry about. Bacterial viruses only infect bacteria. Generally speaking, any particular disease, whether caused by bacteria or by viruses, only infects a closely related group of organisms.

Viruses – Living or Non-Living? That is the Question.

Viruses are packages of genes in protein coats. The essential features of a virus are shown in Figure 2.11. On their own, viruses are inert. Only when they come into contact with a suitable host to infect do they actually have the chance to multiply. Viruses are the smallest organisms known, much smaller than bacteria. Viruses are not true biological cells since they do not grow by dividing in two, generating energy, making protein, or carrying out any of the normal processes of living.

Some scientists claim that a virus is alive because it can multiply and because it contains all the genes necessary for its own replication. Others reply that a virus is not a genuine cell and cannot multiply on its own; it needs a host cell to provide nutrients and energy. Viruses are on the borderline between living and non-living. They are particles in suspended animation, waiting for a genuine living cell to come along so they can infect it and replicate themselves. Viruses are the ultimate parasites, totally dependent on other life forms for their energy, materials and even the equipment to manufacture their own components.

2.11 STRUCTURE OF A VIRUS

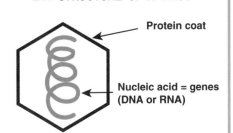

Protein coat

Nucleic acid = genes
(DNA or RNA)

Human Viral Diseases

Many common childhood diseases such as measles, mumps and chickenpox, are caused by viruses. So are the common cold and flu. More dangerous viral diseases include polio, smallpox, herpes, Lassa fever, Ebola

Most human diseases are limited to humans only or to humans and closely related animals. Conversely, most diseases of other animals are not dangerous to people. You cannot catch feline leukemia virus, nor can you give your cat smallpox.

The smallest viruses have only three genes. The largest have 200 to 300 genes. Most viruses have somewhere in between, say around 50.

and AIDS. Do viruses do anything useful? There's good news and bad news. Infection by a mild virus can provide resistance against a related but more dangerous virus (see Ch. 21). Again, viruses may be put to good use by genetic engineers as described later in this book (see Ch. 19). All the same, about the best that can be said for the natural role of viruses is that some of them do relatively little damage.

Treatment of Viral Infections

immunization process of preparing the immune system for future infection by treating the patient with weak or killed versions of an infectious agent

Viral diseases usually cannot be cured after you have caught them. Either your body fights off the infection or it does not. However, viral diseases can often be prevented by immunization, if you are vaccinated before catching the virus. In this case the invading virus will be killed by your immune system (see Ch. 22) which has been put on alert by the vaccine, and the disease will be prevented.

Why are Antibiotics Often Prescribed for Viral Diseases?

Antibiotics only kill bacteria; they are of no use against viruses. So why do doctors often prescribe antibiotics for viral diseases like flu or colds? There are two reasons. One reason is that giving antibiotics may help combat secondary or opportunistic infections caused by bacteria, especially in patients who are in poor health. Secondly, many patients would be upset if they were told the truth. They would rather be given medicine, even if it is of no use, than face the facts that there is no cure. Many doctors go along with this rather than lose their patients and their profits.

Read more about the genetics of molecular biology's guinea pigs, the bacteria, in Chapter 8 and more about viruses in Chapters 19 and 21.

Additional Reading

Biology of Microorganisms by Brock TD, Madigan MT, Martinko JM, & Parker J. 8th ed, 1997. Prentice Hall, Englewood Cliffs, New Jersey.

Microbiology: Principles & Applications by Black JG. 3rd edition, 1996. Prentice Hall, Upper Saddle River, New Jersey.

Basic Genetics

From very ancient times people have realized in a vague way that children look like their fathers and mothers, and that the young of animals and even plants resemble their own parents (Fig. 3.1).

3.1 WE RESEMBLE OUR PARENTS

I think this one's yours.

The birth of modern genetics was due to the discoveries of Gregor Mendel whose greatest insight was to consider inheritance one characteristic at a time. Rather than asking questions like, "Does the little darling look more like its mummy or its daddy?" Mendel analyzed specific, clearcut properties. He actually used pea plants and followed the inheritance of such characteristics as whether the seeds were smooth or wrinkled, whether the flowers were red or white and whether the pods were yellow or green, etc. Whether or not any particular offspring inherited such a characteristic from its parents can be answered by a simple "yes" or "no," but never "maybe."

Today we would say that each of the characteristics examined by Mendel is determined by a single gene. Genes are units of genetic information and each gene provides the instructions for some property of the organism in question. Each gene may exist in alternative forms - for example, red or white flowers - which are called alleles. Note that different alleles of the same gene are related but have minor variations that produce different outcomes (red versus white) for the same characteristic (flower color).

Late last century a young and beautiful actress proposed marriage to the playwright, George Bernard Shaw.

"We will have children who are not only pretty but clever as well!" the actress argued.

George Bernard Shaw replied, "But, my dear, what if they get my face and your brains?"

Gregor Mendel discovered the basic laws of genetics by crossing pea plants

gene   a unit of genetic information

allele   one particular version of a gene

The overall nature of an organism is due to the sum of the effects of all of its genes. Remember that in lower organisms like bacteria there may be 2,000 to 4,000 genes, whereas in higher organisms like plants and animals, there may be 50,000 to 100,000.

Genes and Biochemical Pathways

Until very recently, genetics was a rather abstract matter, since no one knew what genes were actually made of, or how they operated. The first great leap forward came when biochemists proposed that each step in a pathway was controlled by a single gene. A gene determining whether flowers are red or white would then be responsible for a step in the biosynthetic pathway for red pigment. If the gene were defective, no red pigment would be made and the flowers would remain white.

protein polymers made from amino acids; they make up most of the structures in the cell and also do most of the work

Each biosynthetic reaction is carried out by a special **protein** known as an **enzyme**. Each enzyme has the ability to mediate one particular chemical reaction and so the "*one gene - one enzyme*" model of genetics was put forward by G. W. Beadle and E. L. Tatum who won a Nobel prize for this not-so-complex scheme.

enzyme a protein which carries out a chemical reaction

A biochemical pathway may have many steps; we will just consider one for now. Red pigment is made from some precursor in a single step (Fig. 3.2).

3.2 ONE GENE – ONE ENZYME

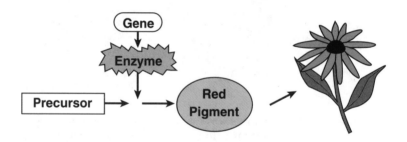

Mutants and Wild Types

mutation an alteration in the genetic information carried by a gene

Genetic alterations are known as **mutations**. Our flowers are normally red when everything is working properly. So, in this case, the white version of the gene is defective and is obviously a **mutant** allele.

mutant organism carrying a mutated gene

The properly functioning, red version, of this gene is referred to as the **wild-type** allele. As the name implies, the wild-type is supposedly the original version as found in the wild, before domestication and/or mutation messed up the beauties of nature. In fact, there are frequent genetic variants in wild populations and it is not always obvious which version of a gene is the wild-type.

wild-type the original or "natural" version of a gene or organism

Geneticists refer to the red allele as R and the white allele as r (not "W"). Although this may seem strange, the idea is that the "r"-allele is merely a defective version of the gene for red pigment. The "r"-allele is NOT a separate gene for making white color. In this example, there is no enzyme that makes white pigment, simply a failure to make red pigment. Our red and white flowers can now be reformulated as follows (Fig. 3.3):

3.3 WILD TYPE AND MUTANT FLOWERS

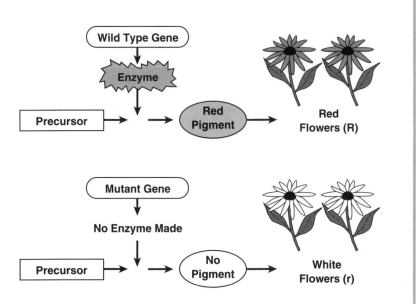

Originally it was thought that each enzyme was either present or absent; that is, there were two alleles corresponding to Mendel's "yes" and "no" situations. In fact, things are often more complicated. An enzyme may be only partially active and genes may actually have dozens of alleles, a matter to be discussed later. A mutant allele that results in the complete absence of an enzyme is known as a **null allele**.

null allele mutant version of a gene which completely lacks any activity

Phenotypes and Genotypes

In real life most biochemical pathways have several steps, not just one. So let's be a little bit more realistic. We will extend our pathway so it has two steps (Fig 3.4).

Our original red/white gene has become gene II. If gene II is working properly it will produce enzyme II which will make red pigment. If gene II is defective, there will be no enzyme II and precursor B will not be converted to red pigment. The flowers will remain white.

Here gene II is responsible for the final step of pigment production. "But what about gene I and enzyme I?" you cleverly ask. If

3.4 TWO STEP BIOCHEMICAL PATHWAY

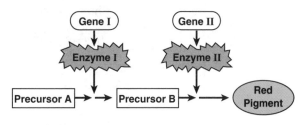

gene I is defective, enzyme I will be missing and you will never make precursor B, let alone any red pigment. True enough. In the case of a pathway such as this, a defect in gene I will have the same apparent effect as a defect in gene II; both defects will produce white flowers (Fig. 3.5).

phenotype the visible effect of the genotype

genotype the total genetic make-up of an organism

We refer to the outward characteristics - the flower color - as the **phenotype** and the genetic make-up as the **genotype**. Obviously, the phenotype "white flowers" may be due to several possible genotypes, including defects in gene I, gene II (as in Fig. 3.5) or even in genes not mentioned here, but which are responsible for producing precursor A in the first place. If we have white flowers, we will not know which gene is defective until we carry out some further analysis. This might involve assaying the biochemical reactions or mapping the genetic defects.

In fact, if gene I is defective, it no longer matters whether gene II is O.K. or not. A defect near the beginning of a pathway will make the later reactions irrelevant. This is known in genetic terminology as **epistasis**. Gene I is epistatic on gene II, that is, it masks the effects of gene II. From a practical viewpoint, this means that we cannot tell whether gene II is defective or not, if we already have a defect in gene I.

Although we have four different genotypes, three of these look the same, that is, they have identical phenotypes - white flowers. Boy, such a simple system begins to get complex in a hurry!

3.5 PATHWAY BLOCKAGE AND FLOWER COLOR

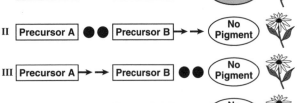

→ → = wild-type gene and functional enzyme

●● = mutant gene and no enzyme; reaction blocked

epistasis when a mutation in one gene masks the effect of alterations in another gene

DNA Nucleic acid polymer of which the genes are made

chromosome structure bearing the genes of a cell and made of a single strand of DNA

Chromosomes Carry the Genes

The second major insight into how genes work was the discovery that genes consist of **DNA** (see Ch. 4 for details). The genes belonging to each cell are arranged on **chromosomes**, which are simply giant molecules of DNA (see Fig. 3.6). Lower organisms fit all their genes onto a single chromosome, whereas higher organisms need several chromosomes to accommodate all their genes. Genes are often drawn on a bar representing a portion of a chromosome as shown.

3.6 GENES ON A CHROMOSOME

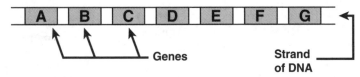

haploid having one copy of each gene

Haploid and Diploid Organisms

Lower organisms like bacteria have only a single copy of each gene; they are called **haploid** organisms. If you are haploid and one of your genes is defective, you are in trouble. To avoid this predicament, higher organisms have an insurance policy. They usually have duplicate copies of each gene and so they are diploid. If one copy of the gene is defective, the other copy will save their more evolutionarily advanced butts. Occasionally we find living cells with more than two copies of each gene. If you have three copies you are triploid, if you have four copies, tetraploid, and so on.

Because genes are parts of chromosomes, diploid cells have two copies of each chromosome. In Figure 3.7, identical capital letters indicate sites where alleles of the same gene can be located.

diploid, triploid and tetraploid having two, three or four copies of each gene, respectively

3.7 ONE PAIR OF CHROMOSOMES CARRYING DUPLICATE GENES

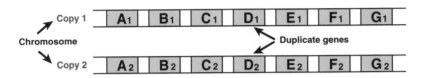

Suppose you are diploid and have two copies of a gene which is involved in making red pigment for flowers. If both copies are wild-type, R-alleles (genotype, RR), then the flowers will obviously be red. If both copies are mutant r-alleles (genotype, rr) then the flowers will certainly be white.

But what if one copy is "red" and the other copy is "white" (genotype, Rr or rR)? Our enzyme model predicts that one copy of the gene produces enzyme and the other does not. Overall we have some enzyme, so we will have red flowers. This assumes that half the usual amount of enzyme can get the job done. In fact, most of the time this is true, as most enzymes are present in a cell in excess levels. From outside, a flower which is Rr will look red, just like the RR version. When two different alleles are present, the one that dominates the situation is known as the dominant allele. The other one is the recessive allele. In this case the R allele is dominant and the r allele recessive.

dominant allele the allele whose properties are expressed as the phenotype

recessive allele the allele whose properties are not observed because they are masked by the dominant allele

homozygous having two identical alleles of the same gene

heterozygous having two different alleles of the same gene

We have four possible types of individual plant from a genetic viewpoint; that is, we have four genotypes: RR, Rr, rR and rr. The genotypes Rr and rR only differ depending on which of the pair of chromosomes carries r or R (see Fig. 3.8). Three of these, RR, Rr and rR, have red flowers and share the same phenotype while only rr plants have white flowers. When you have two identical alleles you are homozygous for that gene (either RR or rr), but if you have two different alleles you are heterozygous (Rr or rR).

3.8 THE FOUR GENOTYPES

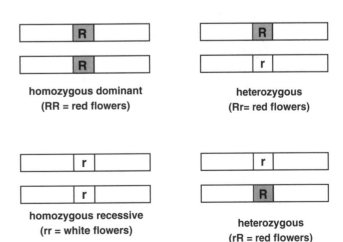

homozygous dominant
(RR = red flowers)

heterozygous
(Rr= red flowers)

homozygous recessive
(rr = white flowers)

heterozygous
(rR = red flowers)

Reproduction and the Number of Genes

As humans, we are told that we have dominion over nature, so we can interfere at will with the sex life of plants for our own vicarious amusement. Now, if we cross two parents who are both homozygous and red, *i.e.*, RR, we will obviously get descendants who are all red - there is no alternative. So also for two rr parents; all their offspring will be white. But what happens when we cross RR with rr or Rr with rR parents?

To understand what happens here, we need to consider how many copies of each gene are passed from parent to offspring. If both copies of the parents' genes were passed on to all their descendants, the offspring would have four copies of each gene, two from their mother and two from their father. The next generation would end up with eight copies and so on. Clearly, something has to be done!

gametes cells specialized for sexual reproduction and which are haploid (have one set of genes)

somatic cells cells making up the body and which are usually diploid (have two sets of genes)

3.9 REPRODUCTION: EGGS AND SPERM

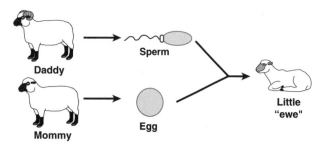

Sperm

Daddy

Mommy

Egg

Little "ewe"

When two animals or plants reproduce, the mother makes egg cells and the father makes sperm cells. These reproductive cells are called germ-line cells or **gametes**, as opposed to the **somatic cells** which make up the body. When an egg fuses with a sperm a new individual is created (Fig 3.9).

Although the somatic cells of higher organisms are diploid, the egg and sperm cells only have a single copy of each gene and are haploid. During the formation of egg or sperm cells, the diploid set of chromosomes must be divided to give only a single set. This is achieved by a process known as **meiosis**. Figure 3.11 ignores the technical details of meiosis and just illustrates its genetic consequences.

meiosis formation of haploid gametes from diploid parent cells

Genetic Crosses and Mendelian Ratios

Because the egg and sperm cells only have a single set of genes, each parent passes on only a single allele of each gene. Which of the original pair of copies gets passed to any particular descendant is purely a matter of chance.

For example, when we cross an RR parent with an rr parent, each offspring gets a single

deep thought: the ability of a sperm to swim backward will never be passed on to future generations.

Egg

One-way street

3.11 PRINCIPLE OF MEIOSIS

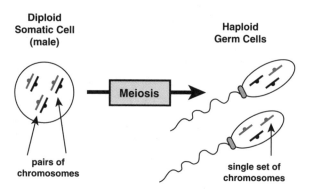

Diploid
Somatic Cell
(male)

Meiosis

Haploid
Germ Cells

pairs of
chromosomes

single set of
chromosomes

R-allele from the first parent and a single r-allele from the second parent. The offspring will therefore all be Rr. Thus, by crossing a plant that has red flowers with a plant that has white flowers, we end up with offspring that all have red flowers. Note however, that the offspring are not genetically identical to either parent, they are heterozygous. Genetic crosses like this are often shown in diagrammatic form (Fig. 3.12):

3.12 GENETIC CROSS OF RR x rr

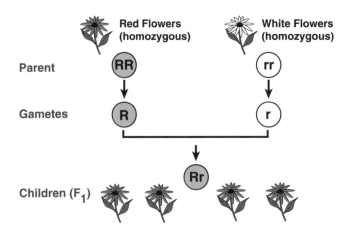

Green Eggs and Sperm

Green plants have green eggs and sperm!
They have! I tell you, don't you squirm!
And when you mix these tiny motes,
Then they fuse and make zygotes.

Zygotes grow and give you seeds -
This is how a green plant breeds.
Seeds are what a farmer needs -
To grow more plants, he plants the seeds!

And you can plant them in the rain.
And in the dark. And on a train.
And on a ship. And in the sea.
They are so good, so good, you see!

And I will plant them here and there.
In fact! I'll plant them EVERYWHERE!
I do so like green eggs and sperm!
And those little seeds so firm.

with apologies to Dr. Seuss and *Sam-I-Am*

And now for the part that always puzzled people before Mendel explained it. Let's continue by crossing two Rr plants taken from the first, or F_1 generation of offspring as shown in Figure 3.13. Successive generations of descendants are labeled F_1, F_2, F_3, etc., this stands for first, second, third, **filial generation**.

filial generations successive generations of descendants from a genetic cross which are numbered F_1, F_2, F_3, etc., to help keep track of them

At random, each parent contributes one copy of the gene which may be R or r. We have labeled these to show which parent they came from, but remember, there is no actual genetic difference between R_1 and R_2 or between r_1 and r_2. The offspring get one copy from each parent. Half those who get R_1 from parent No. 1 will get R_2 from parent No. 2 and the other half will get r_2 - and similarly for r_1.

3.13 POSSIBLE DISTRIBUTION OF GENES IN AN Rr x Rr Cross

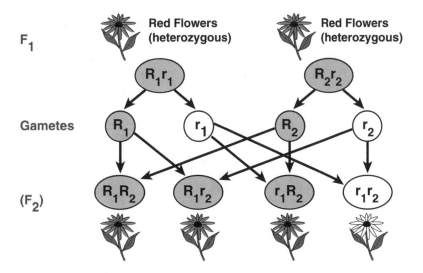

This can be seen by using a checkerboard diagram (Fig. 3.14):
Overall we get:
one RR - homozygous red
two Rr - heterozygous red
one rr - homozygous white

3.14 CHECKERBOARD DIAGRAM FOR Rr x Rr

Parent 1

gametes	**R**	**r**
R	**RR**	**Rr**
r	**Rr**	**rr**

Parent 2 / Progeny

Thus, crossing two red flowers gave a total of three red to one white. The relative numbers of each type of progeny are often referred to as Mendelian ratios. Note that white flowers have reappeared after skipping a generation. This is because the parents were both heterozygous for the r allele which is *recessive* and so was masked by the R allele. This explains how by perfectly legitimate means, two people who both have brown eyes can produce a child who has blue eyes. The allele for blue eyes is recessive to brown. It also explains why inherited diseases do not afflict all members of a family and often skip a generation. Most inherited diseases are caused by recessive genes as discussed in Chapter 13.

Sex Determination and Sex Linked Genes

The sex of a diploid organism, such as an animal or a plant, is determined by its sex chromosomes. Among mammals, possession of two X-chromosomes makes you female and possession of one X-chromosome and one Y-chromosome makes you male. When a diploid mother produces egg cells, each one has a single X-chromosome. When a diploid father makes sperm cells, half the sperm carry a single X-chromosome and the other half carry a single Y-chromosome. So the chromosome from the father is the one determining the sex of the offspring. The checkerboard diagram for sex determination is shown in Figure 3.15.

3.15 SEX DETERMINATION CHECKERBOARD

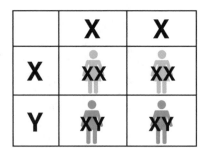

Hence, half of the children will be male and half will be female. Other genes, which have nothing to do with sex, are also carried on the sex chromosomes. Which allele of these genes you inherit will correlate with your sex and so they are called sex-linked genes. Some hereditary diseases show sex-linkage and these are discussed further in Chapter 13.

Partial Dominance and Co-Dominance

So far we have assumed that one wild-type allele of our flower color gene will produce sufficient red pigment to give red flowers since the R-allele is dominant. Sometimes this is not so, and we have a situation where a single good copy cannot get the job done. We then end up with the situation shown in Figure 3.16, where R is partially dominant:

partial dominance when a functional allele only partly masks a defective allele

3.16 PARTIAL DOMINANCE

RR =	Rr =	rr =
Red Flowers	**Pink Flowers**	**White Flowers**

Another possibility is that we have, in addition to the null allele, more than one functional allele. For example let's suppose:

R = makes red pigment

r = null mutant - the enzyme is missing - therefore no pigment of any color is made

B = mutant with an altered enzyme which converts the precursor to a blue pigment instead of a red one

Both R or B are able to make pigment and so both are dominant over r (absence of pigment). The combination of R with B gives both red and blue pigment in the same flower which will look purple, and so they are said to be co-dominant. We now have six possible genotypes and four possible colors (phenotypes) of flowers (Fig. 3.17).

co-dominance when two functional alleles both contribute to the observed properties

3.17 CO - DOMINANCE

RR =	Rr =	BB =	Br =	rr =	BR =
Red Flowers	**Red Flowers**	**Blue Flowers**	**Blue Flowers**	**White Flowers**	**Purple Flowers**

Linkage and Recombination

Real animals and plants have thousands of genes. Suppose we consider just a few genes - A, B, C, D, etc., which have corresponding mutant alleles a, b, c, d, etc. These genes may be on the same chromosome or they may be on different chromosomes. Let's pretend that A, B and C are on one chromosome and D is on another chromosome as shown in Figure 3.18.

3.18 GENES ON DIFFERENT CHROMOSOMES

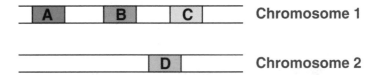

Now let's remember that higher organisms are diploid and have duplicate sets of chromosomes. Let's also make our cell heterozygous with the genotype Aa, Bb, Cc, Dd. We have (Fig. 3.19):

3.19 MULTIPLE HETEROZYGOUS GENES

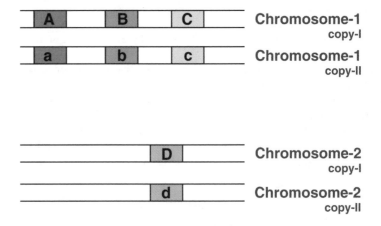

Suppose that we now carry out a genetic cross using this individual as one of the parents. Because the three alleles A, B, and C are on the same chromosome, that is, the same molecule of DNA, they will tend to stay together, as will a, b, and c. So if you inherit allele A from this parent you will usually get alleles B and C with it rather than alleles b and c. The genes A, B and C are said to be **linked**. In contrast, gene D is on a separate strand of DNA and there is as much chance of getting allele d accompanying allele A during inheritance as allele D. Gene D is unlinked to any of A, B, or C.

The alleles A, B, and C (or a, b, and c) do not stay together forever. What happens is that during the process of meiosis in the testis and ovary, the DNA strand carrying A, B and C lines up next to the one carrying a, b and c. Swapping of segments of the chromosomes can now occur by breaking and rejoining of the DNA strands, a process known as **crossing over**.

linkage two genes are linked when they are on the same DNA molecule (that is, on the same chromosome)

crossing over when two different strands of DNA are broken and are then joined to one another

The genetic result of this, the shuffling of different alleles between chromosomes, is called **recombination** (see Fig. 3.20).

recombination mixing of genetic information from two chromosomes as a result of crossing over

3.20 RECOMBINATION

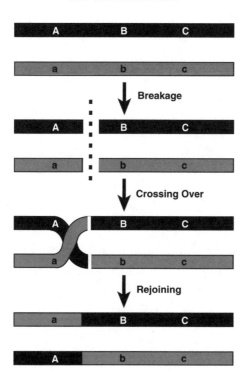

The farther apart two genes are on the chromosome the more likely a crossover will form between them and the higher will be their frequency of recombination. Geneticists measure the recombination frequencies of genes in order to estimate how far apart they are on the chromosome. As we shall see in later chapters, it may be important to know the precise location of a gene. This is true, for example, when we are attempting to identify and clone genes responsible for hereditary defects (see Ch. 13).

Now that we have looked at the basic ideas of genetics let's move on to consider the genetic material, the DNA molecules themselves, in more detail. In the next chapter we'll examine the structure of DNA and in Chapter 5 we'll discuss its replication.

Additional Reading

Genetics by Weaver RF & Hedrick PW. 1989. Wm. C. Brown Publishers, Dubuque, Iowa.

Understanding Genetics: A Molecular Approach by Rothwell NV. 1993. Wiley-Liss, New York.

Principles of Genetics by Tamarin RH. 1982. Willard Grant Press, Boston.

Required Reading: The Molecular Basis of Heredity

4

The fundamentals of modern genetics were laid when Mendel found that hereditary information is made up of discrete fundamental units which we now call genes. The discovery that atoms are made of subatomic particles ushered in the nuclear age. Similarly, the realization that genes are made up of molecules that obey the laws of chemistry has opened the way both to a deeper understanding of life and to its artificial alteration by genetic engineering.

gene a unit of information within the chromosomes that can be inherited

The Gene is the Fundamental Unit of Heredity

The unit of heredity is known as a gene. Each gene is responsible for a single inherited property or characteristic of the organism.

 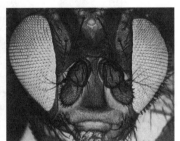 

WHITE EYES   **RED EYES**

MIXED EYES

Fig. 4.1. Example: Fruit flies inherit either red or white eyes. The eye color of a fruit fly depends on a single gene. If this gene is damaged by a mutation, instead of the normal red eye color, a fly with white eyes results. White-eyed flies pass on this characteristic to some of their children, as an indivisible genetic unit. Each individual descendant gets either white eyes or it gets red eyes. It does not get eyes which are half white and half red, nor does it get one red eye and one white eye!

Certain properties of higher organisms, such as height or skin color, are due to the combined action of multiple genes. Consequently, in these

mutation an alteration or defect in the genetic information

25

cases there is a gradation of the property. Such multi-gene characteristics at first caused a lot of confusion and they are still difficult to analyze, especially if more than two or three genes are involved.

Genes are made of **DNA** or deoxyribonucleic acid. Each gene is a linear segment of a long DNA molecule. DNA is a **polymer** made up of a linear arrangement of subunits known as **nucleotides.** There are four types of nucleotides as we will see below.

The information in each gene is due to the order of the different types of nucleotides, just as the information in this sentence is due to the order of the 26 possible letters of the alphabet which make it up. This chapter provides you with the "alphabet" of molecular biology. If you do not know the genetic alphabet, it will not be easy to understand how the alphabet can be used to compose the "words" that DNA or RNA represent as they function to store and communicate information. So like the many hours you spent during childhood learning and repeating the alphabet or mastering the musical scale, so must you spend time with this chapter in mastering the basics. Practice your ABCs over and over and the remaining chapters will be gravy (Fig. 4.2).

DNA deoxyribonucleic acid

polymer a molecule of similar repeating units which are linked together by a common bonding mechanism

nucleotides the subunits from which DNA is built; a nucleotide = phosphate + sugar + base

simple organisms like bacteria have 2,000 to 3,000 genes while higher organisms like flies or humans have 50,000 to 100,000 genes

a typical gene has about 1,000 nucleotides

4.2 PRACTICE MAKES PERFECT

To read a sentence of heredity, just read the nucleotide (N) sequence as shown in Figure 4.3.

4.3 THE ORDER OF THE NUCLEOTIDES "TELLS THE STORY"

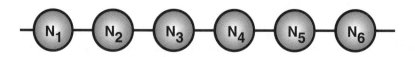

What is the Chemical Composition of a Nucleotide?

Genes are segments of DNA molecules which consist of a linear sequence of subunits known as nucleotides. Each nucleotide has three components: a phosphate group, a sugar and a nitrogen containing base (Fig. 4.4).

In DNA the sugar is always deoxyribose. The different types of nucleotide differ only in the nature of the nitrogen-containing base. In DNA there are four alternative bases: adenine, thymine, guanine, and cytosine. These are the only four "letters" of the DNA genetic "alphabet." Aren't we lucky?

When writing out genetic information these bases are abbreviated by convention to: A, T, G and C. The phosphate groups and the deoxyribose sugars form the backbone of each strand of DNA. The bases are joined to the deoxyribose and stick out sideways. A single strand of DNA is shown in Figure 4.5 and Figure 4.6.

4.4 THE COMPONENTS OF A NUCLEOTIDE

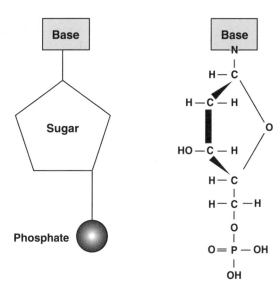

A Simple View **A Chemist's View**

base   alkaline chemical substance, in particular the cyclic nitrogen compounds found in DNA and RNA

deoxyribose   a sugar with five carbons

adenine, thymine, guanine, cytosine   the four bases found in DNA

4.5 THE DNA TEAM

4.6 HOOKING NUCLEOTIDES TOGETHER

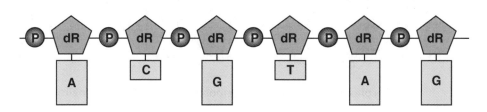

How are Nucleotides Joined?

To understand how nucleotides are joined, we must clarify the situation by numbering the carbon atoms of the sugar molecule. Figure 4.7 to the right shows the convention for numbering nucleotides.

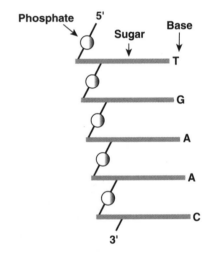

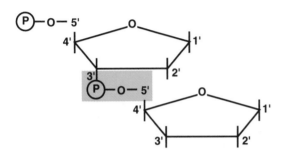

Nucleotides are joined by linking the phosphate on the 5' end of the deoxyribose of the first one to the 3' position of the next as shown on the left (Fig. 4.8).

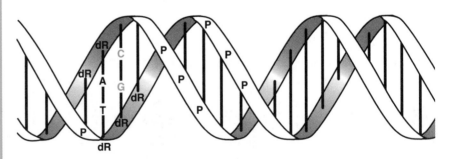

To indicate the linkages that build the DNA molecule, we can represent a strand of DNA as shown in the figure on the right (Fig. 4.9). Here, the bases have been abbreviated to a single letter.

In practice, DNA is normally found as a double stranded molecule. Not only is DNA double stranded but the two separate strands are wound around each other in a helical arrangement. This is the famous double helix (Fig. 4.10).

helix or helical   a helix is a three-dimensional spiral with a constant width

double helix   structure in which two strands of DNA are twisted spirally around each other

The DNA helix is a right handed helix which means that the coils turn clockwise if you look down the helix as it points away from you

4.10 **DOUBLE HELIX**

P = phosphate,   dR = deoxyribose,   A, T, C, G = base pairs

Base Pairs

In double stranded DNA, the bases of one strand are paired with the bases in the other strand. Adenine (A) in one strand is always paired with thymine (T) in the other and guanine (G) is always paired with cytosine (C). The bases A and G are referred to as the purine bases as they contain a double ring structure known as a purine ring. The other two bases, C and T are the pyrimidine bases, since they contain a single, pyrimidine ring. Each base pair consists of one double size purine base paired with a smaller pyrimidine base. So, although the bases themselves differ in size, all of the allowed base pairs are the same width. This is necessary to allow them to fit neatly into the double helix.

Watson and Crick were awarded a Nobel prize for discovering the structure of DNA

base pair two bases held together by hydrogen bonds

4.11 HYDROGEN BONDING BETWEEN A & T AND C & G

Purine   Pyrimidine

Pyrimidine   Purine

What Holds Base Pairs Together?

In double stranded DNA each base pair is held together by linkages known as hydrogen bonds. The A-T base pair has two hydrogen bonds and the G-C base pair is held together by three as in Figure 4.11. Hydrogen bonds are very weak, but since a molecule of DNA usually contains millions of base pairs, the added effect of millions of weak bonds is strong enough to keep the two strands together (Fig. 4.12).

hydrogen bonds bonds resulting from the simultaneous attraction of a positive hydrogen atom to both of two other atoms with negative charges

complementary base pairing because base pairs are always A with T or G with C it is possible to deduce the other partner if given one base of a pair

4.12 HYDROGEN BONDS ARE STRONG WHEN TOGETHER

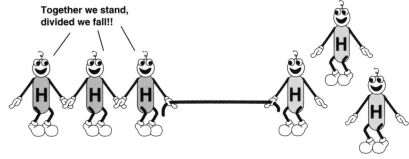

Together we stand, divided we fall!!

The hydrogen bonding in DNA base pairs uses either oxygen (O) or nitrogen (N) giving three alternative arrangements. In each case the hydrogen (H) is held between the other two atoms and serves to link them together (Fig. 4.13).

If you know one of the bases in a base pair of double stranded DNA you can figure out the other. If one strand has an A then the other will have a T and vice versa. Similarly, G is always paired with C. This kind of pairing is termed complementary base pairing. The significance of this arrangement is that if you know the base sequence of either one of the strands of a DNA molecule, you can deduce the sequence of the other strand. Such mutually deducible sequences are known as complementary sequences.

4.13 THREE POSSIBLE TYPES OF HYDROGEN BONDS

O – H – N

O – H – O

N – H – N

complementary sequences two sequences which can be deduced from each other because an adenine base is always paired with a thymine and a cytosine base is always paired with a guanine

Hydrogen Bond Formation

Before hydrogen bonds form and the bases pair off, the hydrogen atom is found attached to one or other of the two bases (shown by the complete lines in Fig. 4.14). During base pairing, the hydrogen binds to an atom of the second base (shown by the dashed lines).

4.14 HYDROGEN BOND FORMATION

last bonds to form are represented by colored dotted lines

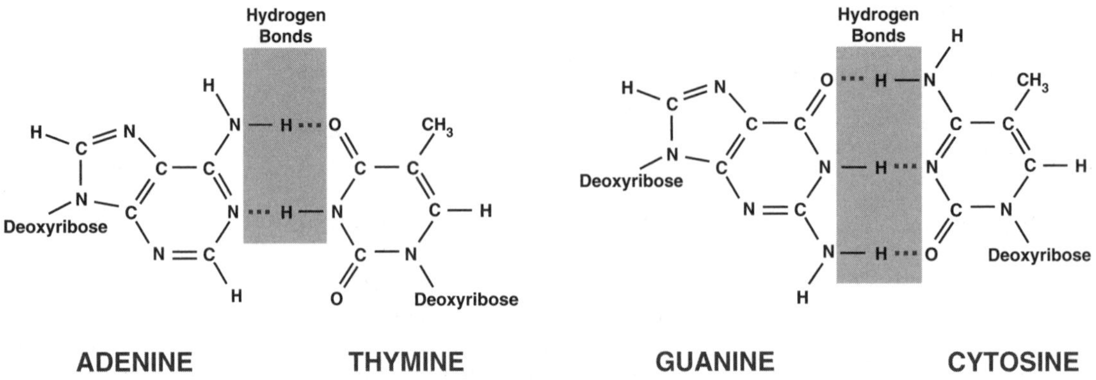

ADENINE   **THYMINE**   **GUANINE**   **CYTOSINE**

Genes are Located on Chromosomes

chromosome structures made of DNA that bear the genes of a cell

proteins polymeric molecules made of sub-units known as amino acids. They are described in Chapter 7

regulatory region DNA sequence appearing in front of a gene. It is used for regulation rather than to encode a protein

intergenic region DNA sequence between genes

The genes are found in linear order and are a major component of structures known as **chromosomes**. Each chromosome is an exceedingly long single molecule of DNA.

In addition to the DNA, which comprises the genes themselves, the chromosome has some accessory **protein** molecules which help maintain its structure.

The DNA of the chromosome is divided into segments (Fig. 4.15). Many of these segments are actual genes. In front of each gene is a **regulatory region** of DNA involved in switching the gene on or off. Between genes are spacer regions of DNA often referred to as **intergenic regions**.

4.15 COMMONLY FOUND SEQUENCE OF INFORMATION ON A CHROMOSOME

About Chromosomes

Chromosomes from bacteria are circular molecules and replicate as shown (Fig. 4.16). Since bacteria generally have only around 2,000 to

3,000 genes, and their intergenic regions are very short, one chromosome is sufficient to accommodate all of their genes.

Chromosomes from higher organisms are linear molecules. Higher organisms with 50,000 to 100,000 genes need several chromosomes to accommodate all of their genes.

Humans have two duplicate sets of 23 different chromosomes, making a total of 46 chromosomes (Fig. 4.17).

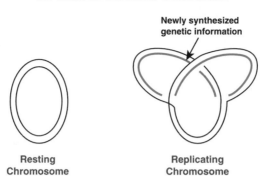

4.16 CIRCULAR BACTERIAL CHROMOSOME

Newly synthesized genetic information

Resting Chromosome

Replicating Chromosome

The record number of chromosomes is held by certain plants that have several hundred.

Yeast, one of the lowest of the "higher organisms" has approximately 12,000 genes spread over 17 different chromosomes of various sizes.

An old one:
Q. How do you tell a boy chromosome from a girl chromosome?

A. You pull down its genes.

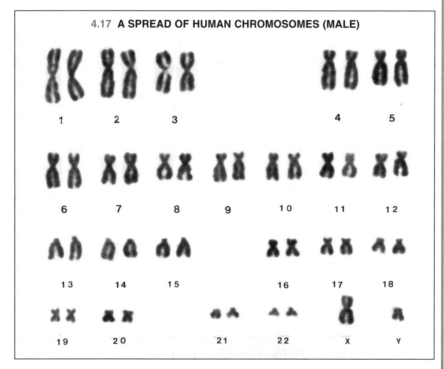

4.17 A SPREAD OF HUMAN CHROMOSOMES (MALE)

1 2 3 4 5
6 7 8 9 10 11 12
13 14 15 16 17 18
19 20 21 22 X Y

cell a cell is the basic unit of life

nucleus the nucleus of a biological cell is an internal compartment containing genetic information. Only the cells of higher organisms possess a nucleus. The nucleus containing the chromosomes is surrounded by the **nuclear membrane**. In primitive organisms such as bacteria, there is no separate nucleus and all the cell components are present in one compartment.

mRNA or messenger RNA the molecule that carries genetic information from the genes to the rest of the cell

chromatin the visible component of the nucleus containing DNA and associated proteins

More about Chromosomes

In higher organisms such as animals and plants, the chromosomes are found in a separate compartment of the **cell**, the **nucleus**. The nucleus is divided off from the rest of the cell by the **nuclear membrane** (see Fig 4.18 next page). The genes communicate with the rest of the cell by dispatching genetic information in the form of special messenger molecules, the **messenger RNA**, through pores in the nuclear membrane. The genes are bound to proteins in the nucleus and together they form visible material called **chromatin**.

Large and complex living organisms like animals and plants are divided into many cells, whereas simple and primitive creatures like bacteria consist of only a single cell. Each cell is surrounded by a membrane and contains all of the essential chemical and structural components necessary for life. In particular, each cell has a full set of genes providing it with the genetic information necessary to operate itself as a living organism.

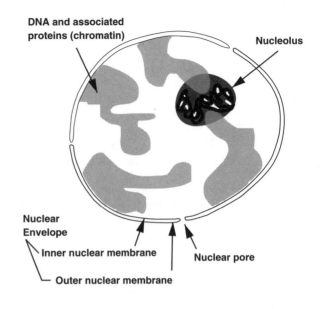

nucleolus a structure in the nucleus where some other kinds of RNA molecules, not messenger RNA, are synthesized

eukaryotic cells advanced cells of higher organisms that have several chromosomes within a compartment called the nucleus

prokaryotic cells primitive type of cells such as bacteria, with a single chromosome and no nuclear membrane

Cells which have a separate nucleus are known as eukaryotic cells to distinguish them from the more primitive prokaryotic cells of bacteria which lack a separate nucleus. The two types are compared in Figure 4.19 and show the processes taking place within each that will be covered in future chapters. The primitive eukaryotes usually consist only of a single cell and no subcompartments, so their single chromosome is free to interact directly with other cell components.

4.19 **PROCESSES TAKING PLACE IN PROKARYOTES AND EUKARYOTES**

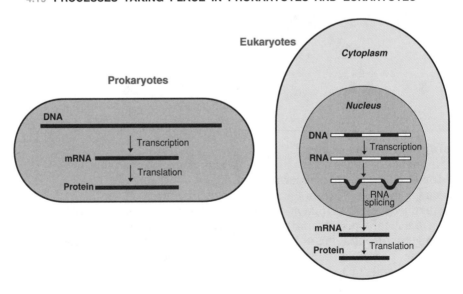

Supercoiling of DNA

An average bacterial cell is about one millionth of a meter long. The length of a single DNA molecule which carries the 3,000 or so genes needed by our bacterial cell comes to about one millimeter! Thus a bacterial chromosome is a thousand times longer than a bacterial cell. How does it fit? The answer is that the DNA inside a cell is coiled to make it more compact. This is referred to as supercoiling.

We take the DNA which is already a double helix and twist it up again as in Figure 4.20. The original double helix has a right-handed twist but the supercoils twist in the opposite sense, that is, they are left-handed or "negative supercoils." There is roughly one supercoil every 200 nucleotides in typical bacterial DNA. The bacterial chromosome contains approximately 50 giant loops of supercoiled DNA arranged around a protein scaffold. In Figure 4.21 the line represents a double helix and the helixes are the supercoils.

4.20 SUPERCOILING OF DNA

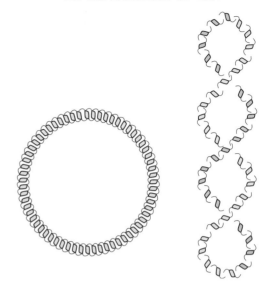

Circular DNA with zero supercoiling

Long strand of negatively supercoiled DNA

supercoiling the manner in which a long strand of helical DNA is twisted so that it can fit into a cell

4.21 BACTERIAL CHROMOSOME WITH SUPERCOILED DNA

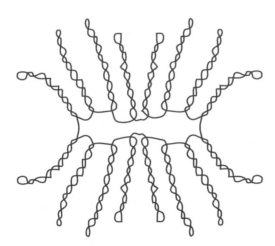

Plants and animals have much more DNA than do bacteria and must fold their DNA in an even more complex way to fit into the cell nucleus. This, and other problems of being a higher organism, are dealt with in Chapter 11.

Summary of Genes, DNA and Bases

Figure 4.22 summarizes the relationships between chromosomes, genes, DNA, nucleotides and bases.

4.22 FROM NUCLEOTIDES TO CHROMOSOME

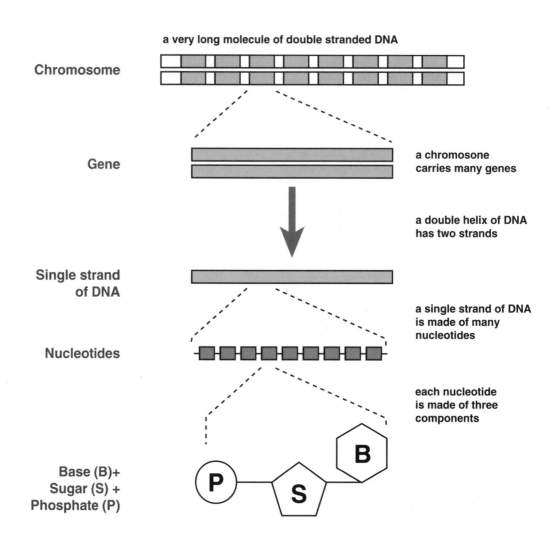

a very long molecule of double stranded DNA

Chromosome

Gene

a chromosone carries many genes

a double helix of DNA has two strands

Single strand of DNA

a single strand of DNA is made of many nucleotides

Nucleotides

each nucleotide is made of three components

Base (B)+ Sugar (S) + Phosphate (P)

P S B

Additional Reading

Essentials of Molecular Biology by Freifelder D & Malacinski GM. 2nd edition, 1993. Jones & Bartlett Publishers, Boston & London

Molecular Biology of the Cell by Alberts B, Bray D, Lewis J, Raff M, Roberts K, & Watson JD. 3rd edition, 1994. Garland Publishing, Inc., New York & London.

Duplicating the DNA: Replication

5

"Be fruitful, and multiply ..."

Genesis 1:22

Like Chapter 4, this chapter is vitally important to understanding the manipulations of DNA that are described in subsequent chapters. So while the DNA is dividing, please give it your undivided attention!

Inheritance of Genetic Information

Since each cell needs a complete set of genes, it is necessary for the original cell to duplicate its genes before dividing. Because the genes are made of DNA and make up the chromosomes, this means that each chromosome must be accurately copied. Upon cell division, both daughter cells will receive identical sets of chromosomes, each with a complete set of genes.

Replication of the DNA

In molecular terms this means that the DNA of the original, or mother, cell is duplicated to give two identical copies. This process is known as replication. Upon cell division each of the descendants gets one complete copy of the DNA. The original genes of the mother cell are on a double stranded DNA molecule so the first step in replication is to separate the two strands of the DNA double helix.

The next step is to build a complementary strand on each of the two original strands. Since A only pairs with T, and since G only pairs with C, the sequence of each strand dictates the sequence of its complementary strand. We now have two double stranded DNA molecules, both with sequences identical to the original one. One of these daughter molecules has the original left strand and the other daughter has the original right strand. This is known as semi-conservative replication since each of the progeny conserves half of the original DNA molecule (Fig. 5.1).

replication duplication of DNA prior to cell division

complementary strand two strands are complementary if A in one is always paired with T in the other and G is always paired with C (or vice versa)

semi-conservative replication replication of DNA in which each daughter molecule gets one of the two original strands and one new complementary strand

5.1 **SEMI - CONSERVATIVE REPLICATION**

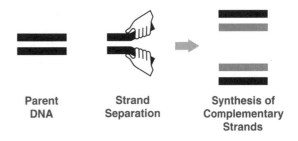

| Parent DNA | Strand Separation | Synthesis of Complementary Strands |

35

enzyme   a protein that carries out a chemical reaction

DNA gyrase   the enzyme that unwinds supercoiled DNA

How Does a Double Helix Separate into Strands?

Because the two strands forming a DNA molecule are held together by hydrogen bonding and twisted around each other to form a double helix, they cannot simply be pulled apart. Worse still, the DNA inside a cell is also supercoiled to pack it into a small space (see Ch. 2). Before separating the strands, both the supercoils and the double helix must be unwound.

This is done in two stages. First the supercoils are unwound by an enzyme known as DNA gyrase. The gyrase cuts both strands of double stranded DNA to give a double stranded break. However, it keeps hold of all of the cut ends. The two halves of the gyrase then rotate relative to each other and the ends are rejoined. This untwists the supercoils (Fig. 5.2). Each rotation costs the cell a small amount of energy.

Once the supercoils have been untwisted, the double helix is unwound by the enzyme DNA helicase (Fig. 5.3). Helicase does not break the DNA chains, it simply disrupts the hydrogen bonds holding the base pairs together.

5.2 DNA GYRASE UNCOILS DNA

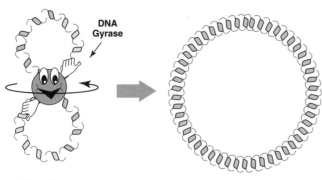

Negatively supercoiled DNA

Circular DNA with zero supercoiling

DNA helicase   the enzyme that unwinds double helical DNA

5.3 OPERATION OF DNA HELICASE

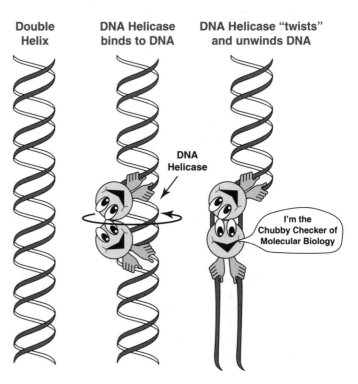

Double Helix

DNA Helicase binds to DNA

DNA Helicase "twists" and unwinds DNA

DNA Helicase

I'm the Chubby Checker of Molecular Biology

How are the Parental Strands of DNA Kept Apart?

The two separated strands of the parental DNA molecule are complementary to each other. Consequently all of their respective bases are capable of pairing off and binding to each other. In order to manufacture the new strands, the two original strands, despite their desire to cling together, must somehow be kept apart. This is done by means of a special "divorce" protein which binds to the unpaired single stranded DNA and prevents the two parental strands from getting back together. This is known as single strand binding protein or SSB (Fig. 5.4).

single strand binding protein (SSB) a protein that keeps separated strands of DNA apart

5.4 SINGLE STRAND BINDING PROTEIN

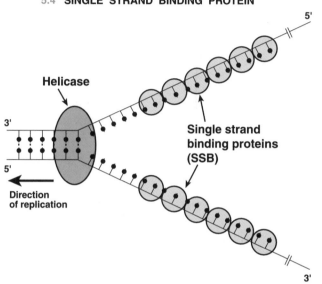

base pairing when two complementary bases (A with T or G with C) recognize each other and are held together by hydrogen bonds

template strand strand of DNA that is used as a guide for synthesizing a new strand by complementary base pairing

Making a New Strand of DNA

The critical issue in replication is the base pairing of A with T and of G with C. Each of the separated parental strands of DNA serves as a template strand for the synthesis of a new complementary strand. The incoming nucleotides for the new strand recognize their partners by base pairing and so are lined up on the template strand (Fig. 5.5).

Actually, things are a bit more complicated. Although hydrogen bonding alone would match bases correctly 99 percent of the time this is not good enough. The enzyme that links

5.5 INCOMING BASES LINE UP ON TEMPLATE STRAND

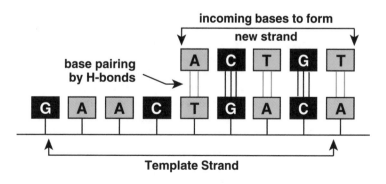

37

5.6 DNA POLYMERASE III MAKING DNA

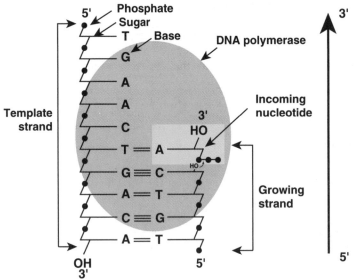

the nucleotides, known as **DNA polymerase III** or **Pol III** (Fig. 5.6) can also sense if bases are correctly paired. If not, the mismatched base pair is rejected.

The nucleotides are then joined together by the enzyme. This DNA polymerase has two subunits. One of these is the synthetic subunit and is responsible for manufacturing new DNA. The other subunit is shaped like a doughnut and slides up and down like a curtain ring on the template strand of DNA (Fig. 5.7). This "sliding clamp" subunit binds the synthetic subunit to the DNA.

DNA polymerase III (Pol III) enzyme that makes most of the DNA when chromosomes are replicated

5.7 DNA POLYMERASE III – SLIDING CLAMP

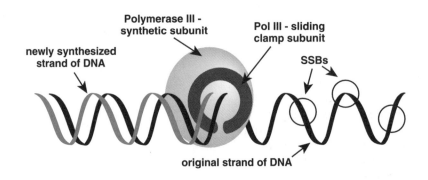

Synthesis Always Goes from 5' to 3'

As you know, nucleotides have three components: a phosphate group, a sugar and the base. In DNA the sugar is deoxyribose, and is joined to the base at position 1' and to the phosphate group at position 5' (Fig. 5.8). (The carbon atoms of the deoxyribose sugar are numbered with prime marks to distinguish them from those of the base which have plain numbers.) When a new nucleotide is added it is joined, via its own phosphate group on position 5', to the 3' position as indicated by the arrow. New DNA strands always start at the 5' end and grow in the 3' direction. In fact, all nucleic acids, whether DNA or RNA, are always made in the 5' to 3' direction.

5.8 NUMBERING OF DEOXYRIBOSE

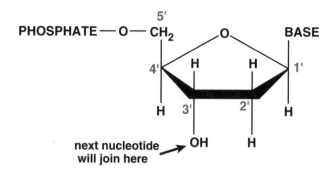

However, DNA is normally double stranded, and it happens that the two strands run in opposite directions, that is, if one goes 5' to 3' then its complementary partner will run from 3' to 5'. The strands are said to be **antiparallel** (Fig. 5.9).

antiparallel parallel but running in opposite directions

5.9 DOUBLE STRANDED DNA IS ANTIPARALLEL

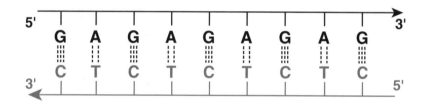

The Replication Fork Is Where the Action Is!

The **replication fork** is the total structure in the region where the DNA molecule is being duplicated. It includes the swivel where the DNA is being untwisted by DNA gyrase, the helicase following right behind, and the stretches of single stranded DNA held apart by the single strand binding protein. It also has two molecules of DNA polymerase III which are busy making two new strands of DNA (Fig. 5.10, next page).

Since DNA is always made in the 5' to 3' direction, and since the two strands of double helical DNA are antiparallel, this means that during DNA replication the two new strands must be synthesized in opposite directions. Because of this, one strand is made continuously and is referred to as the **leading strand**. However, the other strand can only be made in short segments and is known as the **lagging strand**.

replication fork region where the enzymes replicating a DNA molecule are bound to untwisted, single stranded DNA

leading strand the new strand of DNA that is synthesized continuously during replication

lagging strand the new strand of DNA that is synthesized in short pieces and joined together later

5.10 STRUCTURE OF A REPLICATION FORK

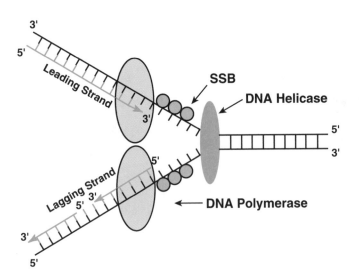

Completing the Lagging Strand

Although the leading strand just keeps getting longer and longer, the lagging strand is handicapped. After the replication fork has passed by, the lagging strand is left as a series of short pieces with gaps between. These short, newly made pieces of DNA are known as Okazaki fragments after their discoverer and must be joined together to give a complete strand of DNA. This is accomplished by two enzymes working in succession: DNA polymerase I and DNA ligase. DNA polymerase I fills in the gaps and DNA ligase joins the ends (Fig. 5.11). DNA polymerase I was discovered before DNA polymerase III, hence the numbering. Both DNA polymerase I and DNA ligase have important uses in genetic engineering.

Okazaki fragments
the short pieces of DNA that make up the lagging strand

DNA polymerase I
enzyme that makes small stretches of DNA to fill in gaps between Okazaki fragments or during repair of damaged DNA

DNA ligase   an enzyme that joins DNA fragments end to end

Nick or Gap?   when dealing with DNA, a "gap" is where bases are missing, whereas a "nick" means that there is a break in the DNA backbone although no bases are missing

5.11 JOINING OF OKAZAKI FRAGMENTS

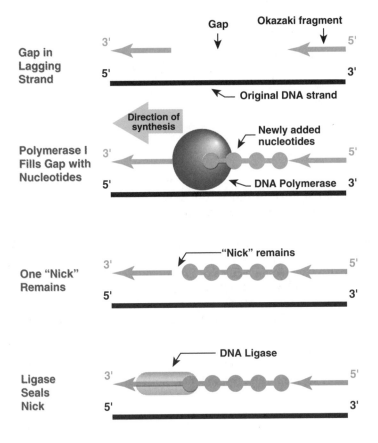

Starting a New Strand

Up to now we have assumed that we have strands of DNA with free ends that can be elongated by DNA polymerase. But how do we get a new strand started? Although the leading strand only needs to be started once, the lagging strand is made in short sections and we need to start again every time we make a new Okazaki fragment.

Curiously, DNA polymerase cannot start a new strand by itself, it can only elongate! Curiouser and curiouser is that new strands are started with a short stretch, not of DNA itself, but of RNA! These short RNA pieces are known as **primers** and the enzyme that starts synthesis of new chains by making the RNA primers is called **primase**. So every time a new fragment of DNA is made, primase sneaks in and lays down a short RNA primer to get things going. Only then can DNA polymerase get to work elongating the strand. For more information see Chapter 11.

Recoiling the DNA into a Helix

As the two new strands of DNA are synthesized, two double helical DNA molecules are produced, each with one old and one new strand. Once the replication fork has moved past, the double stranded DNA molecule automatically rewinds into a helix (Fig. 5.12).

primer short stretch of nucleic acid (RNA in this case) needed to get synthesis of a new strand of DNA started

primase enzyme that starts a new chain of DNA by making an RNA primer

5.12 RESULT OF DNA REPLICATION

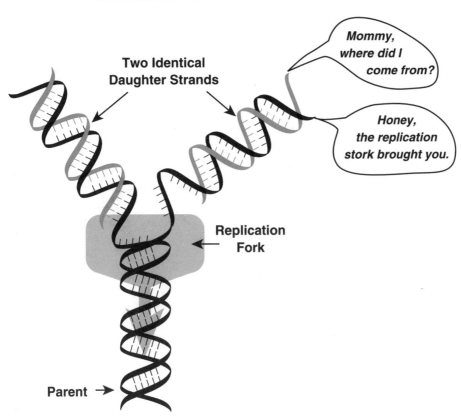

How are New Chromosomes Shared between Daughter Cells?

In the case of bacteria, cell division is relatively simple. There is a single chromosome and the cell is a single compartment (there is no nucleus). The bacterial chromosome is circular and replication proceeds in both directions at once around the circle (Fig. 5.13).

5.13 DIVISION OF CIRCULAR BACTERIAL CHROMOSOMES

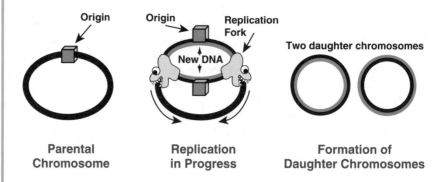

Origin

Origin

Replication Fork

New DNA

Two daughter chromosomes

Parental Chromosome

Replication in Progress

Formation of Daughter Chromosomes

Eventually the two replication forks meet and merge. This yields two new circular chromosomes. These are attached to the wall of the cell and as the bacterial cell elongates, the chromosomes are pulled apart. By the time the cell divides, by building a cross-wall, there is one chromosome in each new cell (Fig. 5.14).

5.14 DIVISION OF BACTERIAL CELL

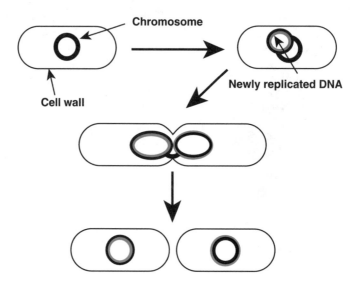

Chromosome

Newly replicated DNA

Cell wall

How Do Cells of Higher Organisms Divide?

Since cells of higher organisms (eukaryotes) are more complex, they handle cell division differently. Not only do they have multiple chromosomes but these chromosomes are all inside the nucleus separated from the rest of the cell by the nuclear membrane. Consequently a much more elaborate process is needed to replicate the chromosomes and to partition them among the daughter cells. This process is called mitosis and involves several phases:

1) disassembly of the nucleus of the mother cell
2) division of the chromosomes
3) partition of the chromosomes between daughter cells
4) division of the mother cell
5) building a new nucleus around the chromosomes in each daughter cell.

Division of Eukaryotic Chromosomes

Eukaryotic chromosomes are linear rather than circular as are bacterial chromosomes. They are often very long. There are several replication forks scattered along the length of each chromosome. Each pair of replication forks starts at a separate origin of replication and then moves in opposite directions. Figure 5.15 shows several sites of replication in a eukaryotic chromosome. The sites where bubble-like DNA is dividing are often called replication bubbles.

mitosis division of mother cell into two daughter cells with identical sets of chromosomes

despite politically correct censorship of hurricanes and other winds, we still refer to parental cells as mothers and their descendants as daughters

origin of replication site on a DNA molecule where replication begins

5.15 REPLICATION OF A EUKARYOTIC CHROMOSOME

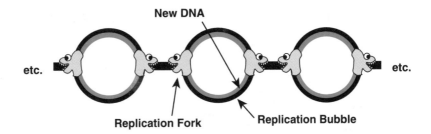

New DNA

etc.   etc.

Replication Fork   Replication Bubble

5.16 THE EUKARYOTIC CELL CYCLE CAN BE COMPARED TO A CLOCK

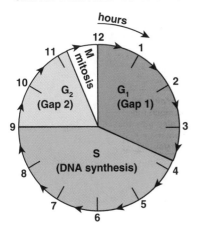

hours

The Eukaryotic Cell Cycle

The cell cycle of eukaryotes may be divided into several phases as shown in Figure 5.16. The process of DNA replication described above takes place in the synthetic or S-phase of the cell cycle. The S-phase is separated from the actual physical process of cell division (mitosis) by two gap or G-phases in which nothing much appears to happen.

Mitosis itself is subdivided into four phases: prophase, metaphase, anaphase and telophase. These are illustrated in Figure 5.17.

prophase condensed chromosomes become visible and the nuclear membrane dissolves

metaphase chromosomes move to the cell equator where they align themselves in pairs

anaphase separate halves of each chromosome are drawn apart by the spindle fibers towards the poles of the cell

telophase a new nuclear membrane is made to surround each set of newly divided chromosomes

5.17 DIVISION OF A EUKARYOTIC CELL BY MITOSIS

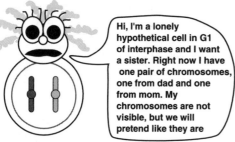

A. G-1 phase

B. S-phase & G-2

C. Prophase

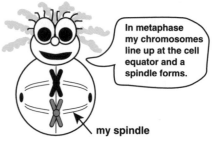

D. Metaphase

E. Anaphase

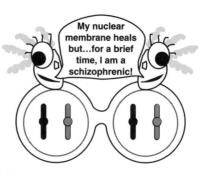

F. Telophase

G. G-1 phase

Phew! Now that we have finished replicating, we can move on to Chapters 6 and 7 and see how the genes oversee the cell's day-to-day existence.

Additional Reading

Essentials of Molecular Biology by Freifelder D & Malacinski GM. 2nd edition, 1993. Jones & Bartlett Publishers, Boston & London.

Genetics and Molecular Biology by Schlief R. 2nd edition, 1993. The Johns Hopkins University Press, Baltimore & London.

Getting The Message Out:

Transcription of Genes to Produce Messenger RNA

6

How is the Genetic Information Used?

During the day-to-day life of a cell, working copies of the genes are used. The DNA molecule that carries the original copy of the genetic information is regarded as sacred and is not used as a direct source of instructions to run the cell. Genetic information can be carried by two kinds of nucleic acid molecules, DNA or RNA. The working copies of genes are made of RNA or ribonucleic acid, which is very similar in chemical structure to DNA. The particular type of RNA molecule that carries genetic information from the genes into the rest of the cell is known as messenger RNA, usually abbreviated to mRNA. The transfer of information from DNA to messenger RNA is known as transcription.

For a gene to be transcribed, the DNA, which is double stranded must first be pulled apart temporarily, as shown in Figure 6.1. Then a molecule of single stranded RNA is made. This is the messenger RNA and it has a base sequence which is complementary to that of the DNA strand used as a template.

What is the Chemical Difference Between DNA and RNA?

There are two related kinds of nucleic acid, deoxyribonucleic acid (DNA) and ribonucleic acid (RNA). (See Chapter 4 for the structure of DNA.) The first difference between them is that in DNA the sugar is always deoxyribose, whereas in RNA

nucleic acid polymeric molecule that carries genetic information as a sequence of bases

RNA or **ribonucleic acid** another nucleic acid that differs from DNA in having the sugar ribose in place of deoxyribose

messenger RNA class of RNA molecule that DNA uses as messengers to carry orders to the rest of the cell

transcription process by which information from DNA is converted into its RNA equivalent

6.1 TRANSCRIPTION: THE PRINCIPLE

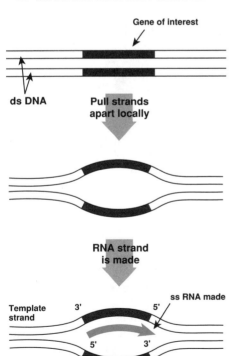

Gene of interest

ds DNA

Pull strands apart locally

RNA strand is made

Template strand   3'   5'

ss RNA made

5'   3'

5'   3'

the sugar is ribose (Fig. 6.2). As its name suggests, *deoxy*ribose has one less oxygen atom than ribose. It is this initial difference which gives the D in DNA versus the R in RNA!

6.2 RIBOSE VERSUS DEOXYRIBOSE

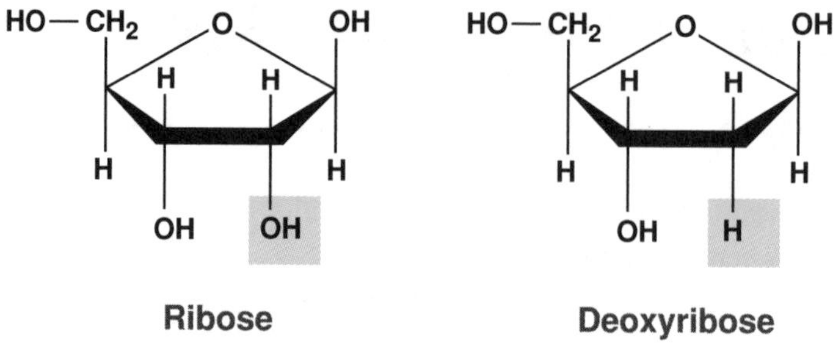

Ribose   **Deoxyribose**

uracil (U)   base that replaces thymine in the RNA molecule and can pair with adenine

The second difference is that in RNA, the base thymine (T) is replaced by the closely related base uracil (U) (Fig. 6.3). Wherever you find thymine in DNA, you get uracil in RNA. Hence, uracil in RNA and thymine in DNA, convey the same genetic information. So, if you include RNA with DNA, the genetic alphabet has five letters (A, C, G, T and U).

6.3 THYMINE VERSUS URACIL

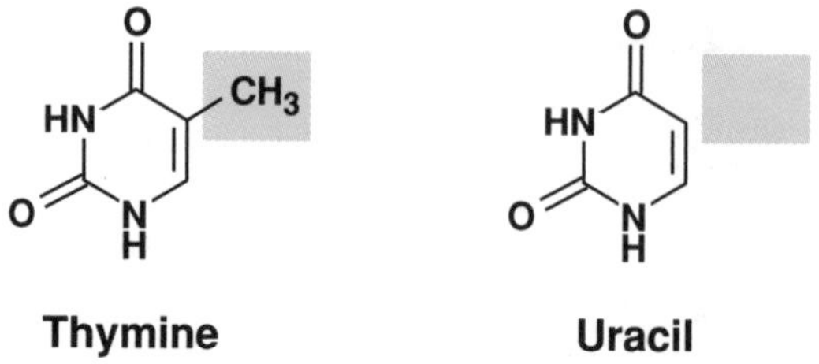

Thymine   **Uracil**

template strand   strand of DNA that is read during transcription

coding strand   the strand of DNA equivalent in sequence to the messenger RNA; sometimes referred to as the non-template strand, the strand of the DNA which is not read during transcription

The third and final difference between DNA and RNA is that DNA is double stranded (ds), whereas RNA is normally single stranded (ss). Thus, when a gene made of dsDNA is transcribed into an RNA message, only one of the strands of DNA is copied. The sequence of the RNA message is complementary to the template strand of the DNA upon which it is synthesized. Apart from the replacement of thymine in DNA with uracil in RNA, this means that the sequence of the new RNA molecule is identical to the sequence of the coding strand of DNA, the one not actually used during transcription.

48

Short Segments of the Chromosome are Turned into Messages

Although a chromosome carries hundreds or thousands of genes, only a fraction of these are used at any given time. In a typical bacterial cell, about 30 percent of the genes are in use at any particular time. In the cells of higher organisms having many more genes, the proportion in use at the same time is much smaller. During cell growth, each gene or small group of related genes, is used to generate a separate RNA copy when, and if, it is needed. Consequently there are many different messenger RNA molecules. Each of these mRNA molecules carries the information from a short segment of a chromosome.

Messenger RNA is Made by RNA Polymerase

RNA is made by an enzyme called **RNA polymerase**. This enzyme binds to the DNA at the start of a gene and opens the double helix. It then goes on to manufacture an RNA message (Fig. 6.4).

6.4 BINDING OF RNA POLYMERASE

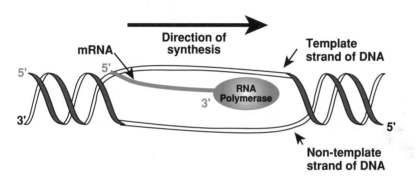

How is the Beginning of a Gene Recognized?

Here we'll talk about bacteria because they're much simpler. The principles of transcription are similar in higher organisms (for the details see Ch. 11.) RNA polymerase is made up of several protein subunits with different roles. A special subunit of bacterial RNA polymerase, the **sigma subunit**, recognizes two special sequences of bases in the coding (non-template) strand of the DNA. These sequences are known as the **-10 region** and the **-35 region** because they are found by counting backwards 10 or 35 bases from the first base of the gene. The stretch of DNA in front of a gene is often referred to as the **upstream region** and the region where RNA polymerase binds is known as the **promoter** (Fig. 6.5).

RNA polymerase enzyme that synthesizes RNA using a DNA template

sigma subunit subunit of bacterial RNA polymerase that recognizes and binds to the promoter sequence

-10 region region of promoter 10 bases back from the start of transcription and which is recognized by RNA polymerase

-35 region region of promoter 35 bases back from the start of transcription and which is recognized by RNA polymerase

upstream region region of DNA in front of a gene; its bases are numbered negatively counting backwards from the start of transcription

promoter DNA sequence in front of a gene which RNA polymerase binds to

6.5 WHAT'S UPSTREAM AND WHAT'S DOWNSTREAM ?

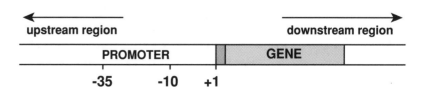

To bind the sigma subunit properly, the base sequence needed at -10 is TATAAT and the sequence at -35 is TTGACA (Fig. 6.6). Such theoretically perfect sequences (in this case TATAAT and TTGACA) are known as **consensus sequences**. Consensus sequences are found by comparing many real life sequences and taking the average. In real life, a few highly expressed genes do have these exact sequences in their promoter.

However, in practice, the -10 and -35 region sequences are rarely perfect, but as long as they are only wrong by one or two bases the sigma subunit will still recognize them. The strength of a promoter depends partly on how close it matches the ideal consensus sequence.

6.6 RECOGNITION OF -10 AND -35 REGIONS BY SIGMA

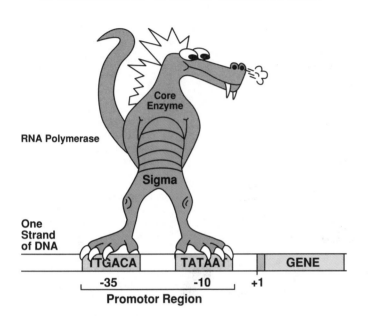

Manufacturing the Message

Once sigma has found a promoter and the RNA polymerase has successfully bound to it, the sigma subunit drops off. The remaining part of bacterial RNA polymerase, known as the **core enzyme**, then makes the mRNA (Fig. 6.7). The DNA double helix is opened up and a single strand of RNA is generated using one of the DNA strands as a template for matching up the bases.

6.7 SYNTHESIZING THE MESSAGE

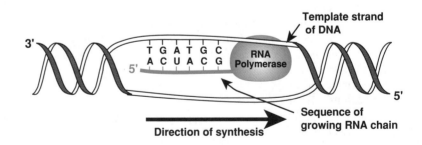

How Does RNA Polymerase Know Where to Stop?

Just as there is a special recognition site at the front of each gene so there is a special sequence at the end. A **terminator sequence** consists of two **inverted repeats** separated by half a dozen bases, followed by a run of As (in the template strand of the DNA). Figure 6.8 will show you what is meant by this. Note that the two inverted repeat sequences are actually on opposite strands of the DNA. The sequence of the mRNA will be the same as the non-template strand of DNA except for the substitution of U for T.

6.8 INVERTED REPEATS IN TERMINATOR

```
                        inverted                     repeats
DNA:                    >>>>>>>>>>
Non-Template Strand   ATTA - TAGCGGCCATC - ACTGTTACA - GATGGCCGCTA - TTTTTTT
Template Strand       TAAT - ATCGCCGGTAG - TGACAATGT - CTACCGGCGAT - AAAAAAA
                                                     <<<<<<<<<<
```

Transcription

```
                 5'     >>>>>>>>>>           <<<<<<<<<<          3'
Messenger RNA        AUUA - UAGCGGCCAUC - ACUGUUACA - GAUGGCCGCUA - UUUUUUU
(remember U is
substituted for T)
```

Although we often talk as if the corresponding single stranded mRNA had "inverted repeats," its second "repeat" is actually the complement of the inverse of the first. Because of this, such inverted repeat sequences on the same strand of an RNA molecule can pair up to generate a stem and loop or "hairpin" structure (Fig. 6.9). The string of As in the DNA gives

stem and loop or "hairpin" structure generated by folding up an inverted repeat sequence

6.9 INVERTED REPEATS MAKE A STEM AND A LOOP

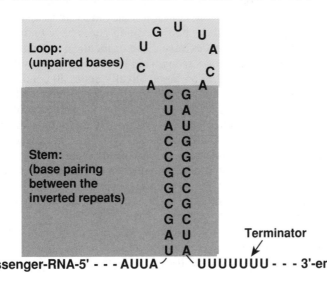

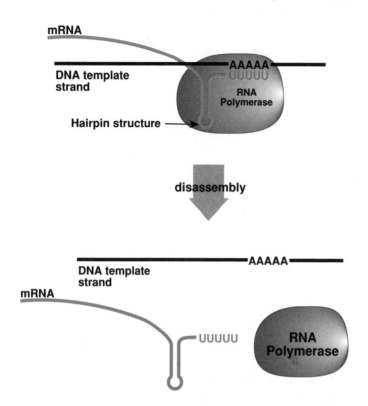

rise to a run of Us at the 3' end of the mRNA.

Once the RNA polymerase reaches the stem and loop it pauses. Long RNA molecules contain lots of possible hairpin structures which cause RNA polymerase to slow down or stop briefly, depending on the size of the hairpin. This provides an opportunity for termination but, if there is no string of Us, the RNA polymerase will start off again. However, a string of Us paired with a string of As in the template strand of DNA is a very weak structure and the RNA and DNA just fall apart while the RNA polymerase is idling (Fig. 6.10). Once the DNA and RNA have separated at the terminator structure the RNA polymerase falls off and wanders away to find another gene.

How Does the Cell Know Which Genes to Turn On?

Although each gene has a promoter and a terminator for starting and finishing the synthesis of messenger RNA, this still does not tell us when to turn on a gene. Some genes are switched on all the time. They are sometimes known as housekeeping genes and they are said to be expressed "constitutively." Most of these housekeeping genes have both their -10 and -35 region promoter sequences very close or identical to consensus. Consequently they are always recognized by the sigma subunit of RNA polymerase and are switched on automatically under all conditions.

Genes which are only needed under certain conditions usually have poor recognition sequences in the -10 and -35 regions of their promoters. In such cases the promoter sequence is not recognized by the sigma subunit unless another accessory protein is there to help (Fig. 6.11). These accessory proteins are known as gene activator proteins and are different for different genes. Each activator protein may recognize one or more genes. A group of genes which are all recognized by the same activator protein will be expressed together under similar conditions, even if the genes are at different places on the DNA. Higher organisms have many genes which are often expressed differently in different tissues. As a result, eukaryotic genes are often controlled by multiple activator proteins also known as transcription factors (Ch. 11). So for now we'll stick to bacterial genes as examples.

housekeeping genes genes that are in constant use to maintain basic cell functions

male chauvinists think that housekeeping genes are the genes found on the X-chromosome of human females

gene activator proteins switch on genes by binding to DNA and helping RNA polymerase to bind

6.11 ACTIVATOR PROTEIN HELPS RNA POLYMERASE BIND

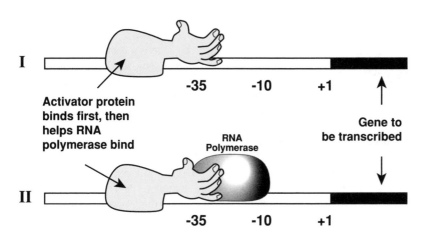

I

-35 -10 +1

Activator protein binds first, then helps RNA polymerase bind

RNA Polymerase

Gene to be transcribed

II

-35 -10 +1

What Activates the Activator?

Long ago, the Greek philosopher Plato pondered the political version of this question: "Quis ipsos custodes custodiet?" meaning, "Who will guard the guardians?" In living cells, especially in more complex higher organisms, there may indeed be a series of regulators, each regulating the next. Ultimately, however, the cell must respond to some outside influence.

As a simple example of an activator, let's consider the use of maltose by the bacterium *Escherichia coli*. Maltose is a sugar made originally from the starch in malt. *E. coli* can grow using this sugar to satisfy all of its needs for energy. An activator protein, called MalT, detects maltose by binding to it (Fig. 6.12). The MalT protein changes shape when it binds maltose. The original "empty" form of MalT cannot bind to DNA. The active form (MalT + maltose) can bind to DNA and it finds the genes needed for growth on maltose and activates them. The result of this is that the genes intended for using maltose are only expressed when this particular sugar is available. The same general principle applies to most nutrients although the details of the regulation often vary from case to case.

maltose a type of sugar consisting of two glucose units and found in malt where it is derived from starch breakdown

6.12 MalT PROTEIN DETECTS MALTOSE

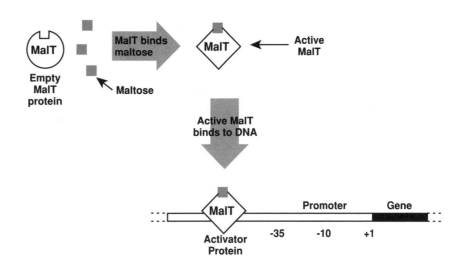

MalT

Empty MalT protein

MalT binds maltose

Maltose

MalT

Active MalT

Active MalT binds to DNA

MalT

Activator Protein

Promoter

Gene

-35 -10 +1

Negative Regulation

repressor proteins switch off genes by binding to DNA and blocking the action of RNA polymerase

lactose a type of sugar found in milk and made of glucose plus galactose

operator site on DNA to which a repressor protein binds

Just as there are activator proteins which help turn genes on, there are also proteins that can turn genes off. Historically, these negative regulators were actually discovered first. They are known as **repressors** and they work in a similar way to activators except they have the opposite effect.

The best known example is the **lactose** repressor, the LacI protein (Fig. 6.13). Lactose is another sugar, found in milk, which bacteria like *E. coli* can grow on. When no lactose is available the LacI protein binds to the stretch of DNA between the promoter and the genes for using lactose. The site where a repressor binds is called the **operator** sequence. The repressor blocks the binding of RNA polymerase, simply by getting in the way. When lactose is present it will bind to the LacI protein. The LacI protein then changes shape and falls off the DNA. Now the RNA polymerase can bind, and the genes for using lactose are switched on. The overall result is the same as for maltose: when lactose is available, the genes for using it are switched on and when there is no lactose, the genes are turned off.

6.13 LAC I REPRESSOR PROTEIN DETECTS LACTOSE

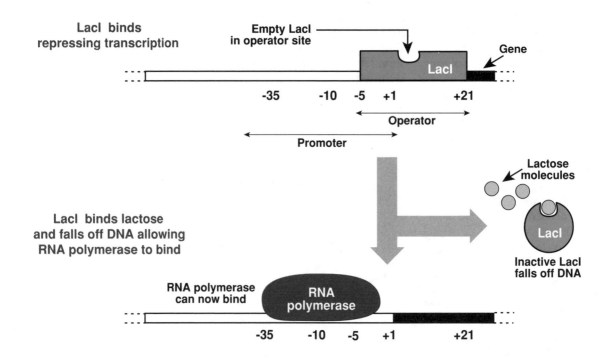

Most Regulator Proteins Bind Small Molecules

Whether our regulator protein is an activator or a repressor, we need to provide it with a signal of some sort. The most common way to do this is by using some small molecule which fits into a binding site on the regulatory protein (Fig. 6.14). This is called the signal molecule. In the case of using a nutrient for growth, the obvious choice is the nutrient molecule itself. As we have seen, this is true for the lactose repressor and the maltose activator.

Most Regulator Proteins Change Shape

When a regulator protein binds its signal molecule it changes shape (Fig. 6.14). Regulator proteins have two alternative forms, the DNA binding form and the non-binding form. Binding, or loss of the small signal molecule, causes the larger protein to flip-flop between its two alternative shapes. Proteins that operate by changing shape in this manner are called allosteric proteins.

Most Regulator Proteins Have Two or Four Subunits

Almost all real regulator proteins act as pairs or in groups of four (Fig. 6.15). All of the subunits bind the signal molecule and then they all change shape together. Because there is an even number of protein subunits bound to the DNA, the recognition site on the DNA is also duplicated - well sort of. Actually, the recognition site is not a direct repeat but an inverted repeat. This is because the subunits of the regulator protein bind to each other head to head rather than head to tail. Consequently, the two protein molecules are pointing in opposite directions. Because they have identical binding sites for DNA, they recognize the same sequence of bases but in opposite directions.

Therefore, one protein subunit binds to the recognition sequence on the template strand of the double helical DNA, and its partner binds to the same sequence but on the non-template strand of the DNA pointing in the opposite direction. This is actually simpler in practice than it sounds (trust us!), precisely because the DNA molecule is a double helix, and twists around to accommodate the proteins

signal molecule the sneaky little molecule that activates an activator protein or deactivates a repressor protein

allosteric proteins proteins that change shape when they bind a small molecule of some sort

6.14 REGULATOR PROTEIN WITH BINDING SITE

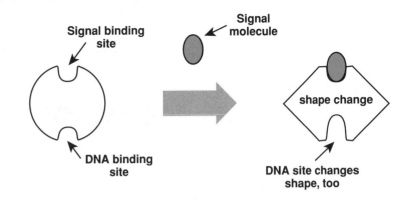

6.15 SUBUNITS OF A DNA BINDING PROTEIN

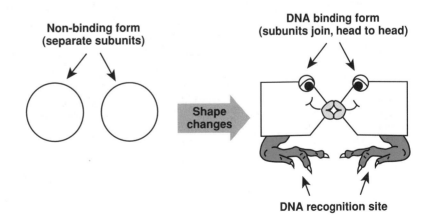

most easily this way (Fig. 6.16). Although the two recognition sequences are on different strands they end up on the same side of the DNA molecule due to its helical twisting.

6.16 INVERTED REPEAT BINDING A PROTEIN

I

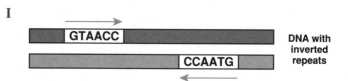

DNA with
inverted
repeats

II

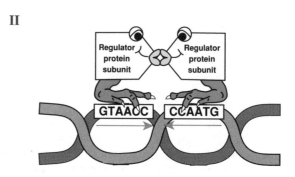

Crp Protein - An Example of a Global Control Protein

So far we have considered how to control genes for single functions such as using a particular sugar for bacterial growth. We must now consider the coordinated control of large groups of genes. This is known as **global regulation** and the proteins in charge of it are called **global regulators**. The Crp global control protein is in charge of selecting from the menu which nutrients to use for growth in bacteria like *E. coli*.

Just as those of us with taste prefer red meat to vitamin-contaminated vegetables, so bacteria select their favorite foods when given a choice. Many bacteria can grow on a wide range of possible sugars such as **fructose** (fruit sugar), lactose (milk sugar), maltose (from starch breakdown) as well as **glucose**.

If given a mixture of glucose, fructose, lactose and maltose, *E. coli* will use the glucose and ignore the others. In molecular terms this means switching off the genes for using all of the other sugars when glucose is available.

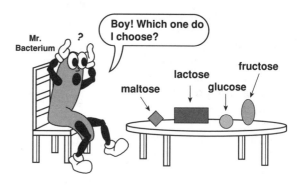

global regulation regulation of a large group of genes in response to the same environmental stimulus

global regulator protein protein that controls expression of many genes in response to the same signal

fructose a type of simple sugar commonly found in fruits

glucose a type of simple sugar usually found making up polymers such as starch and cellulose

56

The **Crp protein** is a global activator that is required for switching on the genes for using maltose, for lactose and for all of the alternative nutrients to glucose. The Crp protein is allosteric, like the MalT and LacI proteins. In order to bind DNA and activate genes, the Crp protein must first bind to a small signal molecule known as **cyclic AMP** (Fig. 6.18). Maybe you have been wondering what Crp stands for. Now we can reveal the truth: Crp = Cyclic AMP Receptor Protein. Cyclic AMP is a global signal that the cell has run out of glucose, its favorite energy source. Only when this has occurred can the genes for using less favored nutrients be switched on (Fig. 6.18).

Crp protein a global activator for the control of the use of alternative sugars to glucose

cyclic AMP (cyclic adenosine monophosphate) a signal molecule used in global regulation

6.18 Crp BINDING CYCLIC AMP

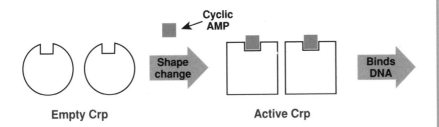

Cyclic AMP

Shape change

Binds DNA

Empty Crp

Active Crp

Consequently, in order to switch on genes for using any individual sugar, say, lactose, we need both an individual signal, the availability of lactose, and a global signal, cyclic AMP which signals the need for nutrition (Fig. 6.19). In practice, most genes respond to two signals, sometimes more. Usually one is a specific signal and the other is a more general signal that applies to many genes.

6.19 LAC GENES – DUAL REGULATION

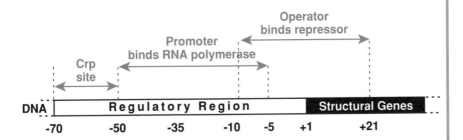

Operator binds repressor

Promoter binds RNA polymerase

Crp site

DNA | Regulatory Region | Structural Genes

-70   -50   -35   -10   -5   +1   +21

Regulatory Nucleotides

Cyclic AMP is a cyclic version of adenosine monophosphate in which the phosphate group is bent around and attached to both the 5' and 3' positions of the ribose sugar. Although it is not used as a building block when making nucleic acids, cyclic AMP is nonetheless a nucleotide of sorts. A variety of modified nucleotides are used by cells as signal molecules and are consequently called **regulatory nucleotides**. Like cyclic AMP, they are mostly used as global signals. Another example is isopentenyl adenosine, found in plants where it acts to control cell division (see Ch. 15).

regulatory nucleotide signal molecule made using same chemical components as nucleic acids, *i.e.*, bases, phosphate groups and ribose or deoxyribose

The Operon Model for Gene Regulation

The above scheme for regulating bacterial genes was first proposed by Francois Jacob ("Fronswa Zhakob") and Jaques Monod ("Zhak Mono"), using the lactose genes as an example. Since then, a vast number of bacterial genes have been fitted to this model or slight variants of it. Jacob and Monod named the various components of this scheme, the repressor, the operator, etc.

Up till now we have talked as if each gene had its own promoter and regulatory sites. In fact, many bacterial genes are found in groups that are transcribed together from the same starting point to give a single messenger RNA. A cluster of genes all switched on together by being transcribed from the same promoter is known as an operon (Fig. 6.20). Despite having more genes than bacteria, higher organisms do not have operons; their genes are regulated one at a time. Nonetheless, genes of higher organisms are regulated by the binding of control proteins, both global and specific, in front of the gene (for more, see Ch. 11).

> **operon** a cluster of genes transcribed together to give a single molecule of mRNA

6.20 THE PARTS OF AN OPERON

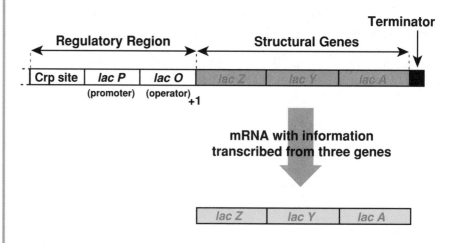

Some operons have only a single gene, most have two to half a dozen and a few have more. Geneticists have an obsession with abbreviations, if possible, of three letters. Another convention is to write gene names in italics. Thus, the lactose operon is generally known as the *lac* operon. The *lac* operon consists of three genes, *lacZ*, *lacY*, and *lacA*.

Whether or not the *lac* operon is switched on or off depends on the two regulator proteins, LacI and Crp. The various possibilities are illustrated in Figure 6.21. Only when the repressor, LacI, is absent and the Crp protein is present to give a helping hand, can the RNA polymerase bind to the promoter and make the messenger RNA.

6.21 THE LAC OPERON: ON OR OFF?

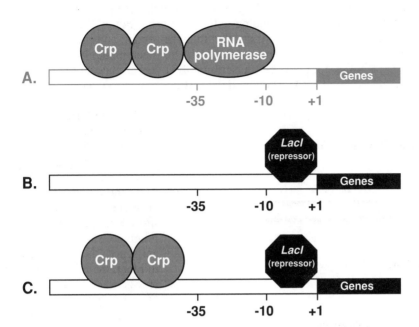

	Crp Protein	Repressor	RNA Polymerase	On or Off
A	Present	Absent	Binds	On
B	Absent	Present	Cannot Bind	Off
C	Present	Present	Cannot Bind	Off
D	Absent	Absent	Cannot Bind	Off

Regulation by Antisense RNA

As we have seen, messenger RNA is transcribed using only one DNA strand as the template strand. The other strand of DNA is not used. But suppose we did use the non-template strand and transcribed RNA from it? We would produce an RNA molecule complementary in sequence to the mRNA. This is known as antisense RNA and can base pair with its complementary mRNA, just as the two strands of DNA in the original gene base pair with each other.

antisense RNA   RNA that is complementary in sequence to messenger RNA and therefore base pairs with it

"antisense" is the form of regulatory message favored by government agencies

ribosome structure composed of RNA and proteins involved in protein synthesis

Antisense RNA is occasionally used in gene regulation by bacteria and higher organisms. If antisense RNA is made, it will base pair with the mRNA and prevent it from binding to the ribosome (Fig. 6.22). Consequently the mRNA cannot be translated to make protein and the gene is effectively switched off, even though mRNA has been made. In practice, antisense RNA is not made by transcribing the non-template strand of the same gene as the mRNA. Another, quite distinct "anti-gene" is used for making the antisense RNA.

6.22 ROLE OF ANTISENSE RNA IN GENE REGULATION

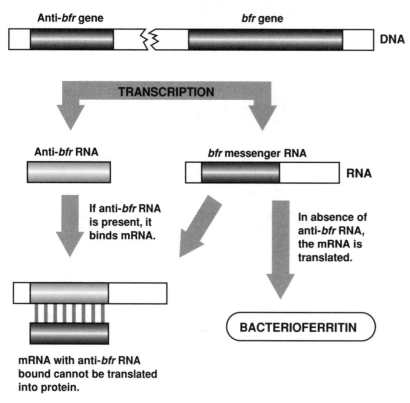

Bacterioferritin is a protein used by bacteria to store surplus iron atoms. The *bfr* gene encodes bacterioferritin itself and the anti-*bfr* gene encodes the antisense RNA. Since only a relatively short piece of antisense RNA is needed to block the mRNA, the anti-gene is similar in sequence but shorter than the original gene. When the iron concentration in the culture medium is low, bacterioferritin is not needed, but it is made if the iron level goes up. The *bfr* gene itself is transcribed to give mRNA in both conditions. However, the anti-*bfr* gene is only transcribed to give antisense RNA in low iron. This prevents synthesis of the bacterioferritin protein when iron is scarce.

So what turns the anti-*bfr* gene on and off? A global regulatory pro-

Ferric uptake regulator (Fur) a global regulator protein that detects and binds iron

tein known as **Fur** (**F**erric **u**ptake **r**egulator) detects and binds iron. When plenty of iron is present, Fur acts as a repressor and turns off the transcription of a dozen or more operons needed for adapting the cell to iron scarcity. In particular, Fur+iron turns off the anti-*bfr* gene which turns on the production of bacterioferritin. So by using antisense RNA we can regulate one gene the opposite way to a bunch of others.

Artificially synthesized antisense RNA will interfere with gene expression, or anything else involving RNA. Antisense RNA is being tested experimentally to suppress cancer by stopping chromosome division (see Ch. 14; Fig. 14.34).

Bacterial Democracy - Quorum Sensing

It's not just the cells of higher organisms that get together. Although bacteria live as single cells, under some circumstances they need to co-operate in communal ventures. Amazingly enough they regulate certain genes by a form of chemical voting known as quorum sensing. The basic idea is quite simple. Because bacteria are so tiny, if only a few are present they will be unable to make much impact. On the other hand, if billions are crowded together, their joint effort may be quite significant. So the bacteria involved all secrete a signal molecule, the auto-inducer, into the medium. If the level of auto-inducer is high enough, this means that enough bacteria are present to have some effect and everybody switches on the genes for communal effort.

The best known example is light emission by sea-faring bacteria. Single bacteria cannot make enough light to be seen, and only if billions cooperate is it worthwhile to indulge in light emission. *Vibrio fischeri* is a marine relative of our tiny friend *Escherichia coli* (see Ch. 2). Just as *E. coli* lives in the guts of animals, so *Vibrio fischeri* wishes to live inside fishes. If a dense enough crowd of *Vibrio fischeri* gathers on organic matter at the bottom of the sea, they all turn on their lights together and the glow attracts a fish which swallows them. Some more daring luminous bacteria provide light for monsters like giant deep sea squid.

The enzyme luciferase is responsible for biological light emission and is widely used for detecting gene expression (see Ch. 16). Luciferase and some accessory proteins are coded for by the *lux* genes. The signal molecule, or auto-inducer, is made by LuxI protein and binds to the LuxR protein (Fig. 6.23). When LuxR has bound auto-inducer it switches on the genes for luciferase.

The key property of auto-inducer is that it drifts freely into and out of the bacteria. If a cell is alone, the auto-inducer simply drifts away into the deep blue sea. If lots of cells are huddled together, auto-inducer from one cell will wander into others instead of being lost. Only if the population density rises above 10 million bacteria per milliliter does enough auto-inducer build-up to turn on the genes for luciferase.

quorum sensing form of regulation in which a gene is expressed in response to population density

auto-inducer signal molecule which can freely exit and enter cells and is involved in quorum sensing

luciferase enzyme involved in a light emission reaction

6.23 LIGHT EMISSION BY SEAFARING BACTERIA

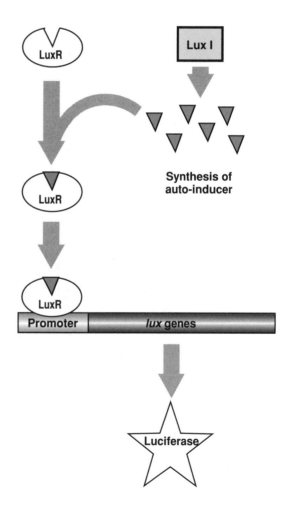

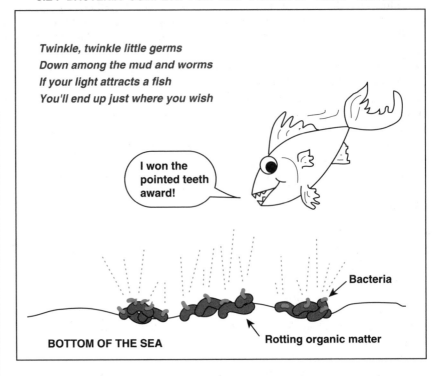

Another, less enlightened, relative of *E. coli*, *Erwinia carotovora*, lives by eating vegetables. Plant cells are many times the size of bacteria and have very thick walls. So breaking these down requires a cooperative effort. When enough bacteria are present on the plant, they all secrete digestive enzymes in unison under control of a quorum sensing system similar to LuxI/LuxR.

Transcription of genes to give mRNA is only half of gene expression. The messenger RNA is just that, a messenger, and does not actually perform any real work. As discussed in the next chapter, for this we need proteins. These genuine working class molecules are made using the information carried by the mRNA.

Additional Reading

Molecular Biology of the Cell by Alberts B, Bray D, Lewis J, Raff M, Roberts K, & Watson JD. 3rd edition, 1994. Garland Publishing, Inc., New York & London.

Essentials of Molecular Biology by Freifelder D & Malacinski GM. 2nd edition, 1993. Jones & Bartlett Publishers, Boston & London

Proteins:
The Buck Stops Here

7

Introduction to Proteins

When it comes to getting the real work of the cell done, the bureaucrats, DNA and RNA, are not much help. All the same, somebody has to get their hands dirty. **Proteins** are biological polymers that carry out most of the cell's day-to-day functions. Some proteins are merely structural or take part in cell movement, others help take up nutrients, others generate energy and yet others carry out biochemical reactions, including the synthesis of nucleotides and their assembly into nucleic acids.

Molecules whose primary role is to carry information (nucleic acids like DNA and messenger RNA) are basically linear molecules with a regular repeating structure. Molecules that form cellular structures or have active roles carrying out reactions are normally folded into three-dimensional (3-D) structures. These include both proteins and certain specialized RNA molecules (rRNA and tRNA, see later in this chapter).

Proteins are made from a linear chain of monomers, known as **amino acids** and are folded into a variety of complex 3-D shapes. A chain of amino acids is called a **polypeptide chain** (Fig. 7.1). So then, what's the difference between a polypeptide chain and a protein? Simply that some proteins consist of more than one polypeptide chain.

7.1 "POLLY" PEPTIDE

Hello! My name is Polly Peptide! I'm not just a pretty face- I have lots and lots of subunits!

Amino Acids

Role of Proteins in the Cell

We can subdivide proteins into four main categories:

1) **structural proteins**
2) **enzymes**
3) **regulatory proteins**
4) **transport proteins**

Structural proteins are found making up many subcellular structures. The flagella with which bacteria swim around, the microtubules used to control traffic flow inside cells of higher organisms, the fibers inside a muscle cell, and the outer coats of viruses (see Ch. 19) are a few examples of structures built using proteins.

protein macromolecule that does most of the cell's work

Proteins are named after Proteus, an ancient Greek god of the sea, who could change himself into any shape he felt like - usually to sleaze out of his obligations.

amino acid monomer from which proteins are built

polypeptide chain polymer chain made of amino acids; one or more such chains make up a protein

structural protein a protein that forms part of a cellular structure

enzyme a protein that carries out a chemical reaction

regulatory protein a protein that controls the expression of a gene or the activity of another protein

transport protein a protein that carries other molecules across membranes or around the body

Enzymes are proteins that carry out chemical reactions. An enzyme first binds another molecule, known as its **substrate**, and then performs some chemical operations with it. Some enzymes bind only a single substrate molecule; others may bind two or more, and react them together to make the final product.

In any case, the enzyme needs an **active site**, a pocket or cleft in the protein, where the substrate binds and the reaction occurs. The active site is produced by folding up the polypeptide chain correctly so that amino acid residues that were spread out at great distances in the linear chain now come together and will cooperate in the enzyme reaction (Fig. 7.2).

7.2 FOLDING POLYPEPTIDE FORMS ACTIVE SITE

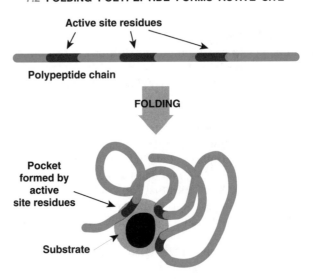

Active site residues

Polypeptide chain

FOLDING

Pocket
formed by
active
site residues

Substrate

The most famous enzyme in molecular biology is β-**galactosidase**, encoded by the *lacZ* **gene** of the bacterium *Escherichia coli*. This enzyme is so easy to assay that it is widely used in genetic analysis (see Ch. 16 for details). The natural substrate of β-galactosidase is the sugar lactose, made by linking together the two simple sugars, glucose and galactose. There is not much else to do with lactose except to split it into these two simpler sugars, so that is exactly what β-galactosidase does (Fig. 7.3).

Analogs are molecules resembling natural substances well enough to fool the enzymes that use them. Some analogs bind but do not react and simply block the active site and inhibit the

7.3 β – GALACTOSIDASE SPLITS LACTOSE

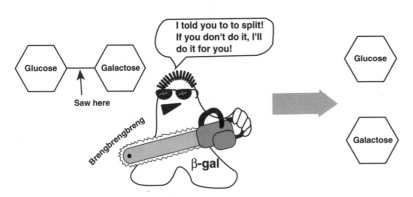

Glucose   Galactose

Saw here

I told you to to split!
If you don't do it, I'll
do it for you!

Brengbrengbreng

β-gal

Glucose

Galactose

enzyme. Such analogs are known as **competitive inhibitors** since they compete with the true substrate for the attention of the enzyme.

Other analogs do react. β-galactosidase splits many molecules in which galactose is linked to something else. We can take advantage of this by giving it **ONPG** which consists of ortho-nitrophenol linked to galactose. When ONPG is split, we get galactose (colorless) and ortho-nitrophenol which is bright yellow. Using ONPG allows us to monitor the level of β-galactosidase by measuring the appearance of the yellow color. Similarly, X-gal is split by β-galactosidase into a blue dye and galactose (see Ch. 16 for applications).

Although regulatory proteins and transport proteins are not enzymes, they also bind other molecules and so they also need "active sites" to accommodate these.

Regulatory proteins vary enormously. Many of them can bind both small signal molecules and DNA. The presence or absence of the signal molecule determines whether or not the gene is switched on (see Ch. 6).

Transport proteins are found mostly in biological membranes, as in Figure 7.4, where they carry material from one side to the other. Nutrients, such as sugars, must be transported into cells of all organisms, whereas waste products are deported. Multicellular organisms also have transport proteins to carry materials around the body, such as hemoglobin which carries oxygen in blood (Fig. 7.5).

To function properly many proteins need extra components, **cofactors** or **prosthetic groups**, which are not themselves proteins. Many proteins use single metal atoms as cofactors, others need more complex molecules. Strictly speaking, prosthetic groups are fixed to a protein, whereas cofactors are free to wander around from protein to protein, however, the terms are often used loosely. A protein without its prosthetic group is referred to as an **apoprotein**.

For example, oxygen carrier proteins such as hemoglobin have a ring-shaped cofactor with a central iron atom, called heme. The heme is bound in the active site of the apoprotein, in this case globin, and so we get hemoglobin (Fig. 7.5). Oxygen binds to the iron atom at the center of the heme and the hemoglobin carries it around

competitive inhibitor chemical substance which inhibits the action of an enzyme by mimicking the true substrate well enough to be mistaken for it

ONPG ortho-nitrophenyl-galactoside - an artificial substrate which releases a yellow color when split by the enzyme β-galactosidase

cofactor or **prosthetic group** extra chemical group attached to a protein. It is not part of the protein chain itself

apoprotein protein without any extra cofactors or prosthetic groups, *i.e.*, just the polypeptide chain(s)

7.4 MEMBRANE TRANSPORT PROTEINS

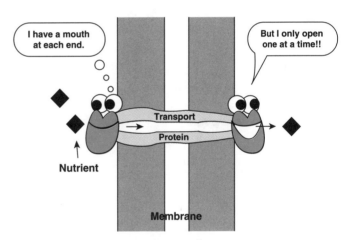

7.5 HEME OF HEMOGLOBIN BINDS OXYGEN

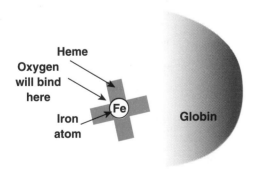

the body. Prosthetic groups are often shared by more than one protein, for example, heme is shared by hemoglobin and by myoglobin, which receives oxygen and distributes it inside muscle cells.

How Are Proteins Constructed?

The monomer or subunit of a protein is known as an **amino acid**. There are 20 different amino acids used in making proteins. They all have a central carbon atom, the alpha carbon, surrounded by the four features as shown in Figure 7.6.

7.6 FOUR COMMON FEATURES OF AMINO ACIDS

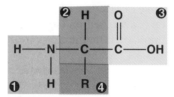

Amino acids possess common features:

❶ NH_2 (amino) group

❷ central hydrocarbon group

❸ COOH (carboxy) group

❹ R or variable group

7.7 FORMATION OF A PEPTIDE BOND

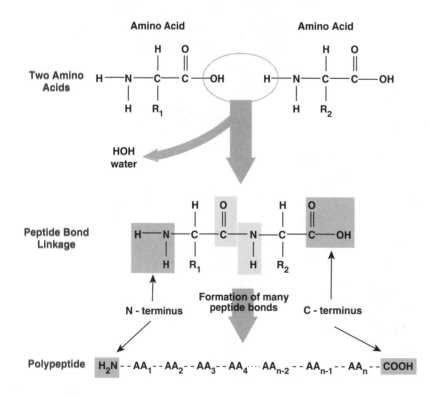

Formation of Polypeptide Chains

Amino acids are joined together by **peptide bonds** (Fig. 7.7) to give a linear polymer called a **polypeptide chain**. The first amino acid in the chain retains its free amino (NH_2) group and this end is often called the **amino-** or **N-terminus** of the polypeptide chain. The last amino acid to be added is left with a free carboxy (COOH) group and this end is often called the **carboxy-** or **C-terminus**. The different side chains of the successive amino acids are labeled R_1, R_2, R_3, etc.

Three Dimensional Structures

To make a complete protein we must next fold the polypeptide chain into the correct 3-D structure. Furthermore, a complete

66

protein may have more than one polypeptide chain. Finally, many proteins have associated cofactors that are not made of amino acids.

The structures of biological polymers, both protein and nucleic acid, are often divided into levels of organization. The first level or **primary structure**, is the order of the monomers - *i.e.*, the sequence of the amino acids for a protein, or of the nucleotides in the case of DNA or RNA. The further levels are as follows:

primary structure the linear order in which the subunits of a polymer are arranged

7.8 HYDROGEN BONDING BETWEEN PEPTIDE GROUPS

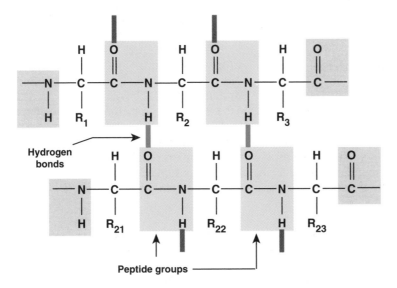

Hydrogen bonds

Peptide groups

secondary structure initial folding of a polymer due to hydrogen bonding

hydrogen bonding bonding resulting from the attraction of a positive hydrogen atom to both of two other atoms with negative charge

Secondary structure: the folding or coiling of the original polymer chains by means of **hydrogen bonding**. In DNA, hydrogen bonding between base pairs forms Watson and Crick's double helix. In proteins there is instead hydrogen bonding between peptide groups (Fig. 7.8).

In proteins there are two alternative secondary structures - the α- (alpha) helix (Fig. 7.9), and the β- (beta) sheet. In the α-helix, a single polypeptide chain is coiled to make the helix and the hydrogen bonds run vertically up and down the helix axis, not sideways across the helix. Actually, the hydrogen bonds in an α-helix are not quite vertical. They are slightly tilted relative to the helix axis because there are 3.6 amino acids per turn rather than a whole number.

The β-sheet also has hydrogen bonding between peptide groups but in this case the polypeptide chain is folded back on itself to

7.9 DIFFERENCES BETWEEN PROTEIN ALPHA - HELIX AND DNA DOUBLE HELIX

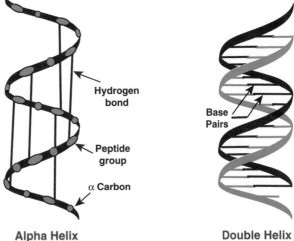

Hydrogen bond

Peptide group

α Carbon

Base Pairs

Alpha Helix
(selected bonds shown)

Double Helix

7.10 HYDROGEN BONDS FORMING A β SHEET

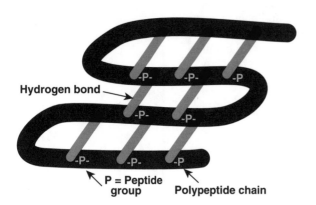

Hydrogen bond

P = Peptide group

Polypeptide chain

give a flattish structure (Fig. 7.10). The hydrogen bonds do go sideways in the β-sheet. From each peptide group, one hydrogen bond goes to one side and a second to the other side.

The next level is the tertiary structure. In a nucleic acid this would be supercoiling. In a protein we fold the polypeptide chain, with its preformed regions of α-helix and β-sheet, to give the final 3-D structure. This level of folding depends on the side chains of the individual amino acids. Since there are 20 different amino acids, a whole variety of final 3-D conformations are possible. Nonetheless, when all is said and done, most proteins are roughly spherical.

tertiary structure   final 3-D folding of a polymer chain

hydrophilic   water loving

hydrophobic   water hating

oil drop model   model of protein structure in which the hydrophobic groups cluster together on the inside, away from the water

This 3-D folding is largely the result of two factors acting in concert. Many of the amino acids have side chains (R-groups) which are very water soluble (hydrophilic). These side chains prefer to be on the surface of the protein so they can dissolve in the water surrounding the protein. In contrast, another set of side chains are water repellent (hydrophobic) and huddle together inside the protein away from the water (Fig. 7.11).

7.11 SOME WATER - LOVING AND SOME WATER - HATING SIDE CHAINS

Water Loving		Water Hating	
Name:	R - Group:	Name:	R - Group:
Aspartic acid	CH_2COOH	Valine	$CH\begin{smallmatrix}CH_3\\CH_3\end{smallmatrix}$
Serine	CH_2OH	Phenylalanine	CH_2 ⟨ ⟩

Since hydrophobic molecules are greasy and insoluble, this arrangement is known as the oil drop model of protein structure (Fig. 7.12, next page).

Quaternary Structure of Proteins

This is the assembly of several individual polypeptide chains to give the final structure. Not all proteins have more than one polypeptide chain, some just have one so they have no quaternary structure. In those having more than one polypeptide chain, the same hydrophilic and hydrophobic forces responsible for tertiary structure are involved. To stick two polypeptide chains together the original chains are designed slightly different.

quaternary structure   aggregation of more than one polymer chain in final structure

68

7.12 OIL DROP MODEL OF PROTEIN STRUCTURE

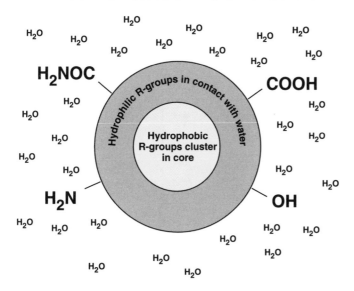

Some of the hydrophobic side chains are left as a cluster exposed to the water at the protein surface (Fig.7.13). This is an unstable arrangement and when two polypeptide chains with exposed hydrophobic patches come into contact with each other, they stick together, rather like velcro.

7.13 GREASY PATCHES STICK POLYPEPTIDES TOGETHER

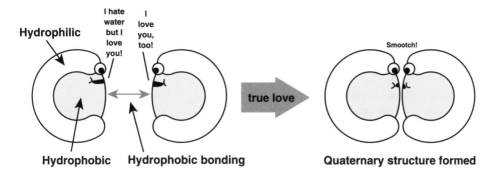

Twenty Different Amino Acids

Unlike nucleic acids that have only four different bases, there are 20 different amino acids in proteins. This allows for a great variety of 3-D structure and of chemical reactivity. The 20 amino acids may be represented by both three letter and one letter abbreviations. The latter are used when writing out protein sequences. Most are obvious, but since some letters of the alphabet have several amino acids, the others need a little imagination.

Amino Acids and their Nicknames

Amino Acid	3 Letter Code	1 Letter Code	Silly Mnemonics
Alanine	Ala	A	
Arginine	Arg	R	arrrrgh!
Asparagine	Asn	N	—
Aspartic acid	Asp	D	—
Cysteine	Cys	C	
Glutamic acid	Glu	E	—
Glutamine	Gln	Q	"Cutamine"
Glycine	Gly	G	
Histidine	His	H	
Isoleucine	Ile	I	
Leucine	Leu	L	
Lysine	Lys	K	—
Methionine	Met	M	
Phenylalanine	Phe	F	Fenylalanine
Proline	Pro	P	
Serine	Ser	S	
Threonine	Thr	T	
Tryptophan	Trp	W	"tWyptophan"
Tyrosine	Tyr	Y	tYrosine
Valine	Val	V	

How are Proteins Made?

Each protein is made using the genetic information stored on the chromosomes. The genetic information is transmitted in two stages. First the information in the DNA is transcribed into messenger RNA (see Ch. 6). The next step uses the information carried by the mRNA to give the sequence of amino acids making up a polypeptide chain. This involves converting the nucleic acid "language," the genetic code, to protein "language," and is therefore known as translation.

An early rule of molecular biology stated that there is one gene for each protein. Although exceptions have been found, it is still usually true that each gene in the DNA gives rise to a single protein.

The overall flow of information in biological cells is known as the central dogma of molecular biology (Fig.7.14) and was first formulated by Sir Francis Crick.

transcription making an RNA copy of a gene

translation making a protein using the information provided by messenger RNA

central dogma a basic plan of genetic information flow in living cells that relates genes (DNA), message (RNA) and proteins

7.14 CENTRAL DOGMA

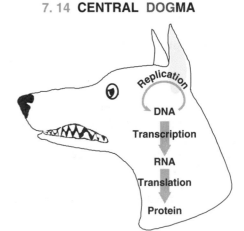

Replication
DNA
Transcription
RNA
Translation
Protein

Decoding the Genetic Code

There are 20 amino acids in proteins but only four different bases in the messenger RNA. So we cannot simply use one base of a nucleic acid to code for a single amino acid when making a protein.

During translation, the bases of mRNA are read off in groups of three, which are known as **codons**. Each codon represents a particular amino acid. Since there are four different bases, there are 64 possible groups of three bases (4^3), that is, 64 different codons in the genetic code. However, there are only 20 different amino acids making up proteins, so some amino acids are encoded by more than one codon. In addition we use a couple of the codons for punctuation. (Hey! The analogy of an alphabet holds up here too!) Figure 7.15 shows nature's genetic code.

codon group of three RNA or DNA bases which encodes a single amino acid

7.15 CODON TABLE

2nd (middle) Base

1st Base	U	C	A	G	3rd Base
U	UUU Phe UUC Phe UUA Leu UUG Leu	UCU Phe UCC Phe UCA Leu UCG Leu	UAU Tyr UAC Tyr UAA STOP UAG STOP	UGU Cys UGC Cys UGA STOP UGG Trp	U C A G
C	CUU Leu CUC Leu CUA Leu CUG Leu	CCU Pro CCC Pro CCA Pro CCG Pro	CAU His CAC His CAA Gln CAG Gln	CGU Arg CGC Arg CGA Arg CGG Arg	U C A G
A	AUU Ile AUC Ile AUA Ile AUG Mat	ACU Thr ACC Thr ACA Thr ACG Thr	AAU Asn AAC Asn AAA Lys AAG Lys	AGU Ser AGC Ser AGA Arg AGG Arg	U C A G
G	GUU Val GUC Val GUA Val GUG Val	GCU Ala GCC Ala GCA Ala GCG Ala	GAU Asp GAC Asp GAA Glu GAG Glu	GGU Gly GGC Gly GGA Gly GGG Gly	U C A G

Let's use this codon table to translate the following RNA message into protein:

RNA CODE

base sequence: GAG - GCC - GUA - AUC - GAA - UGU - UUG - GCA - AGG - AAA

PROTEIN

3-letter code: Glu - Ala - Val - Ile - Glu - Cys - Leu - Ala - Arg - Lys

1-letter code: D - A - V - I - D - C - L - A - R - K

ribosome the cell's
machinery for making
proteins

S-values tell how fast a
particle sediments in an
ultracentrifuge (see Ch.16).
They give a rough indica-
tion of size but you cannot
add them up. A 30S plus a
50S subunit gives a com-
plete ribosome with an S-
value of 70S, not 80S.
(New math, huh!)

Each subunit is made of
protein plus RNA. The
RNA molecules that make
up the ribosome are quite
distinct from messenger
RNA. They are known as -
guess what? - ribosomal
RNA (or rRNA). The
rRNA molecules are NOT
themselves translated into
protein, instead they form
part of the machinery of
the ribosome that
translates the mRNA.

In bacteria the large
subunit has two rRNA
molecules, 5S rRNA and
23S rRNA, and the small
subunit has just the one
16S rRNA. In addition to
the rRNA, there are about
50 different proteins, 30 in
the large subunit and the
other 20 in the small
subunit.

ribosomal RNA RNA
molecules that make up
part of the structure of a
ribosome

Ribosome - The Cell's Decoding Machine

The decoding process is carried out by a submicroscopic machine called a **ribosome** that binds messenger RNA and translates it, so synthesizing a polypeptide chain. Here we'll talk about protein synthesis in bacteria. (The details of protein synthesis differ between bacteria and higher organisms. See Ch. 11 for more on higher organisms.)

The ribosome consists of two subunits, large (50S) and small (30S) and can be pictured as a snowman (Fig. 7.16).

7.16 COMPONENTS OF A RIBOSOME

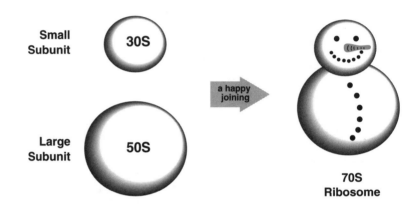

Small Subunit — 30S

Large Subunit — 50S

a happy joining

70S Ribosome

After binding to the mRNA, the ribosome moves along it, adding a new amino acid to the growing polypeptide chain each time it reads a codon from the message (Fig.7.17).

7.17 RIBOSOME, mRNA AND A GROWING POLYPEPTIDE CHAIN

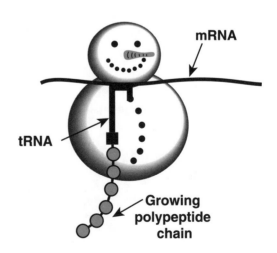

mRNA

tRNA

Growing polypeptide chain

To read the codons we need a set of adapter molecules that recognize the codon on the mRNA at one end and carry the corresponding amino acid attached to their other end. These adapters are a third type of RNA, transfer RNA or tRNA. At one end, the tRNA has an anticodon consisting of three bases that are complementary to the three bases of the codon on the messenger RNA. The codon and anticodon recognize each other by base pairing and are held together by hydrogen bonds (Fig. 7.18). At its other end, each tRNA carries the amino acid corresponding to the codon it recognizes.

transfer RNA RNA molecules that carry amino acids to a ribosome

anticodon group of three complementary bases on tRNA that recognize and bind to a codon on the mRNA

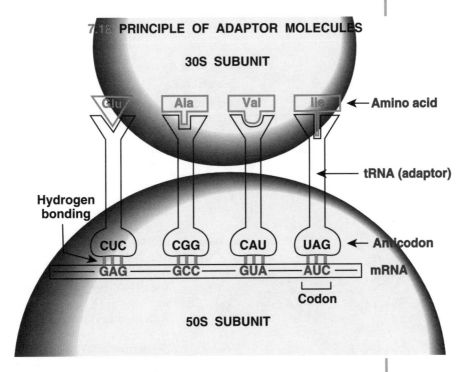

How Many tRNA Molecules are There?

Each transfer RNA carries only a single amino acid so we need at least 20 different tRNAs because there are 20 different amino acids. On the other hand, there are 64 codons to be recognized as some amino acids have more than one codon. In practice, we have a sloppy compromise and the number of different tRNA molecules is somewhere between 20 and 64. Some tRNAs can read more than one codon, though, of course, these must all code for the same amino acid.

Since only complementary bases can pair, how does a tRNA with one anticodon read more than one codon? Easy - it cheats! The rules for dishonesty in base pairing are known as the wobble rules (see Fig. 7.19). The first base of the tRNA anticodon can wobble around a little because it is not squeezed between other bases in a helix structure. For example, if the first anticodon base is G it can pair with C, as usual, or, in wobble mode, with U. Therefore tRNA for histidine, with GUG as anticodon, can recognize both the CAC and CAU codons. Whenever an amino acid is encoded by a pair of codons, the third codon bases are U and C (*e.g.*, histidine, tyrosine)

wobble rules rules allowing less rigid base pairing but only for codon/anticodon pairing

inosine an unusual nucleoside derived from guanosine

or A and G (*e.g.*, lysine, glutamic acid), but never other combinations. Similarly, those privileged amino acids with four or six codons may be regarded as having two or three such pairs.

7.19 WOBBLE RULES

First Base of Anticodon	Pairs with: Third Base of Codon	
	normal	by wobble
G	C	or U
U	A	or U
I	–	C or U or A
C	G only	no wobble
A	U only	no wobble

As originally transcribed, RNA contains only the four bases A, U, G and C. However, some RNA molecules contain bases that are altered chemically after the RNA has been made. This is especially true for tRNA. In fact the anticodon itself may contain the weird base **I** for **inosine**, which is occasionally used as the first anticodon base because it can pair with any of U, C or A (Fig. 7.19).

How Does the tRNA get its Amino Acid?

For each tRNA there is a specific enzyme that recognizes both the tRNA and the corresponding amino acid. The enzymes, known as **amino-acyl tRNA synthetases**, attach the amino acid to the tRNA (Fig.7.20). This is called charging the tRNA. Empty tRNA is known as **uncharged tRNA** while tRNA with its amino acid is **charged tRNA**.

Structure of Transfer RNA

It's time to stop fooling around with silly diagrams and tackle what tRNA actually looks like. A typical tRNA has four short base-paired stems and three loops (Fig. 7.21). This is the cloverleaf structure, intended to reveal details of base-pairing, and shows the tRNA spread out like a coyote flattened by a road roller. The amino acid is bound at the free end of the acceptor stem. The anticodon is at the opposite end in the anticodon loop.

The other two loops of tRNA are named after modified bases. The TψC loop contains "ψ" (spelt "psi" but pronounced "sigh") which stands for pseudouracil and the D-loop has

7.20 CHARGING OF tRNA

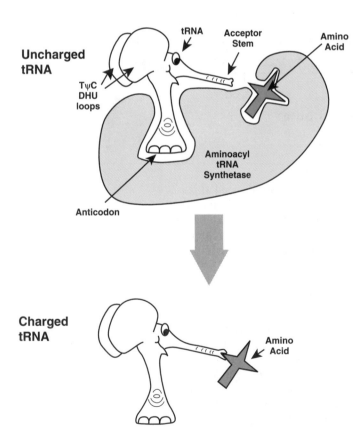

Uncharged tRNA

tRNA — Acceptor Stem — Amino Acid

TψC DHU loops

Aminoacyl tRNA Synthetase

Anticodon

Charged tRNA

Amino Acid

7.21 CLOVERLEAF STRUCTURE OF tRNA

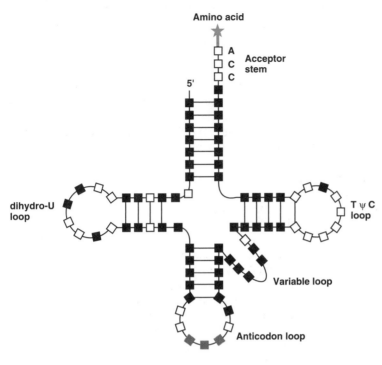

Charged tRNA

amino-acyl tRNA synthetase enzyme that attaches an amino acid to tRNA

uncharged tRNA tRNA without an amino acid attached

charged tRNA tRNA with an amino acid attached

"D" for dihydrouracil. These weird bases are required for proper folding and operation of the tRNA. The TψC loop and the D-loop are needed for binding to the ribosome and for recognizing the enzyme which sticks the amino acids onto the tRNA.

We still haven't told you the true 3-D structure of tRNA. In real life the tRNA cloverleaf is folded up further. The TψC loop and D-loops are pushed together and the molecule is bent into an L-shape, rather as it was shown in Figure 7.20.

Reading Frames

The bases of mRNA are read in groups of three, starting at the 5' end. We always begin with the start codon, AUG. However, consider the following message:

start codon the special AUG codon that signals the start of a protein

GAA<u>AUG</u>U<u>AUG</u>C<u>AUG</u>CCAAAGGAGGCAUCUAAGGA

We have three possible start codons (underlined). Each of these starts at a slightly different point. More disturbingly, each of the three leads us to take quite different groups of three bases as codons. So each gives a translation completely out of step with each of the others. The three alternatives are illustrated below, where bases considered to be part of the same codon have been given the same numbers. These three possibilities are known as "reading frames."

reading frame one of three possible ways to read off the bases of mRNA in groups of three so as to give codons

G A A AUG U AUG C AUG C C A A A G G A G G C A U C U A A G G A
1 1 1 2 2 2 3 3 3 4 4 4 5 5 5 6 6 6 7 7 7 8 8 8
  1 1 1 2 2 2 3 3 3 4 4 4 5 5 5 6 6 6 7 7 7 8 8 8
    1 1 1 2 2 2 3 3 3 4 4 4 5 5 5 6 6 6 7 7 7 8 8

As there are three bases in a codon, changing the reading frame by three (or a multiple of three) gets you back to where you started.

A stretch of RNA, beginning with a start codon, and which can therefore be translated into a protein, is known as an open reading frame, often abbreviated to (and pronounced!) ORF. Any messenger RNA will have several possible ORFs and we have to find the correct one.

Getting Protein Synthesis Started

The first codon is always AUG, which stands for the amino acid methionine (Fig. 7.22). A special tRNA, the initiator tRNA will be charged with chemically tagged methionine (formyl-methionine or fMet) and will bind to the start codon. So all polypeptide chains begin with methionine. (Sometimes the initial methionine is snipped off later, so mature proteins do not always begin with methionine.)

There are also AUG codons in the middle of a message and, consequently, methionines in the middle of proteins. So how does the ribosome know which AUG codon to start with? Near the front (the 5' end) of the messenger RNA is a special sequence, the ribosome binding site. The ribosome binding site is usually called the Shine-Dalgarno or S-D sequence, after its two discoverers (Fig. 7.23). The sequence complementary to this, the anti-Shine-Dalgarno sequence, is found close to the 3' end of the 16S ribosomal RNA and this causes the mRNA to bind to this rRNA. In some cases the S-D is an exact match to the anti-S-D sequence and these mRNAs are translated efficiently. In other cases the match is poorer and translation is less efficient. The start codon is the next AUG codon after the ribosome binding site. It's that simple!

open reading frame (ORF) sequence of mRNA (or corresponding region of DNA) that can be translated to give a protein

fMet or formyl-methionine chemically tagged version of methionine used to start the polypeptide chain in prokaryotic (bacterial) cells

7.22 WHERE DO I START ?

70S Ribosome

ribosome binding site or Shine-Dalgarno (S-D) sequence sequence on mRNA at the front of the message which is recognized by the ribosome

anti-Shine-Dalgarno sequence sequence on 16S mRNA that is complementary to the Shine-Dalgarno sequence

7.23 RIBOSOME BINDING SITE

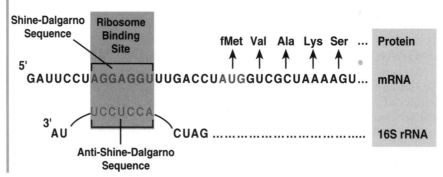

Before protein synthesis starts, the two subunits of the ribosome are floating around separately. Because the 16S rRNA, with the complementary, anti-Shine-Dalgarno sequence, is in the small subunit of the ribosome, the messenger RNA binds to a free small subunit. Next the initiator tRNA, carrying fMet, recognizes the AUG start codon. We also need three proteins, known as initiation factors, that help arrange all the components correctly. Finally the large subunit arrives, and joins its smaller partner as the initiation factors drop off. This sequence of events is shown in Figure 7.24.

Elongation of a Growing Protein

After the large subunit of the ribosome has arrived, the polypeptide chain is made. The ribosome has two sites for tRNA: the A (acceptor) site and the P (peptide) site (Fig. 7.24). We start with the fMet initiator tRNA in the P-site. Another tRNA, carrying the next amino acid, arrives and enters the A-site. The fMet is cut loose from its tRNA and bonded to amino acid No. 2 instead. So tRNA No. 2 now carries two linked amino acids, the beginnings of our growing protein chain.

7.24 FORMATION OF THE INITIATION COMPLEX

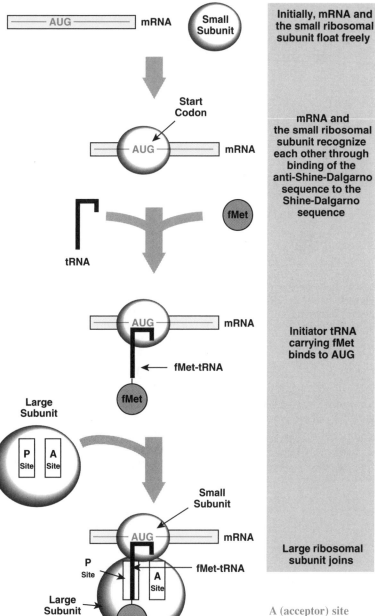

Initially, mRNA and the small ribosomal subunit float freely

mRNA and the small ribosomal subunit recognize each other through binding of the anti-Shine-Dalgarno sequence to the Shine-Dalgarno sequence

Initiator tRNA carrying fMet binds to AUG

Large ribosomal subunit joins

A (acceptor) site binding site on the ribosome for the tRNA which brings in the next amino acid

P (peptide) site binding site on the ribosome for the tRNA that is holding the growing polypeptide chain

Next, another charged tRNA arrives carrying the third amino acid. In order to fit the newcomer into the A-site we must push tRNA No. 2 sideways into the P-site. This in turn pushes the free tRNA in the P-site off the ribosome. It's rather like trying to fit three members of the circumferentially-challenged into the back seat of a small car. As the third gets in, the first one is pushed out the opposite door.

7.25 ELONGATION OF THE POLYPEPTIDE CHAIN

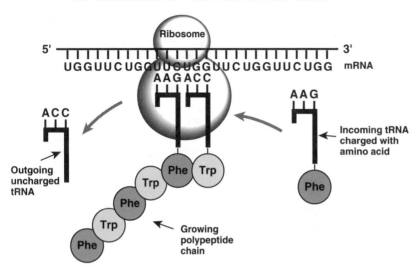

Ribosome

5' UGGUUCUGGUUCUGGUUCUGGUUCUGG 3' mRNA
AAGACC

ACC

AAG

Incoming tRNA charged with amino acid

Outgoing uncharged tRNA

Phe Trp
Trp
Phe
Trp
Phe

Phe

Growing polypeptide chain

As the peptide chain continues to grow (Fig. 7.25), it is constantly cut off from the tRNA holding it (which occupies the P = peptide site) and joined instead to the newest amino acid to be brought by its tRNA into the A-site, hence the name "acceptor" site.

The Elongation Factors

The arrival and sideways shuffling of the tRNAs on the ribosome is supervised by proteins known as **elongation factors**. Both elongation factors require a supply of energy in order to move the tRNA molecules around. The tRNA is delivered to the ribosome and installed into the A-site by elongation factor EF-T (which is actually a pair of proteins, EF-Tu and EF-Ts). The second elongation factor, EF-G, oversees moving everybody sideways at the correct time.

elongation factors proteins that oversee the elongation of a growing polypeptide chain

Termination of Protein Synthesis

Eventually we reach the end of the message. This is marked by a **stop codon**. There are three of these, UGA, UAG, and UAA. As no tRNA exists to read these three codons, the chain can no longer grow. Instead, proteins known as **release factors** read the stop signal and chop the completed polypeptide chain off the final tRNA (Fig. 7.26). This event is so unnerving to the ribosome that it goes all to pieces and falls apart into its separate subunits.

stop codon codon that signals the end of a protein

release factors proteins that supervise the release of a finished polypeptide chain from the ribosome

7.26 TERMINATION OF POLYPEPTIDE SYNTHESIS

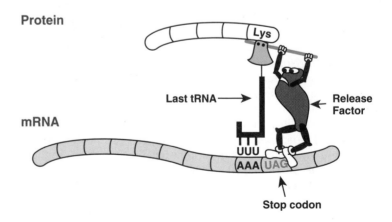

Protein

Lys

Last tRNA →

← Release Factor

mRNA

UUU
AAA UAG

Stop codon

One Messenger RNA Can Code for Several Proteins

In bacteria several proteins may be encoded by the same messenger RNA. As long as each open reading frame has its own Shine-Dalgarno sequence in front of it, the ribosome will bind and start translating. Open reading frames that are translated into proteins are sometimes known as **cistrons**; mRNA which carries several of these is therefore called **polycistronic mRNA** (Fig. 7.27).

In higher organisms this does not happen (see Ch. 11). Instead of a Shine-Dalgarno sequence, the front (5' end) of the messenger RNA molecule is recognized. The first, and only the first, open reading frame is translated.

cistron segment of DNA (or RNA) that encodes a single polypeptide chain

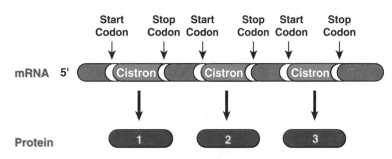

7.27 POLYCISTRONIC mRNA IN BACTERIA

polycistronic mRNA mRNA carrying multiple cistrons and which may be translated to give several different protein molecules; only found in prokaryotic (bacterial) cells

polysome group of ribosomes that bind to and translate the same mRNA

Several Ribosomes Can Read the Same Message at Once

Once the first ribosome has got moving, another can jump onto the same messenger RNA and travel along behind. In practice several ribosomes will move along the same mRNA about a hundred bases apart (Fig. 7.28). This structure is called a **polysome** (short for polyribosome).

7.28 POLYSOME – SEVERAL RIBOSOMES WORK ON ONE mRNA

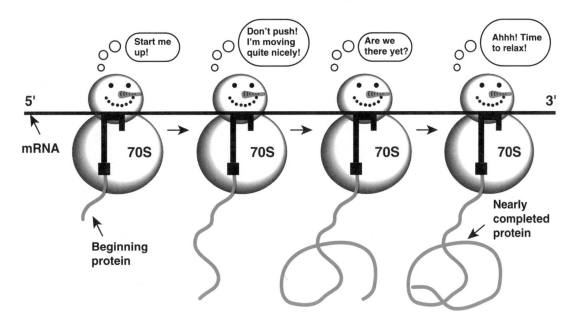

Coupled Translation and Transcription in Bacteria

When mRNA is transcribed from the original DNA template, its synthesis starts at the 5' end. The mRNA is also read by the ribosome starting at the 5' end. What this means is that the ribosome can start translating the message before the synthesis of the messenger RNA molecule has actually been finished. The result is that you find partly finished mRNA, still attached to the bacterial chromosome via RNA polymerase, with a horde of enthusiastic ribosomes already jumping aboard to get started. This is known as coupled transcription-translation (Fig. 7.29). If nature had not invented this system, Henry Ford would have.

coupled transcription-translation   when ribosomes of bacteria start translating an mRNA molecule which is still being transcribed from the genes

7.29 COUPLED TRANSLATION AND TRANSCRIPTION IN BACTERIA

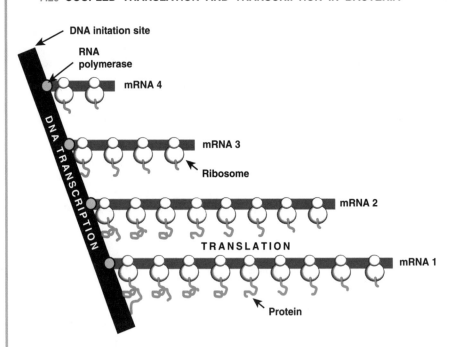

(Note: this is "verboten" in higher, eukaryotic cells, because the DNA is inside the nucleus and the ribosomes are outside.)

Some Proteins Come to a Bad End

Murphy's Law tells us that if something can go wrong, it will. Ribosomes have their problems, too. One snag they sometimes come across is when they receive a defective messenger RNA. Whether the mRNA was never properly finished or whether it was mistakenly snipped short by an over-enthusiastic RNA cutting enzyme, it can cause havoc. A ribosome that is translating a message into protein expects, sooner or later, to come across a stop codon. Even if an mRNA molecule comes to an abrupt end, ribosomes may only be released by release factor and this in turn needs a stop codon. If the mRNA is defective and there is no stop codon, a ribosome that reaches the end will just sit there forever and the ribosomes behind it will all be stalled too (Fig. 7.30).

Murphy's Law - law under which the authors of this book have operated most of their lives

7.30 RIBOSOME PILE-UP AT PREMATURE END OF mRNA

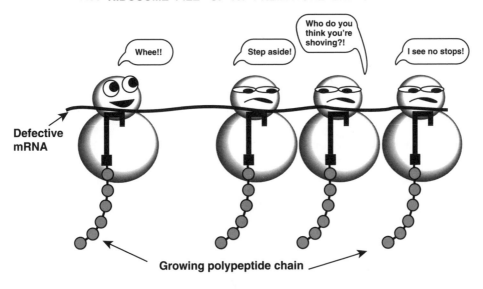

Defective mRNA

Growing polypeptide chain

So what do we do to free these trapped ribosomes? It turns out that bacterial cells contain a small but heroic RNA molecule that rescues stalled ribosomes. This was named tmRNA because it acts partly like transfer RNA and partly like messenger RNA. Like a tRNA, the tmRNA carries an amino acid, actually alanine. When it sees a stalled ribosome it binds beside the defective mRNA (Fig. 7.31). Protein synthesis now continues, first using the alanine carried by tmRNA, and then continuing on to translate the short stretch of message that is also part of the tmRNA. Finally, the tmRNA provides a proper stop codon so that release factor can disassemble the ribosome and free it for its next assignment (Fig. 7.31).

But what about the protein we just made? Clearly, it too is defective. Shouldn't it be destroyed, you intelligently ask. The tmRNA is one step ahead of you! The short stretch of amino acids specified by

tmRNA   a special RNA used to terminate protein synthesis when it finds a ribosome stalled by a bad mRNA

7.31   RIBOSOME RESCUED BY tmRNA – I

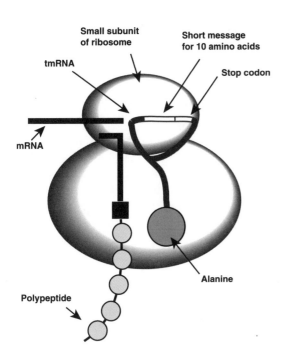

Small subunit of ribosome

Short message for 10 amino acids

tmRNA

Stop codon

mRNA

Alanine

Polypeptide

tail specific protease
enzyme that destroys mis-
made proteins by eating
them tail first

the message part of tmRNA and added to the end of the defective protein act as a signal. This is recognized by **tail specific protease** which munches all proteins carrying this mark of doom (Fig. 7.32). Perhaps we should think of tmRNA as meaning "terminate me" RNA!

7.32 RIBOSOMES RESCUED BY tmRNA - II

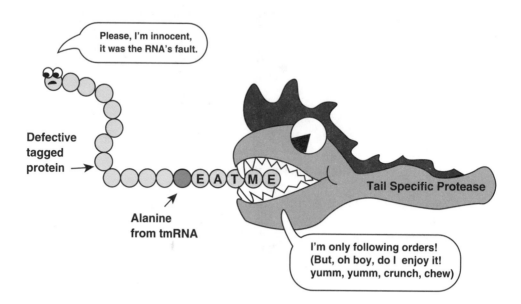

And so we conclude our introduction to the wonderful world of molecular biology. In the following chapters we will discuss the genetics of bacteria and their exploitation by molecular biologists to clone and manipulate genes, including those of higher organisms. That means you! At least we assume that those of you who are capable of reading this book regard yourselves as higher!

Additional Reading

Essentials of Molecular Biology by Freifelder D & Malacinski GM. 2nd edition, 1993. Jones & Bartlett Publishers, Boston & London

Proteins: Structures and Molecular Properties by Creighton TE. 1984. WH Freeman & Co., New York.

Introduction to Protein Structure by Brandon C & Tooze J. 1991. Garland Publishing, Inc., New York & London.

Sex Among the Low-Lifes and Its Exploitation by Molecular Biologists: Gene Transfer in Bacteria

8

No doubt you have spent many a sleepless night wondering whether or not bacteria become romantic. Is there personal fulfillment at the level of the single cell (Fig. 8.1)?

Biologists think that sex serves the purpose of reshuffling genetic information in the hope of producing offspring with combinations of genes superior to those of either parent. Because molecular biologists use bacteria as tools to carry most cloned genes, whether they are originally from corn or cockroaches, we must understand how bacteria transfer genetic information from one to another.

Before we start, it is important to realize that sex and reproduction are not at all the same thing. In animals, reproduction normally involves sex, but in bacteria, and even in plants, these are two distinct processes.

8.1 SEX AMONG BACTERIA

Oprah wants us to reveal our sex lives on slime-time TV!

What consenting adult cells do in a culture dish, is their own business!

Bacteria divide by binary fission. First they replicate their single chromosome and then the cell elongates and divides down the middle (as discussed in Ch. 3). No resorting of the genes between two individuals (that is, no sex) is involved and so this is known as asexual or vegetative reproduction.

binary fission simple form of cell division found among bacteria; accomplished by splitting down the middle

Naked DNA - Transformation

The broadest possible definition of sex means that genetic material is transferred from one partner to the other. The simplest conceivable version of sex would then consist of transferring pure DNA from one cell to another.

transformation changing the properties of a bacterial cell as a result of the uptake of pure DNA

Transformation was first observed by Oswald Avery in 1944 and provided the proof that purified DNA carries genetic information and, therefore, that genes are made of DNA. Avery used DNA extracted from virulent strains of pneumonia. He purified it and added it to harmless strains of the same bacterial species. Some of the harmless bacteria took up the DNA and were changed into virulent strains. So Avery called this transformation.

Warning: Cancer specialists use the same term, "transformation," to refer to the changing of a normal cell to a cancer cell, even though in most cases no extra DNA enters the cell.

competent capable of taking up pure DNA from the external medium

electroporator device used to apply high voltage to cells in order to make them permeable to DNA

Believe it or not, among the bacteria this is possible. Bacterial cells can take up naked DNA molecules and incorporate the genetic information they carry. This is referred to as transformation (Fig. 8.2). But, please note that no actual cell to cell contact is allowed in transformation.

8.2 PRINCIPLE OF TRANSFORMATION

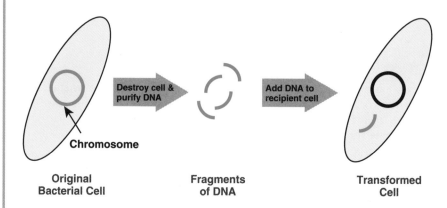

Chromosome

Original Bacterial Cell

Destroy cell & purify DNA

Fragments of DNA

Add DNA to recipient cell

Transformed Cell

Transformation is Used in Genetic Engineering

After genes or other useful segments of DNA have been cloned in the test tube, it is almost always necessary to put them into some bacterial cell for analysis or manipulation. Thus, laboratory transformation techniques are an essential tool in genetic engineering.

Some bacteria readily take up outside DNA. If they can do this, they are said to be "competent." Other bacteria must be brutalized in the laboratory before they will cooperate. There are two ways of doing this, both reminiscent of means used to get information out of political prisoners or early treatments of the mentally ill. The older method is to chill the bacterial cells in the presence of chemicals that damage their cell walls and then to heat shock them briefly. This loosens the structure of the cell walls and allows DNA, a huge molecule, to enter. The modern, high-tech, method is electroshock treatment. Bacteria are placed in a machine called an "electroporator" and zapped with a high voltage discharge. See the poor bacteria in Figure 8.3.

Real Life Transformation

Does transformation actually happen in real life? Yes, probably it does, but only at a very low level. From time to time, bacteria in natural habitats die and disintegrate. In doing so they release DNA which nearby cells may take up. Some sleazy bacteria simply take up any old DNA they find lying around. In practice most bacteria need the kind of "friendly persuasion" described above before they will take up alien DNA.

8.3 ELECTROPORATION OF A HELPLESS BACTERIUM

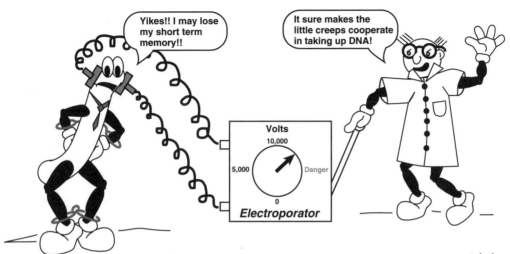

What Happens to the DNA After Uptake?

There are two possibilities, failure or success, known technically as restriction (see Ch. 9) or recombination (see Ch. 5). Both of these processes apply to a wide range of other situations too, so they are dealt with in detail elsewhere. In brief:

Restriction is the destruction of the incoming foreign DNA. Most bacteria assume that foreign DNA is more likely to come from an enemy, such as a virus, than from a friend and they chop it up into small fragments with so-called restriction enzymes. In this case transformation fails. Only DNA that has been modified by closely related bacteria by adding the correct chemical tags is accepted as friendly.

Recombination is the physical incorporation of some of the incoming DNA into the bacterial chromosome (Fig. 8.4). If this happens, some of the host cell's genetic information is replaced by genes from the incoming DNA and the bacteria are permanently transformed. The original version of these genes is lost. If a gene enters a bacterial cell on a fragment of linear DNA, it must be recombined onto the host chromosome in order to survive.

If the incoming DNA is part of a plasmid which can replicate on its own, recombination into the chromosome is not necessary. In practice, it is usually more convenient to avoid recombination.

restriction destruction of incoming foreign DNA by a bacterial cell

recombination merging of genes from two separate molecules of DNA (see Ch. 5 for details)

restriction enzyme enzyme that cuts DNA in the middle of the strand and at a specific site

modified refers here to DNA chemically tagged by adding methyl groups to signal that it is host cell DNA

plasmid circular molecule of double helical DNA which replicates independently of the bacterial cell's chromosome

8.4 RECOMBINATION OF INCOMING DNA

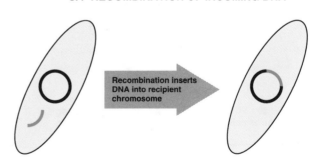

Consequently, molecular biologists normally put the genes they are working with onto plasmids (see Ch. 9 on using plasmids for genetic engineering).

Hitchhiking by Virus - Transduction

When a virus succeeds in infecting a bacterial cell it manufactures more virus particles, each of which should contain a new copy of the viruses own genes. But life is rarely perfect, and sometimes mistakes happen, even to sneaky viruses. Occasionally, instead of packaging virus DNA into the virus particle, fragments of bacterial DNA get packaged. From the viewpoint of the virus, this results in a defective particle. Nonetheless, such a virus, carrying bacterial DNA, may go on to infect another bacterial cell. If so, instead of injecting viral genes, it injects DNA from the previous bacterial victim. This DNA can be destroyed by restriction or incorporated by recombination, as in the case of transformation (see above). If it is successfully incorporated, then **transduction** has occurred.

Bacterial geneticists routinely carry out gene transfer between different bacteria by transduction using bacterial viruses, known as **bacteriophage** (or **phage** for short). If the bacterial strains are closely related, the incoming DNA is accepted as "friendly" and is not destroyed by restriction. In practice, transduction is the simplest way to replace a few genes of one bacterial strain with those of a close relative.

To carry out a transduction, a bacteriophage is grown on a culture of the donor bacterial strain. These bacteria are destroyed by the phage, leaving behind only DNA which carries some of their genes and is now packaged inside the phage particles. This phage sample can be stored in the fridge for weeks or months before use, rather like deep frozen sperm used in artificial insemination. Later, the phage are mixed with a recipient bacterial strain and the DNA is injected. Most recipients get genuine phage DNA and are killed. However, others get donor bacterial DNA and are successfully transduced (Fig. 8.5).

transduction transport of genes from one cell to another inside a virus particle

bacteriophage or **phage** any virus that infects bacteria

8.5 MECHANISM OF GENERAL TRANSDUCTION

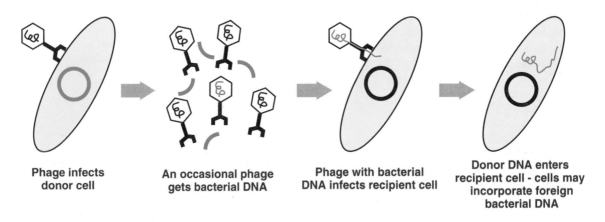

Phage infects donor cell

An occasional phage gets bacterial DNA

Phage with bacterial DNA infects recipient cell

Donor DNA enters recipient cell - cells may incorporate foreign bacterial DNA

The best known examples are the use of phages to transduce the molecular biologists' favorite bacterium, *Escherichia coli*. Different bacteriophages behave differently. The two favorite bacterial phages are **P1** and **Lambda**.

In **generalized transduction**, random fragments of bacterial DNA are picked up by the virus; for example by bacteriophage P1. All bacterial genes have an equal chance of being transferred. P1 makes a mistake by packaging bacterial DNA instead of its own only about once every 10,000 times. (Don't sneer. You can't do anything anywhere near that accurately and neither can I.)

Each P1 particle can carry 90 kb of DNA which is equivalent to about 2 percent of a bacterial chromosome. So any individual gene will be transduced by one in 500,000 of the P1 particles resulting from any particular infection. In practice a typical sample of P1 contains about a thousand million virus particles per milliliter so there is actually plenty of opportunity for transduction to happen.

In **specialized transduction**, certain specific regions of the bacterial chromosome are favored. For example, when bacteriophage Lambda (λ to you classical scholars) infects *E. coli*, it sometimes inserts its DNA into the bacterial chromosome (Fig. 8.6). This occurs at a single specific location known as the **lambda attachment site** (*att* λ).

8.6 INSERTION OF LAMBDA INTO A CHROMOSOME

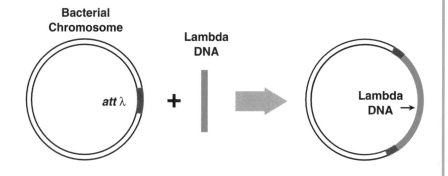

When Lambda multiplies, the original donor cell is destroyed, and several hundred virus particles containing Lambda DNA are produced. However, just as with P1, a small fraction of virus particles end up containing bacterial DNA. There are two differences from the case of P1. First, the **transducing particles** contain a mixture of Lambda DNA and chromosomal DNA. Second, only chromosomal genes next to the Lambda attachment site are transduced by Lambda (Fig. 8.7).

O.K., But Is There Real Sex in Bacteria?
Caution: This section is rated PG13 and should only be read by minors if supervised by a responsible adult!!

We've given you naked DNA and viral transmission of genetic information. But, you ask, what about genuine sexual contact between bacteria?

---

bacteriophage P1 a virus that infects *E. coli* and which is widely used for transduction by bacterial geneticists

bacteriophage Lambda (or λ) a virus of *E. coli* sometimes used for transduction but more often as a cloning vector in genetic engineering

generalized transduction transduction in which the genes transported are picked at random

90 kilobase pairs (90 kb) is 90,000 base pairs of DNA. Since bacterial genes are roughly 1,000 base pairs long, 90 kb of DNA could carry 90 genes.

specialized transduction transduction in which a few specially selected genes are transported

lambda attachment site (*att* λ) specific site on the chromosome of *E. coli* at which λ phage inserts its DNA

transducing particle virus particle containing host cell DNA instead of the viruses own genes

Modern attempts at gene therapy sometimes use a form of specialized transduction. A human virus, engineered so it is no longer dangerous, has the healthy version of a human gene loaded into it. This may then be used to infect the victim of a hereditary disease. See Chapter 15 for some examples.

8.7 SPECIALIZED TRANSDUCTION BY LAMBDA

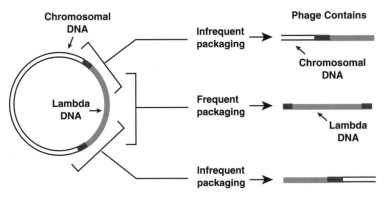

Well, yes, - blush!! - there is and it is known as **conjugation**. This involves two cells, a female or recipient cell, and a male or donor cell. The male has a long, hollow, tubular organ referred to as a **sex pilus**. Shame, shame! But, the pilus is not what you think! The male cell uses the pilus as a grappling hook to grab the female and pull her alongside, rather like a boat hook is used to grab a dinghy (Fig. 8.8). The two cells then form a **conjugation bridge** where they touch (Fig. 8.9), and DNA goes from the male into the female. In practice, mating bacteria snuggle together in groups of five to 10.

conjugation transfer of genes between bacteria involving cell to cell contact

sex pilus long, thin, helical rod of protein that a male cell uses to catch hold of a female cell

conjugation bridge channel that forms where the cell envelopes of mating bacteria touch and fuse together and through which DNA is transferred

8.8 FORMATION OF MATING PAIRS

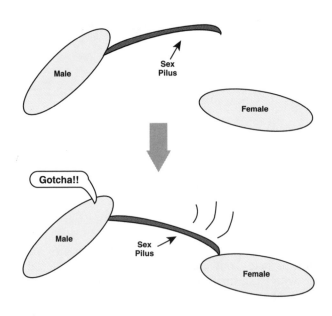

transfer (tra) system cluster of genes that code for the ability of a plasmid to transfer itself from one cell to another

plasmid circular molecule of double stranded helical DNA that replicates independently of the host cell's chromosome

Bacteria are mostly female. To be a male bacterium you need a personal improvement kit. This is known as the *tra* (transfer) system and the genes for it come on a separate DNA molecule known as a **plasmid**. Plasmids are circular molecules of DNA that can replicate in bacterial cells rather like miniature chromosomes. However, they are much smaller than bacterial chromosomes and are not essential for cell growth and survival under normal conditions. Plasmids may carry a variety of genes that confer extra abilities on the bacteria containing them.

8.9 FORMATION OF A CONJUGATION BRIDGE

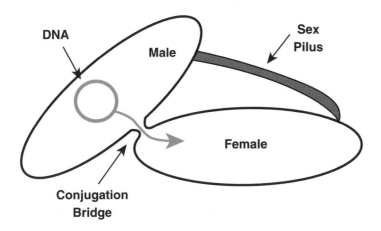

Plasmids which make a cell male are called **fertility plasmids**; the most famous of these is the **F-plasmid** of *E. coli* which is shown in Figure 8.10. Sometimes the male/donor cells are known as F⁺ and the female/recipient cells as F⁻ to indicate that their role in conjugation is determined by the presence or absence of the F-plasmid.

8.10 AN F-PLASMID WITH A *TRA* SYSTEM

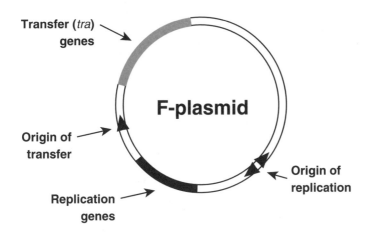

Replication During Plasmid Transfer

We have talked about plasmid transfer as if the whole F-plasmid simply leaves the original male cell and moves into the recipient cell. In fact, only one strand of the F-plasmid DNA is transferred. The details are as follows (Fig. 8.11):

1) One of the two strands of the double stranded DNA of the F-plasmid opens up at the **origin of transfer**.

fertility plasmid type of plasmid that confers ability to mate on its bacterial host

F-plasmid a particular fertility plasmid which confers ability to mate on its bacterial host, *Escherichia coli*

origin of transfer (*oriT*) site on a plasmid where the DNA is nicked just before transfer begins. The origin of transfer enters the recipient cell first

8.11 F-PLASMID REPLICATES DURING CONJUGATION

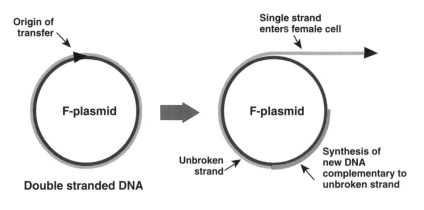

Origin of transfer

F-plasmid

Double stranded DNA

Single strand enters female cell

F-plasmid

Unbroken strand

Synthesis of new DNA complementary to unbroken strand

2) This linearized single strand of DNA moves through the conjugation bridge into the female cell. 3) An unbroken single stranded circle of F-plasmid DNA remains inside the donor cell. This is used as a template for the synthesis of a new second strand to replace the one that just left. 4) As the linearized single strand of F-plasmid DNA enters the female cell, a new complementary strand of DNA is made using the incoming strand as template.

One truly bizarre result of all this is that when the female cell has received the F-plasmid it becomes F^+, in other words, a male! Consequently, bacteria stay together for an even shorter time than most Hollywood marriages!

How Are Chromosomal Genes Transferred?

Although many plasmids allow the cells carrying them to conjugate, usually only the plasmid itself is transferred through the conjugation bridge. Much rarer is the ability of plasmids to carry genes from the host chromosome along with them when they move from one bacterial cell to another.

In order to transfer chromosomal genes, a plasmid must first physically integrate itself into the chromosome of the bacterium. The process of integration needs pairs of identical (or nearly identical) DNA sequences, one on the plasmid and the other on the bacterial chromosome. A variety of different possible **insertion sequences** or **IS-sequences** exist (see Ch. 20 for more on insertion sequences). The chromosome of *E. coli* has seven copies of IS1, 13 copies of IS2 and six copies of IS3 scattered around it more or less at random. The F-plasmid, which is roughly one-fiftieth as big, has three insertion sequences (Fig. 8.12). Two of these are identical copies of IS3 and the third is a single copy of IS2.

Consequently, integration of F can occur at the IS2 or IS3 sites, a total of 19 sites scattered around the chromosome. Integration of the F-plasmid may occur in either orientation (Fig. 8.13).

When an F-plasmid that is integrated into the chromosome is transferred by conjugation, it drags along the chromosomal genes to which it is attached (Fig. 8.14). (Just as before, only a single strand of the DNA moves and the recipient cell has to make the complementary strand itself.)

insertion sequences or **IS-sequences** special mobile chunks of DNA. Integration of a plasmid may occur at matching IS-sequences

8.12 INSERTION SEQUENCES ON F - PLASMID

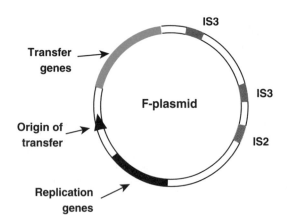

IS3

Transfer genes

F-plasmid

IS3

IS2

Origin of transfer

Replication genes

8.13 F-PLASMID INSERTS INTO BACTERIAL CHROMOSOME

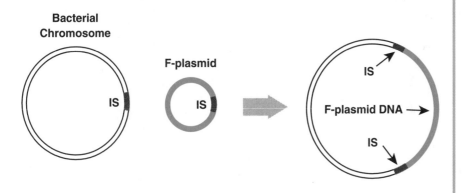

Bacterial Chromosome

F-plasmid

IS

IS

IS

F-plasmid DNA →

IS

Hfr-strain bacterial strain with F-plasmid integrated into the chromosome, allowing high frequency transfer of chromosomal genes

transmissible antibiotic resistance resistance to an antibiotic which is carried on a transferable plasmid

R-plasmids or R-factors plasmids that confer antibiotic resistance on their bacterial hosts

Consequently, bacterial strains with an F-plasmid integrated into the chromosome are known as **Hfr-strains** because they transfer chromosomal genes at high frequency.

A prolonged mating of 90 minutes or so is needed to transfer the whole chromosome. More often, bacteria break off after a shorter period of, say, 15 to 30 minutes, and only part of the chromosome is transferred. Since different Hfr-strains have their F-plasmids inserted at different sites on the bacterial chromosome, they start their transfer of chromosomal genes at different points.

How Were Plasmids Discovered?

Plasmids were discovered in Japanese bacteria just after World War II. They were responsible for the problem known as **transmissible antibiotic resistance.** Dysentery, due to bacteria, was originally treated with sulfonamides, the earliest type of antibiotic. However, it wasn't long before humans were on the run again as bacteria resistant to these antibiotics began to appear! What was far, far, worse was that once resistance had arisen, it was transferred from one strain of bacteria to another at a high frequency. It turned out that the genes for antibiotic resistance were being carried from one bacterium to another on plasmids. Plasmids that confer antibiotic resistance are called **R-plasmids** or **R-factors** (Fig. 8.15).

8.14 CONJUGATIONAL TRANSFER OF CHROMOSOMAL DNA

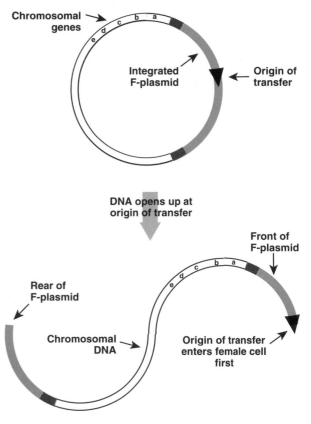

Chromosomal genes

Integrated F-plasmid

Origin of transfer

DNA opens up at origin of transfer

Front of F-plasmid

Rear of F-plasmid

Chromosomal DNA

Origin of transfer enters female cell first

8.15 A TYPICAL R - PLASMID

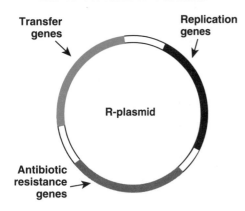

Transfer genes

Replication genes

R-plasmid

Antibiotic resistance genes

Shigella the bacterium which causes one form of dysentery

A major factor in the spread of resistance plasmids among bacteria is the practice of feeding antibiotics to pigs and chickens in order to increase the yield of meat.

multiple antibiotic resistance simultaneous resistance to several antibiotics, often carried by a single plasmid

vegetative replication type of replication which occurs in the absence of plasmid transfer

host range the range of different types of bacteria which a plasmid (or a virus) can infect

enteric bacteria a family of related bacteria often found in the intestines of animals

By 1953, the year Watson and Crick discovered the double helix, 80 percent of the dysentery-causing bacterium, *Shigella*, found in Japan had become resistant to sulfonamides. A single plasmid may carry genes for resistance to more than one antibiotic. By 1969, a third of the *Shigella* in Japan were resistant to four antibiotics: sulfonamides, chloramphenicol, tetracycline and streptomycin.

Today, the transfer of **multiple antibiotic resistance** plasmids between bacteria has become a major clinical problem. Patients with infections after surgery or with severe burns that have become infected are most at risk.

General Properties of Plasmids

Plasmids are circular DNA molecules that can replicate independently of the bacterial chromosome. They can only multiply inside a bacterial host cell. They carry genes for managing their own life cycles and also, usually, genes that affect the properties of the host cell. These properties vary greatly from plasmid to plasmid, the best known being resistance to various antibiotics.

Because of their unique properties, plasmids are invaluable to the molecular biologist and are used to carry genes for genetic engineering. Consequently, a variety of plasmids, modified for different purposes, are widely used in all molecular biology labs.

When the cell it inhabits divides, the plasmid must divide too. The plasmid replicates itself in step with the host chromosome so, at cell division, each daughter cell gets a copy of the plasmid as well as its own chromosome (Fig. 8.16). This **vegetative replication** is quite distinct from the type of replication that happens during plasmid transfer (see above). Vegetative replication starts at *oriV,* the origin of vegetative replication, which is at a different site on the plasmid from *oriT,* the origin used during transfer. All plasmids must have a vegetative origin since they must all divide to survive. But only those plasmids which can transfer themselves have a special transfer origin.

8.16 PLASMIDS REPLICATE IN STEP WITH CELL DIVISION

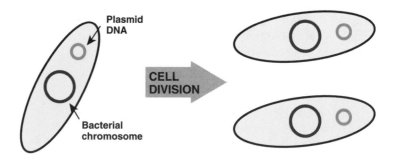

Plasmid DNA

CELL DIVISION

Bacterial chromosome

The **host range** of plasmids varies widely. Some plasmids are restricted to a few closely related bacteria; for example, the F-plasmid only inhabits *E. coli* and related **enteric bacteria** like *Shigella* and

Salmonella. Others have a wide host range; for example, plasmids of the P-family can live in hundreds of different types of bacteria. P-type plasmids were originally discovered in bacteria called *Pseudomonas*, which sometimes infect patients with severe burns. They are often responsible for resistance to multiple antibiotics including penicillins.

When a plasmid settles down to live in a bacterial cell it becomes very possessive of its home. The resident plasmid keeps out other closely related plasmids. Thus, two plasmids belonging to the same family cannot coexist peacefully in the same bacterial cell. This is called incompatibility (Fig. 8.17) and the families are known as incompatibility groups and are designated by letters of the alphabet; *e.g.,* F-type plasmids include the F-plasmid and its relatives. Plasmids of the same incompatibility group have almost identical DNA sequences in their genes for replication, although the genes they carry for optional characteristics may be very different.

incompatibility the inability of two related plasmids to live together in the same bacterial cell

incompatibility group family of related plasmids. Two members of the same family cannot inhabit the same cell simultaneously

8.17 PLASMID INCOMPATIBILITY

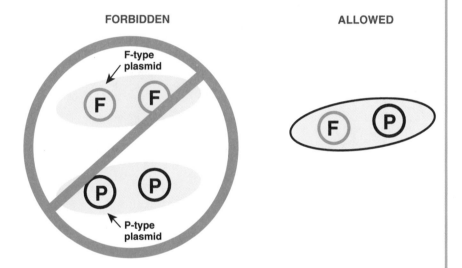

It is perfectly possible to have two, or even more, plasmids in the same cell as long as they belong to different families. So a P-type plasmid will happily share the same cell with an F-type.

The copy number is just what it sounds like, the number of copies of the plasmid in each bacterial cell. It is usually one or two plasmids per chromosome (as with for F- and P-plasmids) but may be as many as 50 or more in certain cases (such as ColE-plasmids). The number of copies has a major effect on the strength of plasmid-borne characters, especially on their antibiotic resistance. The more copies of the plasmid per cell, the more copies there will be of the antibiotic resistance genes and the higher the resulting level of antibiotic resistance.

The size of plasmids varies enormously. The F-plasmid is fairly average in this respect and is about 1 percent the size of the *E. coli* chromosome. Most high copy plasmids are much smaller (ColE-plasmids are about

copy number number of copies of a plasmid in each host cell

ColE-plasmid a small, high copy plasmid which carries genes for a toxin known as colicin E (see Ch. 20)

Plasmids, such as the ColE-plasmid, with very high copy numbers are handy in genetic engineering since they will produce a lot more of whatever it is you have cloned

10 percent the size of the F-plasmid). Very large plasmids, up to 10 percent of the size of a chromosome, are sometimes found but they are difficult to work with and few have been properly characterized.

Movement of Plasmids

transferability ability of a plasmid to move itself from one cell to another

Transferability is the ability of certain plasmids to transfer themselves from one bacterial cell to another. To do this they need to manufacture a sex pilus and form a conjugation bridge with a suitable recipient cell. Many medium size plasmids such as the F-type and P-type plasmids, can do this and are referred to as Tra^+ (transfer positive).

Since plasmid transfer requires the operation of a large number of genes, only medium or large plasmids possess this ability. Very small plasmids such as the *ColE*-plasmids, simply do not have enough DNA to carry the genes needed.

mobilizability ability of a small plasmid to move into another bacterial cell if a self-transferable plasmid assists it

Although small plasmids cannot transfer themselves, they can sometimes hitch a ride with larger plasmids, a property known as mobilizability. For example, the ColEl plasmid can be mobilized by the F-plasmid. Some, but not all, non-self-transferable plasmids can be mobilized.

Now that we can move genes among bacteria, let's move on to some real genetic engineering. In the next chapter we'll actually clone some genes. Yee-hah!!

Additional Reading

Microbial Genetics by Maloy SR, Cronan JE, & Freifelder D. 2nd edition, 1994. Jones & Bartlett Publishers, Boston & London

Bacterial and Bacteriophage Genetics by Birge EA. 3rd edition, 1994. Springer-Verlag, New York, Berlin & Heidelberg

Messing About With DNA

9

Suppose we want to create our very own monster by genetic engineering. How do we go about it? Frankenstein made his monster by sewing a brain into a body and then charging up his creation with a lightning bolt. Genetic engineers make patchwork organisms not by joining organs but by splicing genes together. So let's get started on some basic operations.

First Catch Your Mouse

The first step is to get hold of some DNA. Since all living organisms have genes, they all have DNA which you can extract. If you intend to tackle something large like a tree or a hippopotamus, it is only necessary to hack off a small sample. Nowadays, genetic manipulations using laboratory mice or rats are often done by snipping off a piece from the end of the tail.

However, by far the easiest place to get DNA is from bacteria. A few drops of a bacterial culture will give plenty of DNA for most purposes and since the bacteria are single cells and contain no bone, fat, gristle, etc., the DNA is relatively easy to get out. For animals and plants we must grind our sample into tiny fragments before proceeding.

Next we break open the cells. This may be done mechanically in a blender or it may be done using chemicals to degrade and dissolve the components of the cell walls. For bacteria, we usually use a mixture of lysozyme that digests the outer layer of the cell wall, followed by a detergent that dissolves the greasy cell membrane. For a mouse's tail we use enzymes that degrade the connective tissue and disperse the cells. The DNA is then liberated into solution and is purified by a further series of steps.

Feel free, help yourself

Snip!

lysozyme an enzyme that breaks down the tough, structural layer of bacterial cell walls.

detergent molecule that binds grease at one end and uses its other end to dissolve in water

Purifying the DNA

Two general types of procedure are used here, centrifugation and chemical extraction. First we centrifuge, then we extract.

Centrifugation is a routine technique in molecular biology labs. The basic idea is quite simple: the sample is spun at high speed and the

phenol a corrosive chemical liquid which dissolves proteins very enthusiastically

centrifugal force causes the larger or heavier components to sediment to the bottom of the tube (Fig. 9.2). For example, after destroying the cell wall of bacteria by lysozyme and detergents, we are left with a solution containing the fragments of the wall, which are small, and the DNA, which is a gigantic molecule. So the next step is to centrifuge. The DNA and some other large components are hurled to the bottom of the tube and the garbage from the destruction of the cell wall remains in solution and is poured down the drain.

9.2 PRINCIPLE OF CENTRIFUGATION

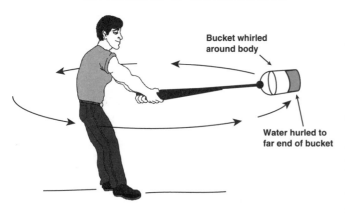

Bucket whirled around body

Water hurled to far end of bucket

The sedimented DNA is then redissolved. However, it still has lots of proteins and RNA mixed in with it. We extract these by chemical means. For example, one step used in almost all DNA purifications uses the chemical **phenol**. Phenol, also known as carbolic acid, is very corrosive and extremely dangerous. The reason for this is that it dissolves and denatures the proteins which make up 60 to 70 percent of all living matter.

If you spill phenol on yourself it will extract and destroy your proteins and cause painful burns which heal very slowly.

If you add phenol to a sample of DNA it will dissolve and remove all of the proteins in your sample, and in doing so, purify the DNA. When phenol is added to water, the two liquids do not mix to form a single solution; instead the denser phenol forms a separate layer below the water. When shaken, the two layers mix temporarily; when the shaking stops, the layers separate again. The phenol dissolves almost all of the protein, but the nucleic acids, DNA and RNA, stay in the water layer (Fig. 9.3). To ensure good separation of the layers we centrifuge briefly. Then we suck off and keep the water layer with the DNA and RNA.

9.3 PHENOL EXTRACTION OF DNA

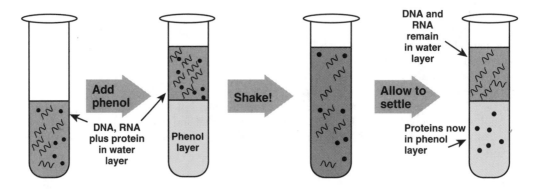

Add phenol

DNA, RNA plus protein in water layer

Phenol layer

Shake!

DNA and RNA remain in water layer

Allow to settle

Proteins now in phenol layer

ribonuclease an enzyme that breaks down RNA

Getting Rid of the RNA

To get rid of the RNA we use an enzyme that degrades RNA, **ribonuclease**. This converts the RNA into tiny fragments but leaves the DNA unchanged as a giant macromolecule.

96

We now add an equal volume of alcohol. The alcohol dissolves in the water so enthusiastically that it occupies all of the water and pushes the larger and less soluble DNA out of solution. However, the small RNA fragments remain dissolved. We now centrifuge again and the DNA is sedimented to the bottom of the tube and we can pour off the solution containing the RNA fragments (Fig. 9.4). The tiny pellet of DNA left at the bottom of the tube is often scarcely visible. Nonetheless, it contains billions of DNA molecules, sufficient for most experiments. This DNA is redissolved and is now ready for use in genetic engineering.

9.4 REMOVAL OF RNA AND SEDIMENTATION OF DNA

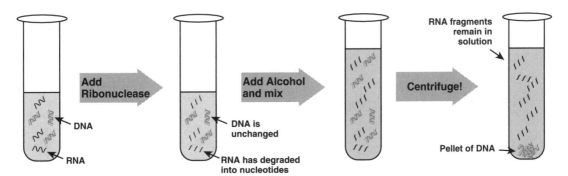

Cutting Up the DNA

The DNA isolated from a typical bacterial cell is a giant circular chromosome with about 3,000 genes on it. In genetic engineering we normally work with one or two genes, or only a fragment of a gene, at a time. So we need to cut the DNA into pieces of a manageable size. This is done by a special type of enzyme known as a restriction enzyme which acts as molecular scissors.

Restriction enzymes all bind to DNA at a specific sequence of bases, the recognition site. This base sequence is usually four, six or eight bases long and the bases form an inverted repeat. In an inverted repeat, the sequence on the top strand of the DNA is the same as the sequence of the bottom strand read in the reverse direction, as shown in Figure 9.5.

Each restriction enzyme has its own specific recognition site. Since any random series of four bases will be found quite frequently, four base-recognizing enzymes cut DNA into lots of very short pieces.

restriction enzyme an enzyme which binds to DNA at a specific base sequence, and then cuts the DNA

recognition site specific base sequence where a restriction enzyme binds

inverted repeat length of DNA which has the same sequence if read in the opposite direction (but on the other strand)

9.5 INVERTED REPEAT

5' -- W - X - Y - Z - z' - y' - x' - w' -- 3'

3' -- w' - x' - y' - z' - Z - Y - X - W -- 5'

Note: w', x' y', and z' refer to whichever bases are complementary to w, x, y, and z respectively.

Conversely, finding a particular eight-base sequence is rare, so eight-base recognizing enzymes cut DNA only at relatively long intervals and generate a few large pieces. The six-base enzymes are the most convenient in practice, as they give an intermediate result. Two examples of widely used restriction enzymes and their sites are shown in Figure 9.6

9.6 SOME REAL RECOGNITION SITES FOR RESTRICTION ENZYMES

5' -- G - G - A - T - C - C -- 3'   5' -- G - A - A - T - T - C -- 3'

3' -- C - C - T - A - G - G -- 5'   3' -- C - T - T - A - A - G -- 5'

Recognition site for BamH1   **Recognition site for EcoR1**

Where is the DNA Cut?

You might think the DNA should be cut at the recognition site where the restriction enzyme binds. This is often true, but not always. There are two classes of restriction enzyme:

Type I restriction enzymes cut the DNA a thousand or more base pairs away from the recognition site. This is done by looping the DNA so the enzyme can get a grip on it both at the recognition site and the cutting site (Fig. 9.7). Since the exact length of the loop is not constant, and since the base sequence at the cut site is not fixed, these enzymes are not of much use to molecular biologists. Even more bizarre is that these enzymes are suicidal. Most enzymes carry out the same reaction over and over again on a continual stream of target molecules. Type I restriction enzymes perform a kamikaze attack and when they cut DNA, the enzymes are destroyed as well!

9.7 TYPE I RESTRICTION ENZYME

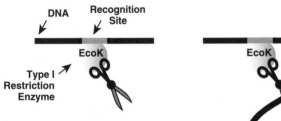

Type II restriction enzymes cut the DNA in the middle of the recognition site. Since the exact position of the cut is known, these are the restriction enzymes which are normally used in genetic engineering. There are two different ways of cutting the recognition site in half (Fig. 9.8).

9.8 TYPE II RESTRICTION ENZYMES

5' -- G - T - T - A - A - C -- 3'   5' -- G - A - A - T - T - C -- 3'

3' -- C - A - A - T - T - G -- 5'   3' -- C - T - T - A - A - G -- 5'

cut by Hpa1   cut by EcoR1

5' -- G - T - T | A - A - C -- 3'   5' -- G   A - A - T - T - C -- 3'

3' -- C - A - A | T - T - G -- 5'   3' -- C - T - T - A - A   G -- 5'

Blunt Ends   **Sticky Ends**

One way is to cut both strands of the double stranded DNA at the same point. This leaves **blunt ends** as shown in Figure 9.8. The alternative is to cut the two strands in different places which generates overhanging ends. Because these overhanging ends are "unmarried," they will base pair with each other and they are known as **sticky ends.**

blunt ends ends of a DNA molecule which are fully base paired

sticky ends ends of a DNA molecule with short single stranded overhangs

Old genetic engineers don't die - they just come to sticky ends!

9.9 STICKY ENDS

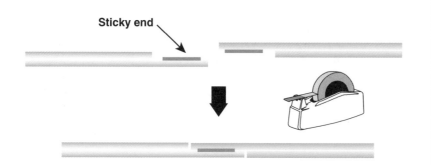

Sticky end

DNA ligase an enzyme that joins DNA fragments end to end

Enzymes that generate sticky ends are far more useful. This is because if two different pieces of DNA were cut with the same restriction enzyme, the same sticky ends would be generated on both fragments. This allows the two different pieces of DNA to be bound together by matching the sticky ends (Fig. 9.9). Such pairing is temporary since the pieces of DNA are only held together by hydrogen bonding between the base pairs, not by permanent covalent bonds. Nonetheless, this gives time for the permanent bonding of the sugar-phosphate backbone.

How are Fragments of DNA Joined Together?

Another type of enzyme is used to join DNA fragments permanently. This is **DNA ligase,** the same enzyme we met in Chapter 5 where it joined up the fragments of the lagging strand during DNA replication. If DNA ligase finds two DNA fragments touching each other, end to end, it will join them together (Fig. 9.10). In practice, pieces of DNA with matching sticky ends will tend to stay attached much

9.10 LIGATION OF STICKY ENDS

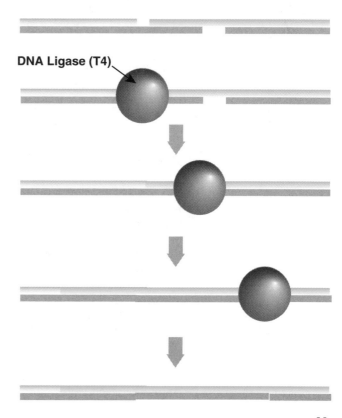

DNA Ligase (T4)

of the time and consequently DNA ligase will join them permanently without much trouble. Since DNA fragments with blunt ends have no way to hold on to each other, they drift apart most of the time. Ligating them takes a very long time and a lot of DNA ligase, and even then it is very inefficient. In fact, bacterial ligase cannot join blunt ends at all. In practice genetic engineers normally use T4 ligase. Originally this came from the T4 bacterial virus and is both easier to use in the test tube and can join blunt ends if need be.

bacterial ligase the DNA ligase found in bacterial cells

T4 ligase the form of DNA ligase coded for by a gene belonging to bacteriophage T4

Where Do Restriction Enzymes Come From?

$$$ You buy them $$$! Several hundred restriction enzymes with different recognition sites are now available commercially. O.K., wise guy, so where did they come from originally?

9.11 RESTRICTION ENZYMES DESTROY VIRUS

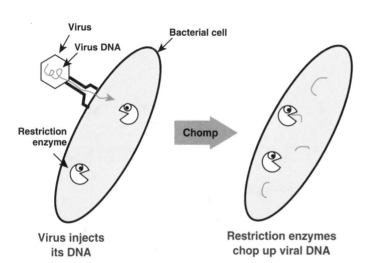

Virus injects its DNA

Restriction enzymes chop up viral DNA

When bacteria are attacked by viruses, the virus coat is left outside and only the virus DNA enters the target cell (Ch. 2; Fig. 2.9). The virus DNA will take over the victim's cellular machinery and use it to manufacture more virus particles unless the bacterial cell fights back. The bacteria usually fight back by chopping up the DNA of the virus, thereby destroying it. This "restricts" the entry of the virus and so the enzymes that chopped up the DNA were called restriction enzymes. In practice most viruses get wasted by the victims they attack (Fig. 9.11). Only a few survive and succeed in conquering the bacteria.

Where Do Specific Restriction Enzymes Get Their Funny Names?

Restriction enzymes have names made up of the initials of the type of bacteria they come from. Let's take EcoRI. EcoR means that the enzyme was found in *Escherichia coli* strain RY13 and the "I" ("one" not "eye") means it was the first restriction enzyme found in this strain.

EcoRI a restriction enzyme found in *E. coli*

Protection of the Cell's Own DNA

Why don't restriction enzymes destroy the DNA of their own bacterial cell? In practice, whenever a bacterial cell makes a restriction enzyme, it also makes a "kinder and gentler" protein known as a modification enzyme. This chemically modifies the DNA of the bacterial cell. Modification enzymes bind to the DNA at the same recognition site as the corresponding restriction enzymes. They then add a molecular tag to one of the bases in the recognition sequence (Fig. 9.12). Once the recognition site has been tagged or "modified," the restriction enzymes can no longer cut it. The result is that all the DNA in a bacterial cell is immune to that

modification enzyme an enzyme that alters a base in the recognition site of a restriction enzyme. This protects the DNA from being cut

9.12 MODIFICATION ENZYMES

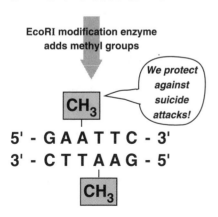

5' - G A A T T C - 3'
3' - C T T A A G - 5'

EcoRI modification enzyme adds methyl groups

We protect against suicide attacks!

CH₃

5' - G A A T T C - 3'
3' - C T T A A G - 5'

CH₃

cell's own restriction enzymes. Only foreign DNA that tries to invade the cell from outside is attacked. Clever, huh?

How Can Individual Fragments of DNA Be Separated Out?

After cutting up a long piece of DNA we will probably need to separate the pieces from one another. This is usually done by **electrophoresis** (Fig. 9.13). This technique will separate molecules only if they are electrically charged. Two electrodes, one positive and the other negative are connected up to a high voltage source. Positively charged molecules move towards the negative electrode and negatively charged molecules move towards the positive electrode.

electrophoresis movement of charged molecules toward an electrode of the opposite charge. Used to separate nucleic acids or proteins or any molecule that has a charge

9.13 ELECTROPHORESIS: THE PRINCIPLE

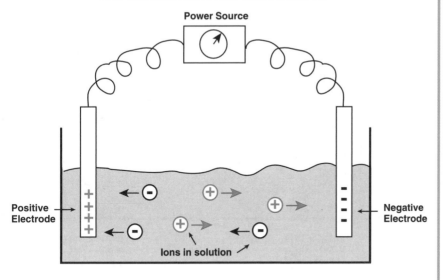

Since DNA carries a negative charge on each of the many phosphate groups making up its backbone (Fig. 9.14), it will move towards the positive electrode during electrophoresis. The bigger a molecule the more force required to move it. However, the bigger a DNA molecule, the more negative charges it has! So these two factors cancel out because all our fragments of DNA have the same number of charges per unit length. And so they all cruise along at the same speed toward the positive electrode.

9.14 DNA HAS MANY NEGATIVE CHARGES

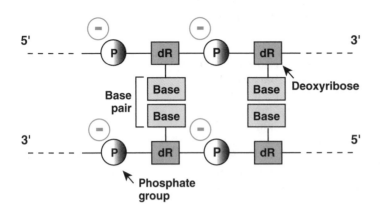

What we want is a way to separate the big fragments from the small ones. To do this we run them through a gel. This is a high-tech version of the familiar gelatin dessert without the artificial flavor and color. **Gelatin** sets due to a microscopic meshwork formed by its own protein fibers.

Gels for DNA work are made of **agarose**. This is a polysaccharide extracted from seaweed. When a hot solution of agarose cools, it congeals to form a meshwork rather like gelatin. The sample of DNA is put into a hole

gelatin a natural protein extracted from the connective tissue of animals by boiling the parts which did not even qualify for hamburgers or hotdogs. Try not to think about this next time you slurp some down

agarose a gel-forming polysaccharide found in some types of seaweed

made at one end of a slab of agarose and electrodes are connected up in contact with each end of the slab (Fig. 9.15).

9.15 AGAROSE GEL ELECTROPHORESIS

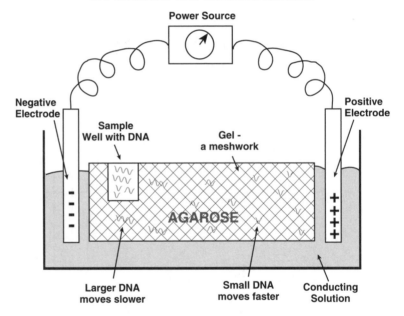

The juice is turned on and the DNA starts to move towards the positive electrode. As the DNA molecules move through the gel they are hindered by the meshwork of agarose fibers. The larger molecules find it more difficult to squeeze through the gaps but the smaller ones are slowed down much less. The result is that the DNA fragments separate in order of size (Fig. 9.16).

Both agarose and DNA are naturally colorless, so we cannot see where the DNA has ended up. To find our DNA fragments we use **ethidium bromide**, a stain which is specific in binding to DNA or RNA. This stains the

ethidium bromide a dye that stains DNA and RNA orange when viewed under UV light

102

9.16 DNA MOLECULES MOVING THROUGH A GEL

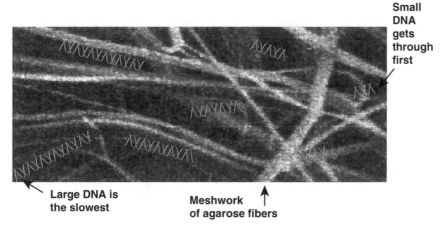

Small DNA gets through first

Large DNA is the slowest

Meshwork of agarose fibers

DNA orange, but only if viewed under ultraviolet light (Fig. 9.17). After locating the orange bands, these are cut out of the agarose slab and the DNA is extracted to yield a pure fragment.

9.17 GEL AFTER RUNNING AND STAINING

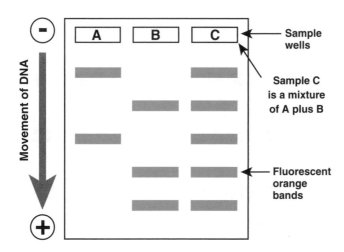

Movement of DNA

A B C

Sample wells

Sample C is a mixture of A plus B

Fluorescent orange bands

Agarose gel electrophoresis can be used to purify DNA for use in genetic engineering or it can be used to measure the sizes of fragments during scientific investigations. To find the size of an unknown piece of DNA we run a set of standard DNA fragments of known sizes alongside, on the same gel (Fig. 9.18).

Making a Restriction Map

A diagram that shows the location of restriction enzyme cut sites on a piece of DNA is known as a restriction map. To generate such a map for

restriction map dia-gram of DNA showing the cut sites for a series of restriction enzymes

103

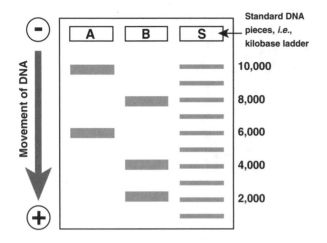

kilobase ladder a set
of standard DNA fragments
with lengths differing by
one kilobase

a piece of DNA we first digest it with a
series of restriction enzymes, one at a
time. The digested DNA is run on an
agarose gel to separate the fragments and
estimate their sizes. This reveals how
many sites each enzyme has in our DNA
and their distances apart. What we do not
yet know is the relative order of the
fragments.

For example, suppose we start with a
5,000bp piece of DNA which is cut into
three fragments of 3,000bp, 1,500bp, and
500bp by the restriction enzyme BamHI.
We have two cut sites but three alternative
arrangements.

You might think there should be six
possible arrangements in Figure 9.19, but
the three that are missing are merely the
first three, turned back to front. We have shown the backward arrangement
just for fragment No. III.

9.19 **POSSIBLE MAPS WITH FIRST ENZYME**

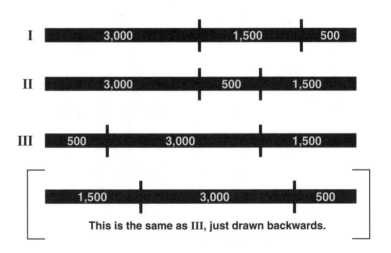

This is the same as III, just drawn backwards.

To decide between these three possibilities, we must use a second
enzyme. Suppose that our second restriction enzyme is EcoRI and that
alone it cuts just once to give two fragments of 4,000bp and 1,000bp.
For EcoRI on its own, there is, of course, only one possible arrangement.
In Figure 9.20 we show a mixed digestion using both enzymes simultaneously. The results of gel electrophoresis for the two single digests and the
double digest are shown.

In the double digest, the largest fragment seen in the BamHI lane has
disappeared. This means that there must be an EcoRI cut site within this
3,000bp BamHI fragment. Since, in this example, there is only one EcoRI

9.20 USING A SECOND RESTRICTION ENZYME

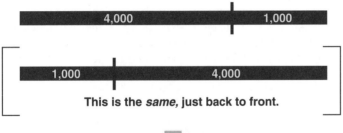

This is the *same*, just back to front.

**Mixed digest
uses both
enzymes simultaneously**

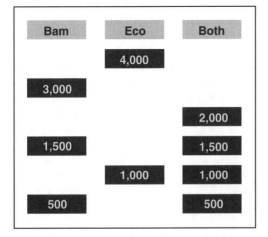

Gel Electrophoresis of Double Digest

cut site, only one of the BamHI fragments disappears in the double digest.
We can now construct the maps shown in Figure 9.21.

9.21 TWO POSSIBLE RESTRICTION MAPS
FROM BamHI and EcoRI DOUBLE DIGESTION

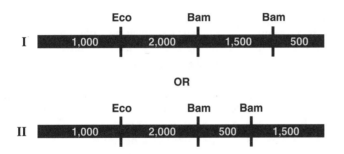

Note that we still have two alternatives left. To sort these out we would need yet a third enzyme. We would prepare double digests with BamHI plus enzyme III and of EcoRI plus enzyme III and analyze the results as above. It is by continuing in this way that clever molecular biologists determine the location of the cut sites for a range of restriction enzymes within a length of DNA. Once we have a map of all the restriction sites, we can cut the DNA as we wish for analysis and for cloning any useful genes it may contain.

Choosing a Vector for Our Own Special Purposes

Now that we have some pieces of DNA containing genes we are interested in, we may need to move them into other cells or manufacture them in bulk quantities for further genetic engineering. For either purpose we mount our cloned genes on what are called vectors.

In principle any molecule of DNA which can replicate itself inside a cell could work as a vector. To make life as easy as possible we consider the following factors:

1) The vector should be a reasonably small and manageable DNA molecule
2) Moving the vector from cell to cell should be easy
3) Growing and purifying large amounts of vector DNA should be straightforward

In practice bacterial plasmids that we learned about in the last chapter come closest to these requirements and are the most widely used vectors. We will discuss the requirements for vectors using plasmids as examples. (Certain viruses are also useful, but we will discuss these later in the book, where we discuss how they are actually used in genetic engineering or gene therapy.)

Multicopy Plasmid Vectors

We have discussed plasmids in Chapter 8, so let's pick one and get going. The ColE1 plasmid of *Escherichia coli* is a small circular DNA molecule that forms the basis of many vectors widely used in molecular biology. It exists in up to 40 copies per cell so obtaining plenty of plasmid DNA is easy and it can be moved from cell to cell by transformation as described in Chapter 8.

Since our basic requirements are satisfied, let's be greedy. Wouldn't it be nice if there was a simple way to:

1) Detect the presence of the vector
2) Directly select for cells that contain the vector
3) Insert genes into the vector
4) Detect the presence of an inserted gene on the vector

Detecting and Selecting Vectors

To satisfy the cravings of the newer generation of pampered and spoiled molecular biologists, it has been necessary to improve the original ColE1 plasmid. This was done in several stages. First we get rid of the genes for colicin E1, a toxic protein for killing bacteria (see Ch. 21) as these are obviously not needed. We next add a gene for resistance to the antibiotic ampicillin, a widely used penicillin derivative (Fig. 9.22). The *ampR* gene is also known as *bla* which refers to β-lactamase, the enzyme encoded by this gene, which degrades penicillins.

---

vector molecule of DNA which can replicate and is used to carry cloned genes or DNA fragments

Vectors do not have any particular gender; they could just as well be Vector as Vectoria!

plasmid circular DNA molecule which is capable of being replicated by the host cell in which it lives - usually a bacterium

ColE1 plasmid a particular plasmid whose derivatives have been widely used as vectors

ampicillin a commonly used antibiotic of the penicillin family

ampR or *bla* gene gene encoding β-lactamase

β-lactamase enzyme which destroys antibiotics of the penicillin class

DNA ligase enzyme which joins ends of DNA strands together

When we transform this new vector into bacterial cells we can directly select those which get the plasmid by incubating in a growth medium containing ampicillin. Those cells containing a plasmid survive, while those which did not get a plasmid are killed. Any time we wish to check that our vector is still present we merely check to see that the cells are still ampicillin resistant. Thus, the antibiotic resistance gene satisfies both requirements 1) and 2) of the preceding section.

Inserting Genes into Vectors

So how about inserting genes into the vector? This is done by cutting out a segment of DNA carrying our gene with a suitable restriction enzyme as described above. We must then cut the plasmid open with the same restriction enzyme so that the vector and the insert have matching ends. When we mix the two together and add DNA ligase, the enzyme which links together DNA strands, we will end up with our gene inserted into our vector as shown in Figure 9.23.

But wait a moment! What if there were more than one cut site in the vector for the restriction enzyme we chose? Then our vector would be chopped into pieces, not merely opened conveniently. This would be bad! Again, we must avoid inserting the cloned gene into any of the genes needed by the plasmid for its own replication and survival within the cell. And, while we are about it, we need flexibility; there are many different restriction enzymes and it would be nice to have a wide choice.

So let's just fix all these questions at one shot. We put into our plasmid a stretch of artificially synthesized DNA about 50 base pairs long which contains cut sites for seven or eight of our favorite restriction

9.22 ColE1 IMPROVEMENT

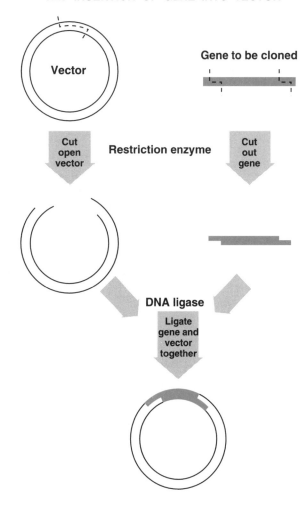

9.23 INSERTION OF GENE INTO VECTOR

107

enzymes. This is known as a multiple cloning site or, more often, just as a
polylinker, and not only allows us a wide choice of restriction enzymes,
but ensures that the insert does not damage the plasmid and goes into more
or less the same location each time (Fig. 9.24).

9.24 PRINCIPLE OF THE POLYLINKER

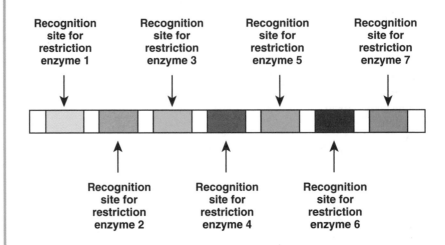

One other thing to do is to make sure that there are no extra cut sites
on the plasmid for any of the enzymes represented in the polylinker.
The couch potato's way to do this is to choose for the polylinker only
enzymes with zero cut sites in the plasmid we started with.

Alternatively, we can get rid of unwanted cut sites by the approach
shown in Figure 9.25. First we grow cells containing the original plasmid
and we make some plasmid DNA. Due to spontaneous mutation (see Ch. 12
to learn about mutations), an occasional plasmid will suffer a base change
within the cut site about which we are concerned. This will abolish recognition of the site by the restriction enzyme. But how do we find this one
rare mutant plasmid? Simple, once someone else has thought of it!
We treat the plasmid sample with the restriction enzyme in question and
then transform the plasmid DNA into fresh bacterial cells without re-ligat-
ing the break. Most plasmids will be destroyed by this procedure, but those
few that have lost the cut site by mutation will remain uncut and survive.

Detecting Insertions into Vectors

Let's suppose we have inserted our cloned gene into a plasmid vector
and transformed it back into a bacterial cell. How do we know if our
cloned gene is actually there? If the cloned gene itself makes something
easy to detect there is no problem. But what if there is no easy test and we
have to detect directly whether our cloned piece of DNA was successfully
inserted in the vector?

There are basically three approaches: A for antibiotics, B for brute
force and ignorance, and C for color. Let's start at the low I.Q. end of the
scale with the brute force method.

We pick a large number of separate bacterial colonies that have received the plasmid vector, hopefully with DNA inserted. We extract plasmid DNA from each of these. We then cut the plasmid DNA with the restriction enzyme used during our original cloning experiment. If there is no insert in the plasmid, this merely converts the plasmid from a circular to a linear molecule of DNA. If the vector contains inserted DNA, we will get two pieces of DNA, one being the original plasmid and the other the inserted DNA fragment. To see how many pieces of DNA we actually have, we separate the cut DNA by gel electrophoresis. If we test enough transformed colonies, sooner or later - perhaps very much later! - we will find one with an inserted DNA fragment in its plasmid.

Somewhat less laborious is to use an antibiotic resistance gene. Actually, we need two antibiotic resistance genes, one to select for cells which have received a plasmid as described above and a second to fool around with to detect inserts (Fig. 9.26). Suppose that the cut site for the restriction enzyme we use is within this second antibiotic resistance gene. Then, when we insert our cloned fragment of DNA, we will disrupt this antibiotic resistance gene. So, cells that receive a plasmid without an insert will be resistant to both antibiotics. Those receiving a plasmid with an insert will be resistant to only the first antibiotic.

Finally, for the truly sophisticated, there is color screening. The best known version is the blue/white screening done using β-galactosidase and X-gal. We start with a plasmid carrying the *lacZ* gene for β-galactosidase and insert a polylinker into the *lacZ* coding sequence, very close to the front of the gene. Luckily, the frontmost part of the β-galactosidase protein is unusual in being inessential for enzyme activity. As long as the polylinker is inserted without disrupting the reading frame we obtain active enzyme. However, if we insert a foreign segment of DNA into the polylinker the *lacZ* gene is inactivated and no β-galactosidase will be made (Fig. 9.27). We can detect β-galactosidase because it turns X-gal blue (see Ch. 16 for details). Plasmids without a DNA insert will produce β-galactosidase and the cell which carries them will turn blue. Plasmids with an insert will be unable to make β-galactosidase and the cells will stay white.

In real life there are more gruesome technical details. The *lacZ* gene is rather large so we actually only put the front part on our plasmid vector. The plasmid encodes the first 146 amino acids of β-galactosidase. On the bacterial chromosome we have a *lacZ* gene missing the front portion but

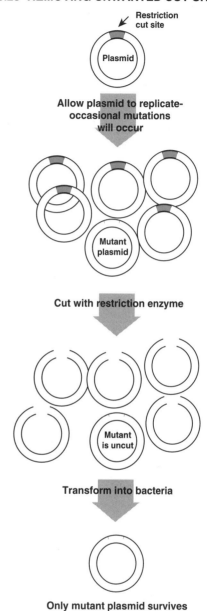

9.25 REMOVING UNWANTED CUT SITES

Restriction cut site

Plasmid

Allow plasmid to replicate-occasional mutations will occur

Mutant plasmid

Cut with restriction enzyme

Mutant is uncut

Transform into bacteria

Only mutant plasmid survives

β-(beta) galactosidase enzyme that splits lactose and related compounds

X-gal substance split by β-galactosidase and yielding a blue dye

lacZ gene the gene which encodes β-galactosidase

109

9.26 USE OF ANTIBIOTIC RESISTANCE GENE

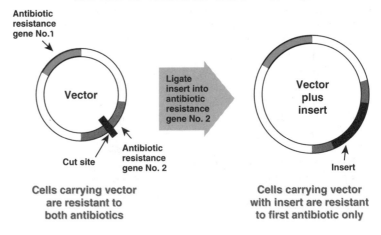

Antibiotic resistance gene No.1

Vector

Cut site

Antibiotic resistance gene No. 2

Ligate insert into antibiotic resistance gene No. 2

Vector plus insert

Insert

Cells carrying vector are resistant to both antibiotics

Cells carrying vector with insert are resistant to first antibiotic only

9.27 BLUE / WHITE SCREENING OF PLASMIDS FOR INSERTS

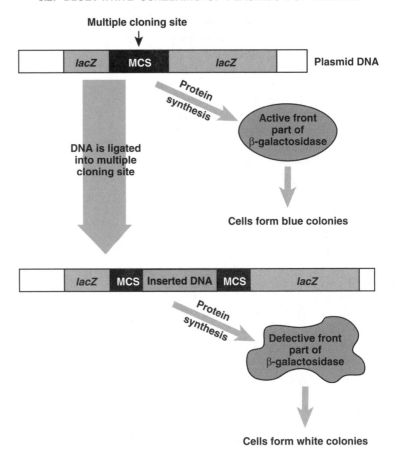

Multiple cloning site

| lacZ | MCS | lacZ | **Plasmid DNA**

DNA is ligated into multiple cloning site

Protein synthesis

Active front part of β-galactosidase

Cells form blue colonies

| lacZ | MCS | Inserted DNA | MCS | lacZ |

Protein synthesis

Defective front part of β-galactosidase

Cells form white colonies

encoding the rest of the β-galactosidase. If both genes are active we get two protein fragments that associate to give an active enzyme. If the plasmid gene is inactivated by cloning DNA into the polylinker, the short front fragment of β-galactosidase is missing and no active enzyme can form. (We should mention that assembling an active protein from fragments made separately is normally not possible and that β-galactosidase is exceptional in this respect.)

Moving Genes Between Organisms: Shuttle Vectors

So we now have plasmid vectors that are easy to use - in bacteria. But what if we want to put the gene we have cloned onto our vector into the cell of some other organism? This may be done using a shuttle vector. As its name implies, this is a vector that can survive in more than one type of host cell.

shuttle vector a vector able to survive in more than one type of host cell

Let's make a shuttle vector for yeast, a simple higher organism (Fig. 9.28). We take our bacterial vector and we add yet another stretch of DNA to it. We need to add:

9.28 SHUTTLE VECTOR FOR YEASTS

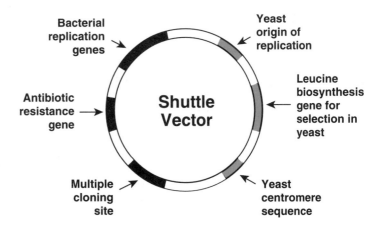

A) An origin of replication that works in yeast. The requirement for an origin of replication is easy to understand. This works pretty much like a bacterial origin of replication except that the DNA sequences needed for recognition differ in different organisms.

B) Genes to allow survival of the plasmid in yeast. Survival of a plasmid in yeast needs more than just replication. When a yeast cell divides, the duplicated chromosomes are pulled apart by microtubules attached to their centromeres, so that each daughter cell gets a full set. Our shuttle vector needs to segregate correctly at cell division also. To achieve this we insert a segment of DNA from the centromere of the yeast chromosomes, the Cen sequence, which is recognized by the microtubules that drag new chromosomes apart.

microtubule tubular structure made of protein that is used for structural support and movement of internal components

C) A gene to select for the plasmid in yeast. Our problem here is that yeast is not affected by most of the antibiotics that kill bacteria. In practice we usually have to settle for something less satisfactory. We use a yeast strain with a defect in a gene for making an amino acid, say leucine, and provide the necessary gene on the vector. If we don't give the yeast leucine it will starve. Only if it obtains the plasmid with the *leu* gene will it survive.

centromere structure used to pull chromosomes apart after division, usually found more or less in the center of the chromosome

Obviously, the detailed requirements for vectors will vary depending on the host organism, but the general ideas will be the same. We will consider ways to get genes into humans, animals and plants later (Chs. 13 and 15).

Yeast Artificial Chromosomes

Genes of higher organisms are often relatively huge. While typical bacterial genes are around a thousand base pairs, the genes for muscular dystrophy or hemophilia may take up a million or more base pairs of DNA. The largest bacterial plasmids that can be conveniently manipulated are around 50 kbp. So how do we carry huge pieces of DNA?

The answer is to continue using yeast, and to construct **yeast artificial chromosomes** or **YACs** (Fig. 9.29). We saw above that for a plasmid to survive in yeast we needed a yeast specific origin of replication and the Cen (centromere recognition sequence). We keep these, but instead of a circular plasmid we make a linear DNA molecule and on both ends stick the correct **telomere** sequence required at the ends of all eukaryotic chromosomes (see Ch. 11). Although artificial, a yeast cell will treat this structure as a chromosome. Of course, for practical use we also need a selectable marker and a suitable multiple cloning site. The great thing about a YAC is that we can insert colossal amounts of cloned DNA and the yeast cell doesn't mind a bit and will happily replicate it.

Because the recognition sequences for replication origins, centromeres and telomeres are so similar among higher organisms, an added bonus is that YACs will survive in mice and are even passed on from parent to child. Admittedly, not every baby mouse inherits the YAC, but all the same, this opens the way for moving around the huge DNA sequences needed for engineering higher animals.

9.29 **YEAST ARTIFICIAL CHROMOSOME**

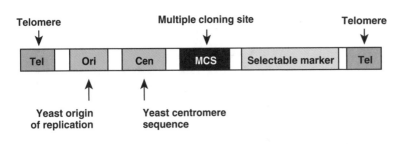

yeast artificial chromosome (YAC) linear DNA which mimics a yeast chromosome by having yeast specific origin, centromere and telomere sequences

telomere specific sequence of DNA found at the end of linear eukaryotic chromosomes

Onward and Upward

So, where is our monster? At the beginning of this chapter we suckered you into thinking we would create a monster. Come on, did you really think it was that easy? If it was really that easy, it would have been done long ago by some unscrupulous biologist.

In the next chapter we will tell you how genetic engineering can be used to make simple products. But before we attempt to improve plants, animals and even people (Ch. 15) we must discuss the genes of higher organisms and the sort of things that can go wrong and need to be repaired. Hang in there all you Frankenstein-wannabe's; chapter 10 will tell you more.

Additional Reading

Molecular Biotechnology: Principles and Applications of Recombinant DNA by Glick BR & Pasternak JJ. 1994. American Society for Microbiology, Washington, DC

Recombinant DNA: A Short Course by Watson JD, Tooze J & Kurtz DT. 1983. WH Freeman & Co., New York

Products from Biotechnology

10

We have entered the era where biotechnology is increasingly used to manufacture foodstuffs, drugs, or chemicals of one sort or another. Many of the small, first generation genetic engineering companies have either folded or been taken over by larger chemical or pharmaceutical corporations. There is a growing array of products, some never before available, others merely "new and improved." These range from relatively simple molecules to complex proteins such as **insulin**. We will illustrate this topic starting with the simple alcohol molecule and moving on to more complex examples including dyes, plastics, antibiotics and hormones.

insulin a protein hormone which controls the sugar level in the blood

Zymomonas a bacterium which ferments glucose to alcohol by a similar pathway to yeast

Did you hear about the genetic engineering company specializing in birth control products which was recently set up in Rome? It's called "Gen-Italia"

A is for Alcohol, Z is for Zymomonas

Neolithic pottery containing a yellowish residue dating back to 5,000 B.C. has been discovered in Iran. Sophisticated chemical analysis showed that it was wine. Perhaps it was chugging down a few jars of this "Chateau Caveman" that provided inspiration for mankind to build its first cities a few hundred years later. The production of alcohol is one of the earliest forms of biotechnology practiced by man. Alcoholic drinks are usually made by fermenting grain of some sort, using yeast. The lone exception is the use, in Mexico, of a bacterium, *Zymomonas*, to ferment sugar from the the sap of the agave plant to give a liquid known as *pulque*. Distillation converts this into tequila.

But humans weren't the first to appreciate alcohol. Elephants, monkeys and other wild animals deliberately consume fruit that has rotted and fermented naturally to yield alcohol. Perhaps the medal should go to a breed of fruit fly from the sherry producing region of Spain. These flies rely on sherry for their sole source of nutrition. They may not fly straight, but they fly happy.

There is little need for genetic engineering in the area of alcoholic drinks. However, conversion of

10.1 THE SHERRY FRUIT FLY

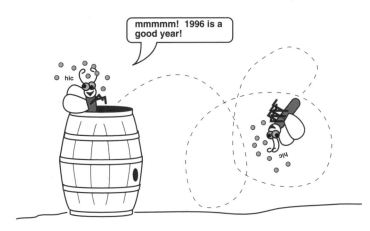

Gasohol is widely used in Brazil, where sugar-rich crops are grown especially for it and yeast is used to make the alcohol

waste biomass to fuel alcohol is another matter. If the United States converted the 100 million tons of waste paper it generates every year into fuel grade alcohol, this would reduce gasoline consumption by around 15 percent. Alcohol may be blended with gasoline to give "gasohol" which works fine in most internal combustion engines.

10.2 SUGAR TO ALCOHOL BY FERMENTATION

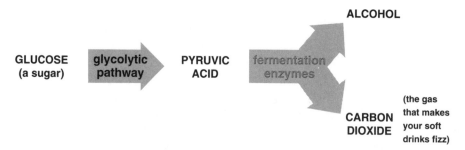

cellulose major carbohydrate of plant cell walls; a polymer of glucose

pathway engineering construction by genetic engineering of a complete biochemical pathway

Alcohol is made from sugar (Fig. 10.2). Sugars are components of the carbohydrates making up much of the bulk of plant matter. So, in principle, fuel alcohol can be made from almost any plant-derived material. *Zymomonas* grows faster than yeast and makes alcohol faster, too. The problem is that it is very fussy in what it eats. In nature *Zymomonas* lives entirely on glucose, and lacks the enzymes to break down other sugars, or carbohydrate polymers like starch and cellulose. Yeast is almost as narrow-minded. Genetic engineering is being used to make improved strains of both yeast and *Zymomonas* containing genes for enzymes capable of breaking down starch, cellulose and others. Building a new biochemical route, step by step, by installing the necessary genes, is known as pathway engineering (Fig. 10.3).

10. 3 PATHWAY ENGINEERING

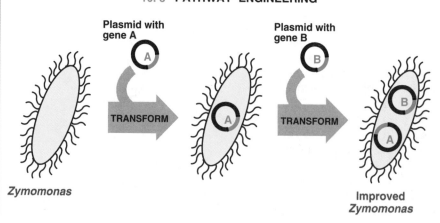

Paper consists almost entirely of cellulose, a linear polymer of the sugar glucose. The polymer chains are packed tightly side by side in a crystalline array with loosely packed, non-crystalline zones at intervals (Fig. 10.4). The challenge is to break down cellulose, yielding glucose that can be turned into alcohol.

Breakdown of cellulose requires several steps (Fig. 10.5), each catalyzed by a separate enzyme, as follows:

1) Endoglucanase snips open the polymer chains in the middle. This enzyme can only get a grip on the polymer chains in the loosely packed zones.
2) Cellobiohydrolase cuts off sections with 10 or more glucose units from the free ends made by endoglucanase.
3.) Exoglucanase chops off chunks of two or three glucose units from the ends of the sections liberated in step step 2.
4) β-glucosidase (also known as cellobiase) converts these dimers and trimers to glucose.

The genes for each of these four enzymes have been cloned from various microorganisms. Since cellulose is too big to enter the cell, the first three enzymes must be secreted and work outside. The dimers and trimers of glucose are released from cellulose outside the cell and are then transported inside. They are next broken into individual glucose molecules and converted to alcohol. So far, pilot projects have degraded cellulose from waste paper to glucose by adding separate enzymes from different sources. Finally, yeast or *Zymomonas* is added to convert the glucose to alcohol. A complete recombinant organism which converts cellulose to alcohol all on its own, as in Figure 10.6, is still to be created.

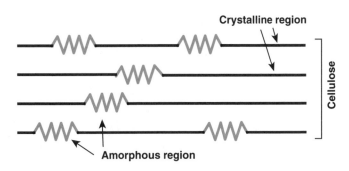

10.4 CELLULOSE STRUCTURE

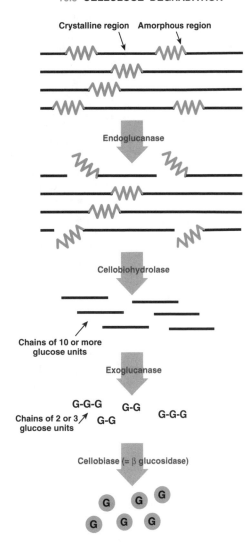

10.5 CELLULOSE DEGRADATION

Zymomonas capable
of converting glucose
to alcohol

a) Clone 4 genes for
cellulose breakdown.

b) Put into plasmid.

c) Transform into
Zymomonas

Zymomonas capable
of converting cellulose
to glucose

Problems from Alcohol Consumption

Genetics not only applies to the production of alcohol, but also to its consumption. In animals, the breakdown of alcohol takes place mostly in the liver. Two biochemical steps are involved, and their speed has a great influence on your susceptibility to alcohol. The two enzymes needed for disposing of alcohol are alcohol dehydrogenase and acetaldehyde dehydrogenase (Fig. 10.7). Susceptibility to alcohol varies greatly between individuals and between different groups of humans. Much of this variation is genetic in origin. Women are more susceptible to alcohol than men. Apart from the obvious factor of lower body weight, they have less alcohol dehydrogenase in the stomach lining. This strategically situated enzyme attacks the alcohol before it reaches the bloodstream (Fig. 10.8). Hence, a greater percentage of the alcohol in a drink enters the bloodstream of women and they get giddy quicker than men.

alcohol dehydrogenase the enzyme that converts alcohol to acetaldehyde

acetaldehyde dehydrogenase the enzyme that converts acetaldehyde to acetate

The acetate produced may be burned as fuel to release carbon dioxide and generate energy, or it may be converted to fat and stored as a beer belly

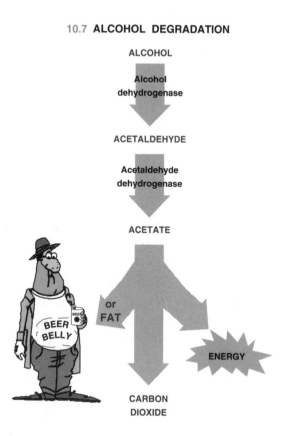

10.7 ALCOHOL DEGRADATION

ALCOHOL

Alcohol
dehydrogenase

ACETALDEHYDE

Acetaldehyde
dehydrogenase

ACETATE

or
FAT

ENERGY

CARBON
DIOXIDE

BEER
BELLY

10.8 ALCOHOL DEHYDROGENASE IN THE STOMACH

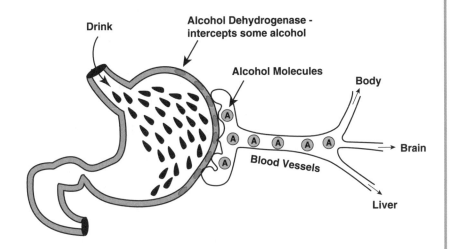

The genes for alcohol dehydrogenase exist in several variant forms. First, there are multiple genes that encode several different structural versions of alcohol dehydrogenase within the same individual. Such variants are called **isoenzymes** and each individual in the population has a full set. Here, as is often the case, the different isoenzymes are expressed preferentially in different tissues of the body (Fig. 10.9).

isoenzyme variant forms of the same enzyme found within a single species; often showing tissue specific distribution

10.9 ALCOLHOL DEHYDROGENASE ISOENZYMES

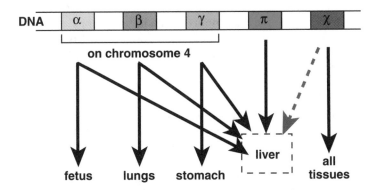

Second, even if we consider a single gene encoding a single isoenzyme, there are still quite a few genetic variants. These are due to slightly different alternative DNA sequences for this gene, *i.e.*, there are multiple alleles. In this case different individuals will possess different alleles. This gives rise to genetic and even racial differences in the ability to metabolize alcohol. Such racial differences are not just stereotypes, as required by politically correct mythology, but are due to the presence of different alleles in different human populations.

Many Northern Europeans have a highly efficient allelic form of liver alcohol dehydrogenase and people of Viking ancestry can often drink most others under the table

Brainwave abnormalities found in adult alcoholics were once thought to be entirely the result of heavy drinking. When heavy drinkers abstain for several months, all defects except the unusually low P3 wave (involved in memory and emotion) return to normal. This defect has been found in pre-teen sons of alcoholics who have not yet started drinking, implying that it is an inherited trait. The genes for alcoholism have not yet been located, but it is probably just a matter of time until they are identified. Alcoholics probably break down alcohol in a different manner from normal. Their blood contains the chemical, **2,3-butanediol**, which is absent in non-alcoholics. In rats in which the normal pathway for alcohol breakdown has been blocked artificially, alcohol travels via an alternative pathway that produces 2,3-butanediol as a by-product.

2,3-butanediol a particular chemical substance with two alcohol groups, made by certain bacteria, but a rare by-product in cells of higher organisms

Ice Forming Bacteria and Frost

Each year, frost causes more than a billion dollars damage to crops in the United States alone. It is not the low temperature itself that does the damage. When water freezes, forming ice, it expands and it is this which bursts water pipes and damages plant tissues. However, ice crystals need a microscopic nucleus or "seed" to form around. The seeding of ice crystals on plants is mostly due to bacteria, especially *Pseudomonas syringae*. If these are absent, ice does not form and water will supercool down to -8°C without solidifying, and the plants will be unharmed.

Those bacteria causing ice formation possess a protein on their cell surfaces known as **ice nucleation factor**, which is encoded by the *inaZ* gene. Like most bacteria, *E. coli* does not normally promote ice formation. But when the *inaZ* gene from *Pseudomonas syringae* is inserted into *E. coli* it gains ice nucleating ability. Conversely, when the *inaZ* gene of *Pseudomonas syringae* is disrupted, ice nucleating ability is lost (Fig. 10.10). The wild, "ice-plus," strains of *Pseudomonas syringae* can be displaced by spraying the "ice-minus" mutants onto crops which are at risk from frost damage. So even when the temperature falls below zero, very few ice crystals form and most of the plants are unharmed - many are cooled, but few are frozen!

ice nucleation factor protein found on the surface of some bacteria. It acts as a seed for ice crystal formation

Blue Genes for Blue Jeans

And the LORD spake unto Moses, saying:

"Speak unto the children of Israel, and bid them that they make them fringes in the borders of their garments throughout their generations, and that they put upon the fringe of the borders a riband of blue: And it shall be unto you for a fringe, that ye may look upon it, and remember all the commandments of the LORD..."

Numbers 15:37-39

The earliest brand of blue garments were made by Levites not Levi's!!

The "blue" dye, hyacinthine purple, is actually a blue/purple mix, containing roughly 50-50 of Tyrian purple and indigo. The secret of this sacred dye was lost about 1,400 years ago when its suppliers, the Phoenicians, were overrun by invading Arabs. Tyrian purple, a reddish purple dye, is also made by a sea snail called the **spiny murex** (Fig. 10.11). Recently it was discovered that one of its relatives, the **banded murex** secretes a mixture of Tyrian purple with indigo, *i.e.*, the lost hyacinthine purple.

spiny murex and **banded murex** two related sea snails which make dyes of the indigo family

10.10 HAVE AN "ICE" DAY

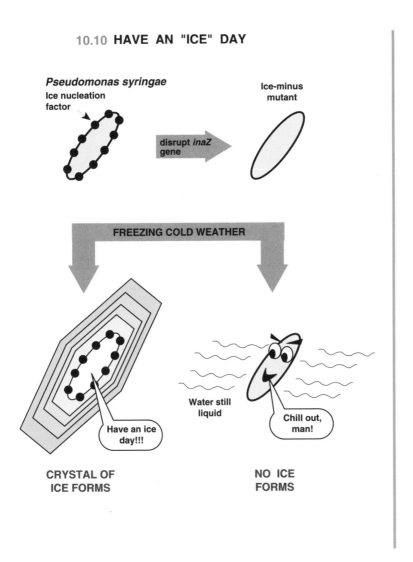

10.11 SEA SNAILS AND DYES

Tyrian purple is itself very closely related to indigo (Fig. 10.12). It has two bromine atoms, extracted from seawater by the sea snail and stuck onto the indigo ring structure. Both dyes are secreted as colorless precursors that turn blue (or purple) by reacting with the oxygen in air. Indigo itself is used for dyeing wool and cotton blue. Yes, blue jeans are made of cotton dyed with indigo.

10.12 CHEMICAL STRUCTURE OF INDIGO

X = hydrogen in indigo itself

X = bromine, from seawater, in Tyrian purple

hydrocarbon molecule made only of carbon and hydrogen; petroleum is a mixture of linear and cyclic hydrocarbons

NAH plasmid bacterial plasmid which carries genes for the breakdown of the hydrocarbon naphthalene

Indigo used to be extracted from plants, but nowadays is synthesized chemically. Recently, genetically engineered bacteria that can make indigo were discovered by accident. Some soil bacteria can degrade cyclic **hydrocarbons** found in oil or produced by chemical factories. The ability to break down naphthalene, the chemical comprising mothballs, is due to a set of extra genes carried on the **NAH plasmid**. When genes from the NAH plasmid were cloned into the laboratory bacterium *Escherichia coli* (Fig. 10.13), some of the bacteria turned blue!

10.13 CLONING THE *NAH* GENE INTO *E. COLI*

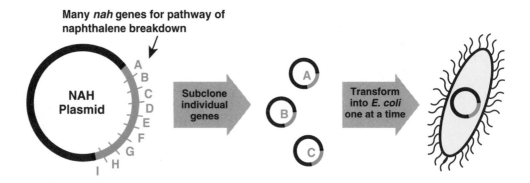

Many *nah* genes for pathway of naphthalene breakdown

NAH Plasmid
A B C D E F G H I

Subclone individual genes

A B C

Transform into *E. coli* one at a time

These blue bacteria turned out to possess the gene for the enzyme naphthalene oxygenase that carries out the first step in breaking down naphthalene. However, this enzyme can carry out similar reactions with other molecules similar in shape to naphthalene. In particular, it converts indole into indoxyl (Fig. 10.14). *Escherichia coli* itself provides the indole from the amino acid tryptophan. Oxygen from the air converts the indoxyl to indigo.

Commercialization of this would involve sticking the recombinant bacteria with the naphthalene oxygenase gene onto a solid support in a bioreactor. Tryptophan would be added at one end and indigo would trickle out at the other (Fig. 10.15).

10.14 THE HYBRID PATHWAY FOR INDIGO

TRYPTOPHAN

A. **tryptophanase**
(already in *E. coli*)

INDOLE

B. **naphthalene oxygenase**
(cloned from NAH plasmid)

DIHYDROXY-INDOLE

C. **spontaneous dehydration**

INDOXYL (colorless)

D. **oxygen in air**

INDIGO (blue!)

Perhaps the indigo genes could be cloned into transgenic sheep and we could get blue wool directly!

10.15 BIOREACTOR FOR INDIGO PRODUCTION

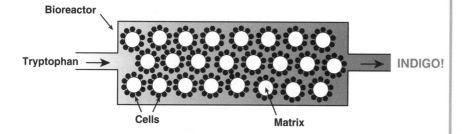

Bioreactor

Tryptophan → → INDIGO!

Cells   Matrix

polyhydroxy-alkanoate (PHA) polymeric plastics made by certain types of bacteria, formed by linking hydroxy-acid subunits together

Biosynthetic Plastics are Also Biodegradable

Plastics are polymers built from chains of monomer subunits, like proteins and nucleic acids. However, most plastics consist of the same monomer mindlessly repeated over and over again. Sometimes two or more closely related monomers may be mixed together and follow each other at random.

Certain bacteria make and store a group of related plastics known as **polyhydroxy-alkanoates (PHAs)** whose composition is shown in

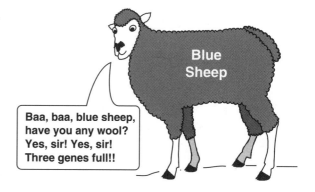

Blue Sheep

Baa, baa, blue sheep, have you any wool? Yes, sir! Yes, sir! Three genes full!!

polyhydroxy-butyrate
(PHB) bioplastic polymer
of hydroxy-butyrate subunits

Figure 10.17. They accumulate PHAs when they run low on essential
nutrients and use them as a source of energy when conditions improve.

10.17 CHEMICAL STRUCTURE OF PHA PLASTICS

Hydroxyacid $HO - CH - CH_2 - COOH$

POLYMERIZATION

Poly-hydroxyacid

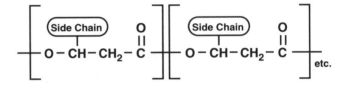

Arabidopsis a small
plant used for genetics
because of its genetic
simplicity; it has only
about 10 times as much
DNA as bacteria

chloroplast organelle
found in plant cells which
traps energy from sunlight
to make sugars by
photosynthesis

The most commonly found PHA has four-carbon (hydroxy-butyrate,
HB) subunits and is therefore called **polyhydroxybutyrate (PHB)**.
However, a plastic made by randomly mixing in 10 to 20 percent of
five-carbon (hydroxy-valerate, HV) subunits has much better physical
properties. Yet other PHAs containing a proportion of eight-carbon or
longer subunits, give materials that are more rubbery.

The poly-HB/HV copolymer is manufactured
by mutant bacteria of the species *Alcaligenes
eutrophus* and is marketed by the Zeneca
Corporation (United Kingdom) under the trade
name of Biopol. It is at present more expensive
than plastics made from oil but is completely
biodegradable. PHAs are restricted at present to
specialized uses. For example, since they break
down slowly inside the body to give natural,
biochemical intermediates, they can be used for
making slow release capsules.

To make PHAs economically competitive they
will need to be produced cheaply and in bulk.
One good way to do this is to insert the genes for
their synthesis into suitable crop plants. This is still
in the experimental stage, but the genes for making
PHA from *Alcaligenes eutrophus* have been successfully inserted into *Arabidopsis,* a plant widely used
for genetic experiments. In fact, the PHA genes were
modified so that the pathway for PHA synthesis was
expressed inside the chloroplasts (Fig. 10.18).

10.18 PHA PLASTICS FROM PLANT CHLOROPLASTS

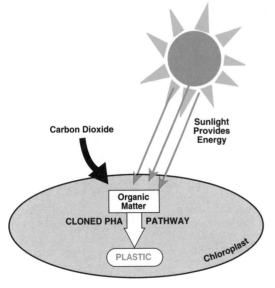

The reason for doing this is that chloroplasts are the sites of photosynthesis and are consequently where newly synthesized organic matter first appears. Locating the PHA pathway in chloroplasts, rather than in the main compartment of the plant cells, gave a 100-fold increase in PHA accumulation.

The final step will be to move the pathway into a genuine crop plant such as rapeseed or soybeans. Eventually we should be able to grow fields full of genuine plastic flowers!

antibiotic chemical substances which kill bacteria selectively, that is, without killing the patient too

Antibiotics

Although antibiotics are molecules of only intermediate complexity, they are the most difficult to tackle by genetic engineering. The reason is that antibiotics are made by microorganisms and their synthetic pathways may have 20 or even more steps. Each step requires a separate enzyme, encoded by its own gene. Cloning all the genes for a single antibiotic is not easy.

The mold that makes penicillin was discovered by Alexander Fleming in the 1920s. Fleming discovered penicillin because he couldn't be bothered to clear up his bacterial cultures, so he just left the petri dishes containing them lying around.

Some of Fleming's bacterial cultures got moldy and he noticed a zone of death around a pretty blue mold (Fig. 10.20). He found that the mold excreted a chemical toxic to bacteria but harmless to animals - penicillin. Fleming called the mold *Penicillium notatum*. It has a cousin, *Penicillium chrysogenum*, which makes a related antibiotic called cephalosporin C. These antibiotics are

FLEMING AND HIS MESSY LABORATORY

Looks like it's time to do the dishes again!

Today, there would be a dozen government agencies lined up to save mankind by shutting down the laboratories of untidy scientists like Fleming. We are lucky he did his work before OSHA, the EPA and the FDA made the world safe for the paranoid

penicillin an antibiotic that kills bacteria by destroying their cell walls and which is produced by certain kinds of mold

cephalosporin C original member of a family of antibiotics closely related to the penicillins

10.20 PENICILLIN FROM MOLD

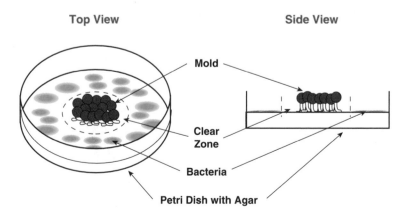

Top View   Side View

Mold
Clear Zone
Bacteria
Petri Dish with Agar

Fleming also discovered lysozyme, an enzyme that breaks down bacterial cell walls. Lysozyme protects against bacterial invasion and is found in blood, sweat and tears, etc. Apparently Fleming's nose dripped into a culture and the bacteria disintegrated. Later, less vulgar but more heroic investigators held raw onions to their eyes and collected the tears. Luckily, lysozyme is found in large amounts in egg white, a substance resembling nasal secretions in many respects but which is more hygienic.

hormone regulatory molecule which carries commands from one tissue to another in the body fluids

made by separate branches of the same biosynthetic pathway (Fig. 10.21). The natural versions are altered chemically to make many different antibiotics.

Although cephalosporin C is pretty feeble itself, it is the starting point for a vast array of broad spectrum antibiotics made by chemical modification. To do this, cephalosporin C must first be converted to 7-ACA (7-amino-cephalosporanic acid) which is not made by any known organism. This step had to be done chemically and gave very low yields. Recently, a mold which makes cephalosporin C was engineered to convert this to 7-ACA. Two extra genes were inserted to create the extended pathway (Fig. 10.22). The gene for D-amino-acid oxidase was taken from a fungus (*Fusarium solani*) and the cephalosporin acylase gene from a bacterium (*Pseudomonas diminuta*).

More and more disease-causing microorganisms are acquiring or mutating to antibiotic resistance (see Ch. 8 for how this happens). This is starting to cause problems in treating patients with infections. We are now in a position where we need to run merely in order to stay in the same place. More and more novel antibiotics are becoming necessary simply to treat infections which used to be susceptible to the original versions of penicillin, cephalosporin and others.

10.21 SYNTHESIS OF PENICILLIN AND CEPHALOSPORIN

Aminoadipic acid + Cysteine + Valine

⬇

Unpronounceable precursor

⬇

Isopenicillin N

↙   ↘

Penicillins   Cephalosporin C

⬇   ⬇

Wide range of antibiotics made by minor chemical modification

Hormones

Hormones are regulatory molecules that travel around the bodies of animals or plants. They are needed to coordinate operations inside large organisms having millions of cells. Animal hormones are made in special tissues, known as glands, and then secreted into bodily fluids, such as the bloodstream (Fig. 10.23).

10.22 THE ENGINEERED PATHWAY TO 7-ACA

Cephalosporin C

D-amino-acid oxidase

H_2O_2

Unpronounceable precursor
7-β-(5-carboxy-5-oxopentanamido)-cephalosporanic acid

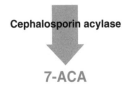

H_2O_2 ➡

Unpronounceable precursor
7-β-(4-carboxy-butanamido)-cephalosporanic acid

Cephalosporin acylase

⬇

7-ACA

Some hormones are proteins, whereas others are smaller, chemically more simple molecules. For example, the sex hormones belong to a class of small molecules known as steroids. Other steroid hormones affect other aspects of development, and are illegally used by weight lifters and other athletes to promote muscle growth.

10.23 HORMONES – THE GENERAL IDEA

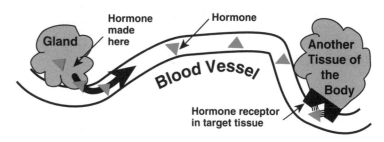

Paradoxically, it is the protein hormones that have been most susceptible to genetic engineering. Once a gene has been cloned, the protein for which it codes can be churned out in large amounts with relative ease. Smaller, non-protein molecules, which seem simpler to an organic chemist, would need half a dozen proteins (enzymes) working in series to synthesize them.

Insulin and Diabetes

Diabetes is a genetic defect due to absence of insulin, a small protein hormone made by the pancreas, which controls the level of sugar in your blood. Lack of insulin results in high blood sugar and this causes a variety of complications. Injections of insulin keep blood sugar levels down to near normal in people with diabetes.

The first commercially available genetically engineered hormone for humans was insulin. Insulin is made by the pancreas and Before Cloning (BC), people with diabetes had to give themselves injections of insulin extracted from the pancreases of animals (Fig. 10.24).

diabetes disease causing inability to control level of blood sugar, due to defect in insulin production

insulin a protein hormone which controls the sugar level in the blood

10.24 INSULIN FROM PANCREAS

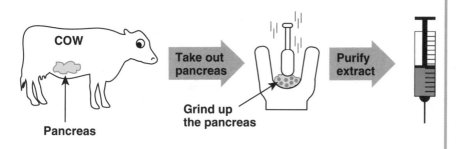

recombinant genetically engineered; may refer to a single product or a whole organism

Recombinant insulin was first marketed by the Eli Lilley Corporation under the registered name of *Humulin* (Human Insulin)

Although this worked well on the whole, occasional allergic reactions occurred, usually to low level contaminants in the extracts. Today, genuine human insulin made by recombinant bacteria is available.

10.25 STRUCTURE OF INSULIN

A chain
(21 amino acids)

B chain
(30 amino acids)

Insulin is a protein made of two separate polypeptide chains (Fig. 10.25). A stroke of amazing originality led to these being called the A and B chains. The two chains are held together by disulfide bonds.

Despite the fact that the final protein has two polypeptide chains, insulin is actually encoded by a single gene. The original gene product, preproinsulin, is a single polypeptide chain, which contains both the A and B chains together with some extra material, the signal sequence and the C- (or connecting) peptide (Fig. 10.26). Preproinsulin does not work as a hormone; it has to be processed via a couple of steps to give insulin. Processing requires several other enzymes, one for each step.

disulfide bond chemical linkage between two cysteine residues binding together two protein chains

preproinsulin insulin as first synthesized with both a signal sequence and the connecting peptide

signal sequence sequence of about 20 amino acids found at the front of proteins marked for export from the cell

proinsulin insulin without the signal sequence but still with the connecting peptide

C-peptide connecting peptide which originally links the A and B chains of insulin but which is absent from the final hormone

10.26 BIOSYNTHESIS OF INSULIN

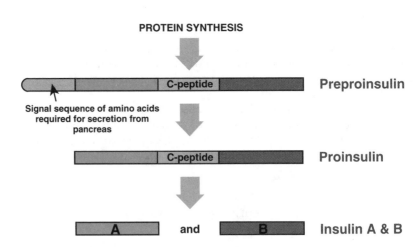

PROTEIN SYNTHESIS

C-peptide — **Preproinsulin**

Signal sequence of amino acids required for secretion from pancreas

C-peptide — **Proinsulin**

A and B — **Insulin A & B**

If we simply cloned the insulin gene and expressed it in bacteria, we would get preproinsulin, not insulin. In practice there are two solutions to this problem. The first is to take the preproinsulin and treat it with enzymes that convert it into insulin. This means we have to manufacture each of the processing enzymes too. Sorry, and no thanks!

The solution was to make two artificial mini-genes, one for the insulin A-chain and the other for the insulin B-chain (Fig. 10. 27). Two pieces of DNA, which coded for the two insulin chains, were synthesized by chemical procedures. These two DNA molecules were then inserted into plasmids which were put into two separate bacterial hosts. The two chains of insulin were then produced separately by two bacterial cultures. They were mixed and treated chemically to generate the disulfide bonds linking the chains together.

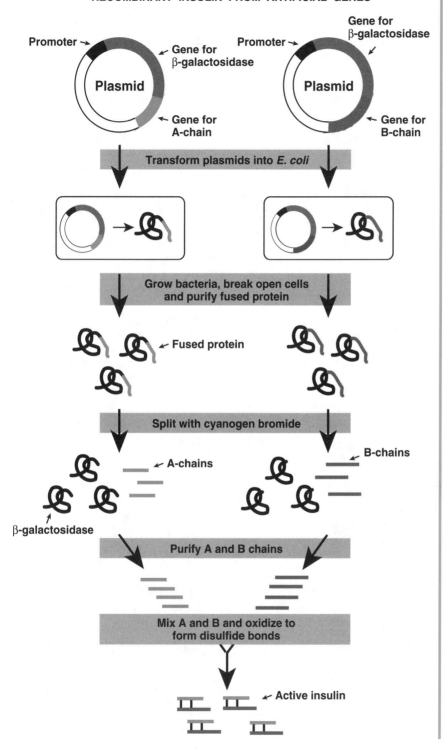

10.27 SYNTHESIS OF TWO CHAINS OF INSULIN FROM RECOMBINANT INSULIN FROM ARTIFICIAL GENES

Promoter

Gene for β-galactosidase

Plasmid

Gene for A-chain

Promoter

Gene for β-galactosidase

Plasmid

Gene for B-chain

Transform plasmids into *E. coli*

Grow bacteria, break open cells and purify fused protein

← Fused protein

Split with cyanogen bromide

← A-chains

B-chains

β-galactosidase

Purify A and B chains

Mix A and B and oxidize to form disulfide bonds

← Active insulin

In reality (Fig. 10.27), the insulin mini-genes were fused to the back of the *lacZ* gene which encodes β-galactosidase. The reason for doing this is that β-galactosidase can be easily purified while the small insulin chains would tend to get lost. After purification, the fusion protein was split with cyanogen bromide, a reagent that splits protein chains at methionine residues. Since the first amino acid of each protein is a methionine, this releases the insulin chains.

The above approach gives insulin which works well. Nonetheless, natural insulin, even natural human insulin, is not perfect. In fact, we can improve upon nature in the case of insulin. The problem is that natural insulin tends to form hexamers, *i.e.*, clumps of six. This clumping covers up the surfaces used by the insulin molecule to bind to the insulin receptors in cells responding to this hormone (Fig. 10.28).

10.28 FORMATION OF INSULIN HEXAMERS

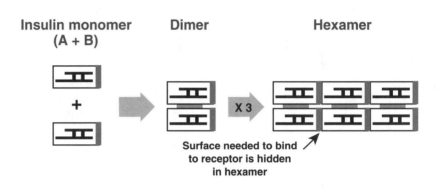

In normal people, insulin is secreted from the pancreas as a monomer and is distributed rapidly by the bloodstream before it gets a chance to clump. However, when insulin is injected, a high concentration of insulin is present in the syringe and clumping occurs. After injection, it takes a while for the hexamers to dissociate and it may take several hours for the patient's blood glucose to drop to normal levels.

10.29 GENETIC ENGINEERING OF INSULIN TO STOP FORMATION OF HEXAMERS

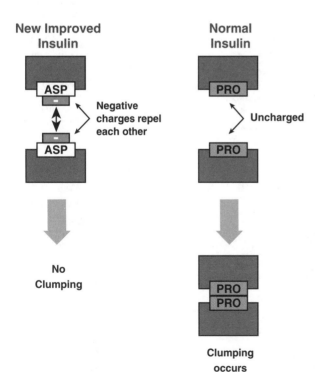

Insulin can be genetically engineered to stop clumping. The DNA sequence of the insulin gene is altered so as to change the amino acid sequence of the resulting protein. A proline located at the surface where the insulin molecules touch each other when forming the hexamer is replaced with an aspartic acid. The side chain of aspartic acid carries a negative charge. So when two genetically altered insulin molecules approach each other, they are mutually repelled by their negative charges and no longer clump together (Fig. 10.29). The [Pro $\Rightarrow$ Asp] insulin causes a faster drop in blood sugar than the native insulin. The Danish pharmaceutical company, Novo, is carrying out clinical trials and the new improved insulin may eventually replace the natural product.

128

Obesity and Cosmetic Biotechnology

Americans spend $12 billion a year on obesity products (diets, slimming machines, liposuction, etc.) yet they remain the most obese people of any industrial nation and are still getting fatter. About 35 million Americans are fat enough (20 percent or more overweight) to significantly affect their health. The increased health costs amount to nearly $40 billion per year, over three times the amount spent on slimming products!

But high-tech help may soon be on the way. The *ob* (for obese) gene of mice encodes a protein hormone called **leptin**. Mice defective in both copies of the *ob* gene lack leptin and weigh up to three times as much as normal mice. When obese mice are treated with leptin, they not only eat much less, but also burn off their fat much faster and lose 30 percent of their body weight (all fat!!). The genes for human leptin and mouse leptin have both been cloned into the bacterium *E. coli* which then makes the hormones in large quantities. Both act pretty much the same on obese mice.

> **leptin** a protein hormone which controls the appetite and the burning of fat by the body

What about fat cats and fat cat owners? When will we be able to buy an anti-fat pill? Unfortunately, things aren't quite as simple as you might have hoped. While some obese humans no doubt lack leptin, others probably lack the receptor for this hormone. Furthermore, leptin defects cause what is technically known as "morbid obesity," gross obesity which appears almost from birth and is relatively rare. In contrast, "fat" mice resemble many people in growing fat as they age. These mice have a genetic defect in the processing of insulin (as we saw earlier, another hormone also involved in how fast you burn sugar and fat). Finally, there are "tubby" mice, defective in the *tub* gene whose function is still unknown.

Presumably, the public will eventually be able to choose between a whole slew of pills labeled "anti-obese," "anti-fat," "anti-tubby," "anti-plump" and, if they are truly lucky, maybe they will live to see "anti-ugly" pills.

10.30 FAT CATS OR "TUBBY TABBIES"

129

A variety of human hormones, growth factors and related protein molecules have been cloned, expressed by bacteria and marketed for clinical use. Here is a brief list of some of these:

Protein	Use
Erythropoietin	Promotes red blood cell formation in the treatment of anemia
Factor VIII	Helping blood clots form in hemophiliacs
Filgrastim and Sargramostim (blood cell stimulating factors)	Used to boost white blood cell counts after radiation therapy or bone marrow transplantation
Insulin	Treatment of diabetes
Interferon (alpha)	Treatment of hepatitis B & C, genital warts, certain leukemias and other cancers
Interferon (beta)	Treatment of multiple sclerosis
Interferon (gamma)	Treatment of chronic granulomatous disease
Interleukin-2	Killing tumor cells
Somatotropin	Treatment of growth hormone deficiency
Tissue plasminogen activator (t-PA)	Dissolving blood clots to prevent heart attacks and lessen their severity

Of course, it would be much more convenient to alter people so they do not need medication. Before we consider this, we need to understand the genetic organization of animals and plants. These higher organisms have many basic features in common with bacteria, but they also differ in many important respects as described in the next chapter.

Additional Reading

Molecular Biotechnology: Principles and Applications of Recombinant DNA by Glick BR & Pasternak JJ. 1994. American Society for Microbiology, Washington, DC

Genetic Organization in *Higher* Organisms

Most of the basic principles of molecular biology were discovered using bacteria. This is because bacteria are relatively simple and easy to use in experiments. However, as the physicist, Sir Isaac Newton once remarked, "If I have seen further than others, it is because I stood on the shoulders of giants!" Investigators working on the molecular biology of higher organisms must make do with standing on the shoulders of microscopic bacteria like *Escherichia coli*. Nonetheless, a great deal has been learned, much of it very recently, about the organization of genetic information in plants and animals. In the near future, genetic manipulation of higher organisms will have a profound impact on both agriculture and health care. Before getting into the practical side of things in later chapters, we must first deal here with the fundamentals.

nucleus compartment of cell enclosed by a membrane which contains the chromosomes

eukaryote organism that shelters its genes inside a nucleus and has multiple linear chromosomes

prokaryote organism without a nucleus and with a single circular chromosome

Definition of Eukaryotic

By definition, possession of a nucleus makes you a eukaryote (see Fig. 11.1). The nucleus is a separate compartment of the cell and is surrounded by a membrane, the nuclear membrane. The nucleus is where higher organisms keep the chromosomes carrying their genetic information.

Lower organisms such as bacteria are known as prokaryotes and typically have two or three thousand genes carried on a single circular chromosome. In contrast, eukaryotes have many thousand genes carried on several linear chromosomes. Humans and fruit flies are both estimated to have between 50,000 and 100,000 genes. The difference between you, dear reader, and some benighted insect, is not so much the number of your genes, but their organization.

11.1 EUKARYOTIC CELL SHOWING ORGANELLES

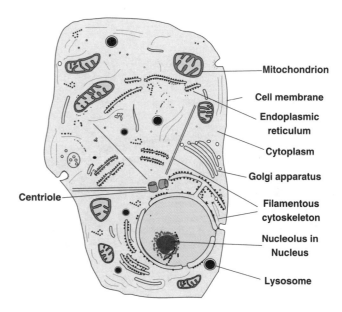

Mitochondrion
Cell membrane
Endoplasmic reticulum
Cytoplasm
Golgi apparatus
Filamentous cytoskeleton
Nucleolus in Nucleus
Lysosome
Centriole

Cell Structure in Eukaryotes

Eukaryotic cells are much bigger than bacterial cells and are divided into separate compartments by membranes. In addition to the nucleus, eukaryotic cells have several other compartments surrounded by membranes. These are known as "membrane bound organelles" since they are like the organs of animals and plants but on a miniature scale.

Mitochondria are Used for Respiration

The most common organelles are the mitochondria (singular, mitochondrion). Almost all eukaryotic cells have mitochondria, which are responsible for generating energy by the oxidation of food molecules (Fig. 11.2). This process is known as **respiration**. You might think you are respiring when your lungs suck in some air, but you are only breathing. To qualify officially as respiration, the oxygen from the air must reach the mitochondria in your cells and be used to burn the food molecules as fuel. Otherwise the air does you no good.

respiration process of releasing energy from food molecules by reacting them with oxygen

photosynthesis process of trapping light energy to form food molecules

11.2 MITOCHONDRION

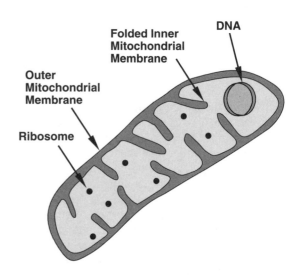

DNA

Folded Inner Mitochondrial Membrane

Outer Mitochondrial Membrane

Ribosome

Chloroplasts Trap Energy from the Sun

Chloroplasts are organelles found only in plants (Fig. 11.3). They carry out **photosynthesis**, a process of trapping the energy in sunlight (the "photo" bit) and using it to make sugars (the "synthesis" part). The green pigment, chlorophyll, which absorbs sunlight, is located on the internal membranes of the chloroplast. These membranes are highly folded so as to pack more light absorbing area into each chloroplast.

11.3 CHLOROPLAST

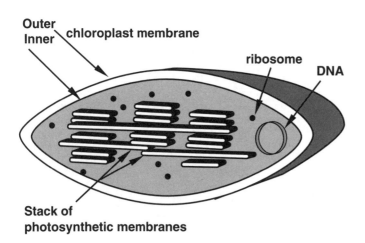

Outer
Inner chloroplast membrane

ribosome
DNA

Stack of photosynthetic membranes

Chromosome Structure in Eukaryotes

Higher organisms have multiple chromosomes, whereas bacteria only have one. Moreover, unlike bacterial chromosomes which are circular, eukaryotic chromosomes are linear. This means they have ends and a middle (Fig. 11.4). Both of these are special structures. The middle is called the **centromere** and although it is often more or less centrally located, it is sometimes closer to one of the ends. At cell division, when the chromosomes are replicated to give two copies, microscopic fibers (microtubules) are attached to the centromeres and drag the two sets of chromosomes apart. The end structures of eukaryotic chromosomes are called **telomeres**.

centromere structure used to pull chromosomes apart after division, usually found more or less in the center

telomere special end structure found on a linear eukaryotic chromosome

11.4 GENERAL EUKARYOTIC CHROMOSOME STRUCTURE

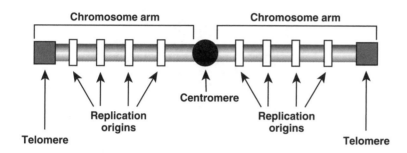

"The End is Nigh" - Telomeres

Whenever a DNA molecule is replicated, the process of making a new chain always starts with a **primer** made of RNA (see Ch. 5 for details). This primer is later removed. If the DNA molecule is circular there is no problem, replication goes around the circle and fills from behind the gap where the primer used to be. If the DNA molecule is linear, there is a problem which must be overcome. As Figure 11.5 shows, a linear chromosome would get shorter, by one primer length, each time it was replicated. Eventually, if nothing was done, it would fade away like the Cheshire cat.

The telomeres at the ends of eukaryotic chromosomes consist of a six base pair sequence repeated about 2,000 times. During each replication cycle the chromosomes are, in fact, shortened due to loss of the RNA primer. However, an enzyme known as **telomerase** cancels this loss out by adding a few of the six base pair chunks each time around (Fig. 11.6).

primer short segment of nucleic acid that binds to the longer template strand and allows the start of synthesis for a new chain of DNA

telomerase enzyme that adds DNA to the end, or telomere, of a chromosome

11.5 THREAT OF SHORTENED EUKARYOTIC CHROMOSOMES

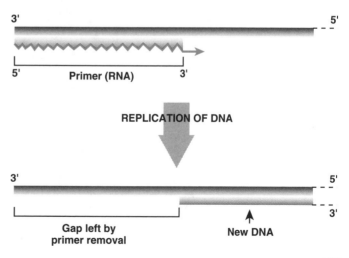

Presumably in honor of the lost RNA primer, telomerase carries around with it a small bit of RNA complementary to the six base pair telomere repeat. This allows it to recognize the telomeres, and reminds it what sequence to make.

11.6 TELOMERASE REPLACES THE ENDS

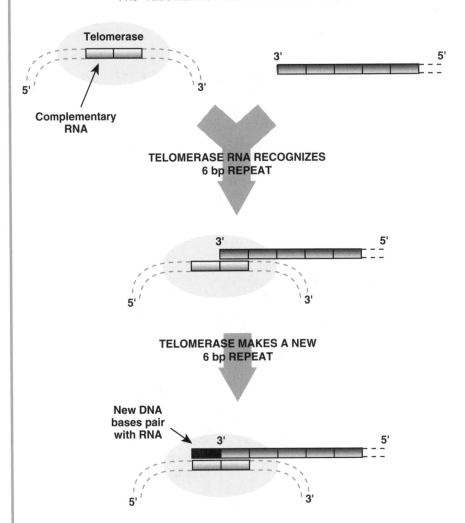

Back up Your Files!!

People often need to make a back-up copy of important documents, especially if they have to submit the original to a government agency! In the same way, most eukaryotic cells have a back-up copy of each gene. Since genes are carried on chromosomes, this means that they have pairs of duplicate chromosomes which are best seen when they line up with their partners at cell division. Human cells have $2 \times 23 = 46$ chromosomes.

A cell with only a single set of chromosomes and hence only a single copy of each of its genes, is known as **haploid**. Cells with duplicate copies of their genes are referred to as **diploid**.

Bacteria have only a single copy of each of their genes. So if a vital gene is damaged, that whole cell containing the defect is doomed. On the other hand, bacteria are small and divide rapidly so they can afford to lose a few of their comrades. Anyway, no one has ever sued for accidental death of a bacterium and won $23 million.

Eukaryotic cells are much larger and divide more slowly. In other words, each cell represents a greater investment of time and resources, so keeping a back-up copy of each gene is a sensible policy.

How is DNA Packaged?

Eukaryotic cells have vastly more DNA than lower organisms. Even though eukaryotic cells have 10 to 20 times as many genes as bacteria, most of their DNA does not even consist of genuine coding sequences. Some of this extra DNA is found between genes, whereas amazingly, other non-coding DNA is actually inserted into the genes (see below). Since as much as 95 percent of the DNA in a eukaryotic cell may be non-coding, this means that it may contain 500 times as much DNA as a bacterial cell.

This eukaryotic DNA is stored on the chromosomes in the nucleus. Each chromosome is a single molecule of DNA. In bacteria, there are approximately 3,000 genes on a single chromosome which is about 1 millimeter long. They are thus 1,000 times longer than the bacterial cell in which they belong (see Ch. 3). Eukaryotic chromosomes may be as much as a centimeter long and must be folded up to fit into the cell nucleus which is 5 **microns** across, a 2,000-fold shortening.

The DNA starts folding by coiling around the **histones**, positively charged proteins that neutralize the negative charge of the DNA itself. DNA with histones bound to it was named **chromatin** when it was first discovered in chromosomes. Each 200 base pairs of DNA is wrapped around nine histone proteins forming a **nucleosome** (Fig. 11.7). Eight of the histones cluster together and 140 base pairs of DNA are coiled around them, so forming a core particle. A linker region consisting of the remaining 60 bp of DNA and the ninth histone molecule join each core particle to the next.

Overall, a chromosome with its DNA twisted into a series of nucleosomes would resemble a string of beads, except that the folding process continues. The chain of nucleosomes is wound helically into a giant solenoid structure with six nucleosomes per turn, the 30 nanometer fiber (Fig. 11.8). In turn these fibers are looped back and forth. There are about 50 of the solenoidal turns per loop and the ends of the loops are attached to a protein scaffold. Eighteen of these loops are then wound radially around the chromosome axis to give a miniband. Roughly a million such minibands form a complete chromosome.

haploid possessing only a single copy of each gene

diploid possessing a double copy of each gene

micron a millionth part of a meter

histones special positively charged proteins that bind to DNA

chromatin complex of DNA plus protein constituting eukaryotic chromosomes

nucleosome subunit of a eukaryotic chromosome consisting of DNA coiled around histone proteins

11.7 NUCLEOSOME

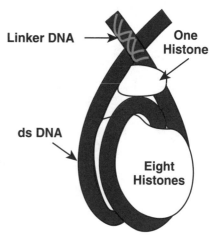

Linker DNA — One Histone

ds DNA

Eight Histones

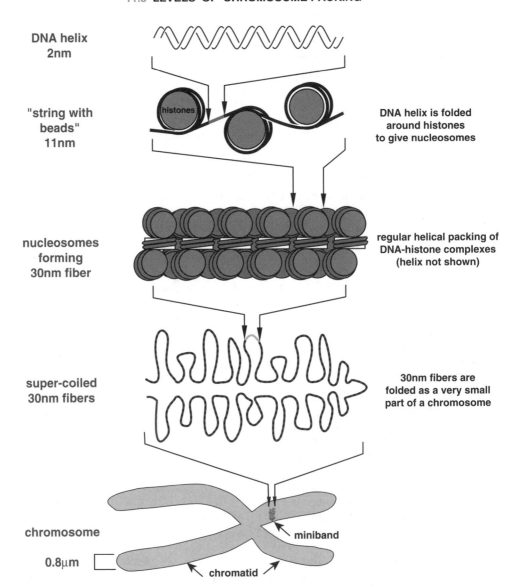

DNA helix
2nm

"string with
beads"
11nm

DNA helix is folded
around histones
to give nucleosomes

nucleosomes
forming
30nm fiber

regular helical packing of
DNA-histone complexes
(helix not shown)

super-coiled
30nm fibers

30nm fibers are
folded as a very small
part of a chromosome

chromosome

0.8μm

miniband

chromatid

SUMMARY OF CHROMOSOME FOLDING

Level of Folding	Consists of	Base Pairs per Turn
DNA double helix	nucleotides	10
Nucleosomes	200 base pairs each	100
30 Nanometer fiber	6 nucleosomes per turn	1,200
Loops	50 solenoid turns per loop	60,000
Miniband	18 loops	1,080,000
Chromatid	1,000,000 minibands	

When chromosomes are visible under the microscope it is because they have been caught in the act of dividing. Before cell division and during normal cell operations, the DNA is spread out for use in transcribing genes. At this point the chromosome consists of a single molecule of double helical DNA. In this state, it does not look at all like typical chromosome pictures. Just before cell division, the DNA condenses and is folded up tightly enough so it becomes much easier to see.

chromatid single double-helical DNA molecule making up the whole, or part, of a chromosome

The typical metaphase chromosome, seen in most pictures, has just replicated its DNA and therefore consists of two identical DNA molecules still held together by the centromere. These duplicate strands of DNA are known as chromatids. This metaphase chromosome is just about to divide into two daughter chromosomes as shown in Figure 11.9. But remember that in a non-dividing cell the chromosomes only have a single chromatid.

11.9 **INTERPHASE AND METAPHASE CHROMOSOMES**

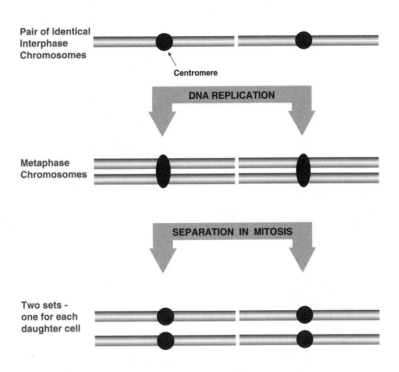

Pair of identical Interphase Chromosomes

Centromere

DNA REPLICATION

Metaphase Chromosomes

SEPARATION IN MITOSIS

Two sets - one for each daughter cell

Repeated Sequences

Most genes are present only as single copies. Such unique sequences account for almost all bacterial DNA. However, in higher organisms, unique sequences may comprise as little as 20 percent of the total DNA. For example, we humans have 65 percent unique DNA, whereas frogs have only 22 percent. The rest of the DNA is made up of repeated sequences of one kind or another.

Since each prokaryotic cell contains 10,000 or more ribosomes, it is not surprising that their DNA usually contains half a dozen copies of the genes for ribosomal RNA and transfer RNA. In the much larger eukaryotic cell, there are hundreds or thousands of copies of the ribosomal RNA and transfer RNA genes.

Sequences present in hundreds or thousands of copies are referred to as "moderately repetitive sequences." Some are multiple copies of highly used genes, like those for rRNA, whereas others are nonfunctional stretches of DNA that merely fill up space in the chromosome. About 25 percent of human DNA falls into this category. Another 10 percent of our human DNA consists of sequences present in hundreds of thousands to millions of copies and which are referred to as "highly repetitive sequences." These are almost all useless as far as is known.

repetitive sequence DNA sequences which exist in many copies

The most famous highly repetitive sequence is the **Alu element**. From 300,000 to 500,000 copies of this 300 base pair sequence are scattered throughout human DNA. Though apparently useless, they make up 6 to 8 percent of our genetic information. They occur singly or in clusters and many are mutated or incomplete. The figure below (Fig. 11.10) shows a chromosome with some of each of these assorted features.

11.10 STYLIZED EUKARYOTIC CHROMOSOME

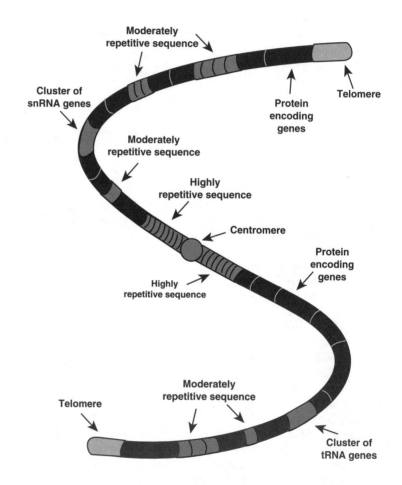

Pseudogenes

In addition to repetitive sequences present in impressive numbers of copies, eukaryotic cells also possess something a whole lot less inspiring, **pseudogenes**. These are defective duplicate copies of genuine genes. These are present in only one or two copies and may be next to the original, functional version of the gene or may be far away, even on a different chromosome. Since they contain various defects which prevent them from being expressed, they are useless.

Gene Structure in Eukaryotes

Not only is useless DNA scattered throughout the eukaryotic chromosomes between the genes, but the actual genes themselves are often interrupted with non-coding DNA. These intervening sequences are known as **introns** and the regions of the DNA which contains coding information are known as **exons**. Most eukaryotic genes consist of exons alternating with introns (Fig. 11.11). In lower, single-celled eukaryotes, introns are relatively rare and often quite short. In contrast, in higher eukaryotes, most genes have introns and they are often longer than the exons. In some genes, the introns may occupy 90 percent or more of the DNA.

intron segment of a gene that does not code for protein

exon segment of a gene that codes for protein

11.11 EXONS AND INTRONS

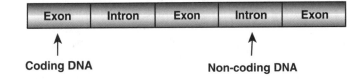

Coding DNA   Non-coding DNA

Transcription in Eukaryotes

Since eukaryotic cells have 10 times as many genes as bacterial cells, deciding which to turn on, and when, is much more complicated. Consequently, the whole business of transcription is more complex. For a start, eukaryotes have three different RNA polymerases, unlike bacteria which have just one. The "union" rules for RNA polymerase are as follows:

RNA Polymerase No.	Genes Transcribed:
I	genes for large ribosomal RNAs
II	genes which code for proteins
III	genes for tRNA, 5S rRNA and a few other tiny RNAs

Since ribosomal RNA and transfer RNA are needed all the time and in all types of cells, the genes encoding them are regarded as **"housekeeping genes"** and RNA polymerases I and III go about their business pretty much automatically. In contrast, RNA polymerase II actually has to think about what it is doing. In a multicellular organism, different cell types produce different types of proteins. For example, red blood cells produce hemoglobin, whereas white blood cells make antibodies. Worse still, protein production often varies during development. Fetal hemoglobin is different from the adult version.

The activity of RNA polymerase II is regulated by a large number of accessory proteins, called **transcription factors**, that bind to and recognize specific sequences on the DNA. These DNA sequences are of two major classes, the **promoter** (Fig. 11.12) and the **enhancers** (Fig. 11.13).

housekeeping genes genes which are switched on all the time because they are needed by all cells for essential life functions

transcription factors proteins which control gene expression by binding both to specific sequences on DNA and to the RNA polymerase

promoter region of DNA in front of a gene which binds RNA polymerase and so promotes gene expression

enhancers regulatory sequences outside the actual promoter which bind transcription factors so increasing gene expression

11.12 PROMOTER LAYOUT IN EUKARYOTES

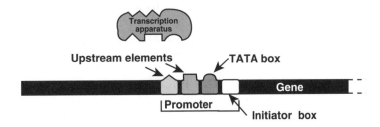

139

Promoters are found in front of all genes, both prokaryotic and eukaryotic. A eukaryotic promoter consists of three regions. The initiator box is a sequence found at the site where transcription starts. About 25 base pairs upstream from this is the TATA box, an AT-rich sequence which is recognized by a protein, imaginatively called the TATA box factor. RNA polymerase II, the TATA box factor and some other proteins stay together as a bulky complex sometimes known as the "transcription apparatus."

The third component of the promoter is the upstream element. There are many different upstream elements. They are about 10 base pairs long and are recognized by specific proteins. In fact, depending on the gene, there may be more than one upstream element in a given promoter. The more upstream elements, the more complex the control of transcription.

Enhancers are sequences which are involved in gene regulation, especially during development or in different cell types. Enhancers do exactly what their name indicates, they enhance the rate of transcription as a result of binding certain specific transcription factors. Although enhancers are sometimes found close to the genes they control, more often they lie at some considerable distance, perhaps thousands of base pairs away. Even odder is that they may be located either upstream or downstream from the promoter. When an enhancer switches a gene on, the DNA between it and the promoter loops out as shown in Figure 11.13.

11.13 ENHANCER ACTION INVOLVES LOOPING OF DNA

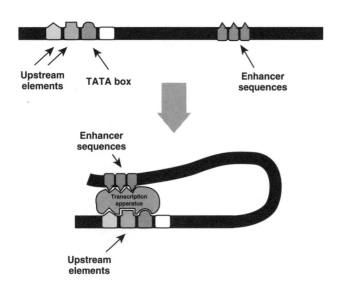

Transcription Factors

These are specialized proteins that regulate gene expression by controlling transcription. They have four domains needed for the following functions:

1) binding to a specific sequence on the DNA
2) binding to the RNA polymerase II complex
3) getting into the nucleus where the genes are kept
4) responding to a stimulus of some sort which signals that the gene should be turned on.

An example of a transcription factor is MyoD (Fig. 11.14) which only appears in those cells destined to become muscle cells. The MyoD factor switches on a wide selection of genes needed in muscle cells, but is not required in other cell types. A special class of transcription factors is needed for the development of embryos into adults in higher organisms.

RNA Processing

You might think now that RNA polymerase has done its work of transcribing the gene, we have our messenger RNA and we can zoom off to the ribosome and get translated into protein. Not so fast. For one thing, transcription has just taken place using the DNA which is inside the nucleus, whereas the ribosomes are outside!

Far far worse, however, is that most eukaryotic genes have those useless intervening sequences, the introns. Thus the DNA sequence of a eukaryotic gene consists of regions which code for part of the final protein, the exons, alternating with the regions of non-coding DNA, the introns.

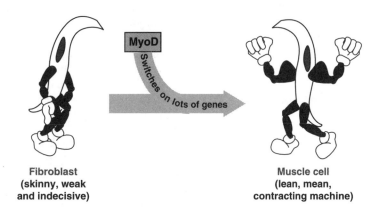

11.14 FIBROBLASTS BECOME A MUSCLE CELL

Fibroblast (skinny, weak and indecisive)

MyoD Switches on lots of genes

Muscle cell (lean, mean, contracting machine)

The RNA molecule resulting from transcription is known as the **primary transcript** (Fig. 11.15). It is not actually genuine certified messenger RNA because it, too, has exons alternating with introns. If we translated the primary transcript we would get a huge dysfunctional protein with lots of extra stretches of garbage.

primary transcript the RNA molecule produced by transcription before it has been processed in any way

11.15 EXPRESSING A EUKARYOTIC GENE

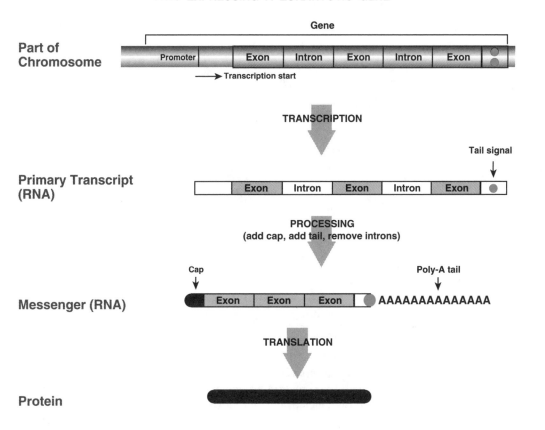

In fact the primary transcript is trapped inside the nucleus until the introns are removed. This is known as **splicing** and involves cutting out the introns and joining the ends to generate an RNA molecule which has only the exons, *i.e.*, it contains an uninterrupted coding sequence.

In order to be recognized as a bona fide messenger RNA molecule, two other modifications must be made. These are the addition of a cap structure to the front and a tail to the rear of the RNA molecule. In fact, as shown in Figure 11.15, these are added before splicing out the introns.

Capping and Tailing

Just as a college graduate receives a cap and gown before leaving the university to earn money in the real world, so too, messenger RNA must be capped and tailed before being allowed to leave the nucleus. RNA molecules destined to become messenger RNA, have a cap added to their 5' ends and a tail added to their 3' ends. This occurs inside the nucleus and before splicing. Shortly after transcription starts, the 5' end of the growing RNA molecule is capped by the addition of a guanosine monophosphate (GMP) residue (Fig. 11.16). This is added in a backwards orientation relative to the rest of the bases in the RNA.

11.16 ADDITION OF CAP TO EUKARYOTIC RNA

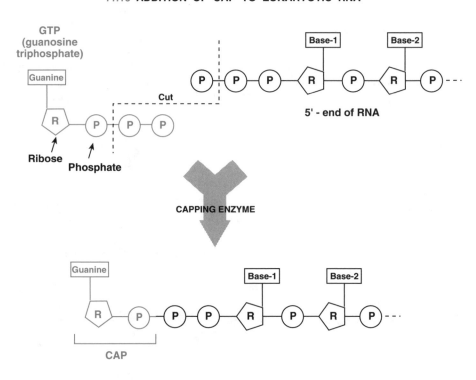

After addition of the GMP, the guanine base has a methyl group attached. Extra methyl groups may be added to the ribose sugars of the first one or two nucleosides of the RNA in some higher organisms.

After being capped, the growing RNA is tailed (Fig. 11.17). There is a recognition sequence - AAUAAA - at the 3' end. The RNA polymerase which is making the RNA molecule, cruises on past this point. However, a specific endonuclease recognizes this sequence and cuts the growing RNA molecule 10 to 30 bases downstream. The enzyme poly(A) polymerase now comes along and adds a run of 100 to 200 adenine residues to form the tail.

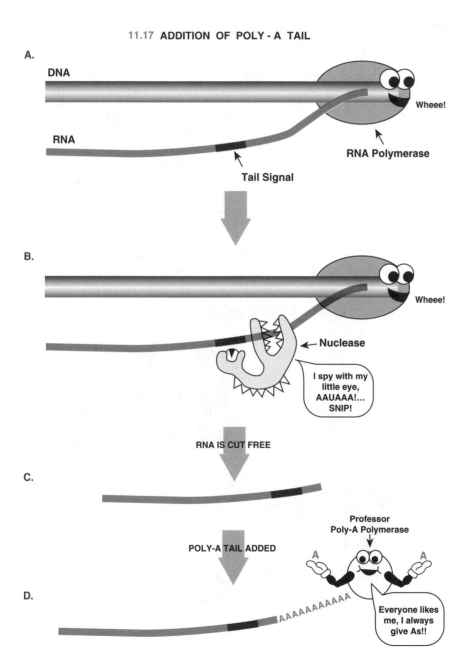

11.17 ADDITION OF POLY - A TAIL

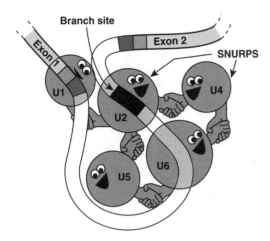

Branch site

Exon 1

Exon 2

SNURPS

U1

U2

U4

U5

U6

spliceosome macromolecular complex which removes introns from RNA

small nuclear RNA (snRNA) small molecules of RNA found only in the nucleus which oversee the splicing of mRNA

snRNP "snurps" - small nuclear ribonucleoproteins contain proteins plus snRNA and are responsible for RNA splicing

RNA Splicing

The next step in the processing of the RNA is the splicing out of the introns. The splicing machinery is known as the **spliceosome** and consists of several proteins and some specialized, small RNA molecules found only in the nucleus. Each **small nuclear RNA** ("snRNA") plus its protein partners forms a small nuclear ribonucleoprotein **(snRNP)**. There are five snRNP's - numbered from U1 to U6 (with U3 missing!). We can picture them as little elves, sorry, I mean snurps, working together to splice the RNA.

During the first stage of splicing, the spliceosome recognizes both ends of the intron and binds to them. This makes the intron DNA loop out as shown in Figure 11.18.

So why do the snurps need both RNA and protein? They use their small RNA molecules for the intellectual task of recognizing the splice and branch sites on the larger RNA molecule they are processing. As you would imagine, this is done by base pairing. The protein part of the snurp then does the manual labor of cutting and sticking. This is illustrated in Figure 11.19 for the U1 snurp which recognizes the 5' splice site.

11.19 SNURP U1 RECOGNIZES SPLICE SITE

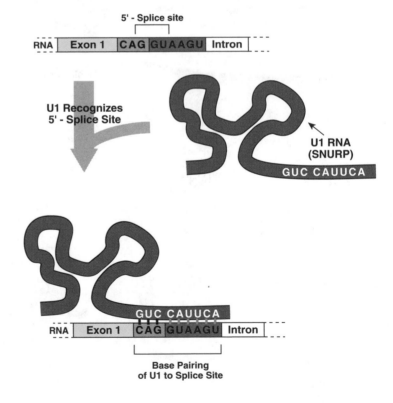

5' - Splice site

RNA | Exon 1 | CAG | GUAAGU | Intron

U1 Recognizes
5' - Splice Site

U1 RNA
(SNURP)

GUC CAUUCA

GUC CAUUCA

RNA | Exon 1 | CAG | GUAAGU | Intron

**Base Pairing
of U1 to Splice Site**

Splicing must be accurate to within a single base since a mistake would throw the whole coding sequence out of register and totally scramble the protein eventually resulting from translation of the mRNA. The overall result of all this cutting and pasting is shown below (Fig. 11.20).

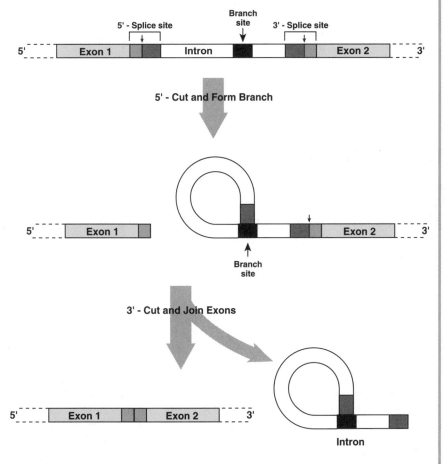

11.20 SPLICING – THE OVERALL SCHEME

Mutation of the proper splice sites within the genes for hemoglobin may cause a defective hemoglobin protein and one form of a hereditary disease known as thalassemia. Depending on the particular mutation, thalassemia may be relatively mild or extremely severe.

In the first stage of splicing, the 5' end of the intron is snipped off from the exon to its right and joined to an adenine in the branch site in the middle of the intron. The branch point is supervised by snRNP U2. Meanwhile the U1 snRNP must hold on to the loose end of the right hand exon. During the second stage of splicing, the intron is cut loose from the left hand exon. Finally, the free 3' end of the right hand exon is joined up to the 5' end of the left hand exon.

The discarded intron is referred to as the "lariat product" because of its shape

Alternative Splicing

Although any particular splice junction must be made with total precision, eukaryotic cells can sometimes choose to use different splice sites within the same gene. Generally, alternative splicing is used by different cell types within the same animal. This allows a single original DNA sequence to be used to make several different proteins which have distinct but overlapping functions. There are four main types of alternative splicing:

alternative splicing variations in processing mRNA which allow more than one possible protein to be made from a single gene

FIRST: Alternative Promoter Selection

In this case, two alternative promoters are possible. The choice between them depends on cell-type specific transcription factors. Note that in this case there are two alternative primary transcripts (Fig. 11.21). If promoter P1 is used, then the sequence containing P2 and exon-2 is

11.21 ALTERNATIVE PROMOTER SELECTION

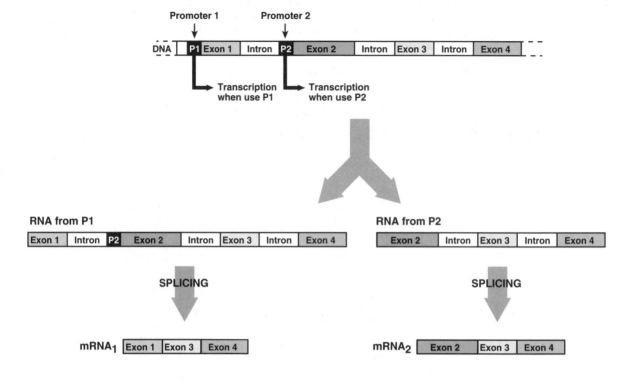

spliced out. If promoter P2 is used, then exon No. 1 is not even part of the primary transcript and is therefore not in the mRNA.

SECOND: Alternative Tail Site Selection

Two alternative sites for adding the poly A tail may be possible. The choice between them again depends on cell type. In this case, cleavage at the earlier poly A site results in loss of the distal exon (Fig. 11.23). If the later poly A site is chosen, then the earlier poly A site and the exon just in front of it are spliced out. This mechanism is used to produce antibodies which recognize the same invading, foreign molecule but which have different rear ends (see Ch. 22). One type of antibody is secreted into the blood, whereas the other type remains attached to the cell surface.

11.22 ALTERNATIVE VERSION OF ALTERNATIVE TAIL SITE SELECTION

Yes, of course, we get to choose our own tails!

THIRD: Alternative Splicing by Exon Cassette Selection

Here we have a genuine choice of the actual splicing sites. Depending on the choice made, a particular exon may or may not appear in the final product as shown in

146

11.23 ALTERNATIVE TAIL SITE SELECTION

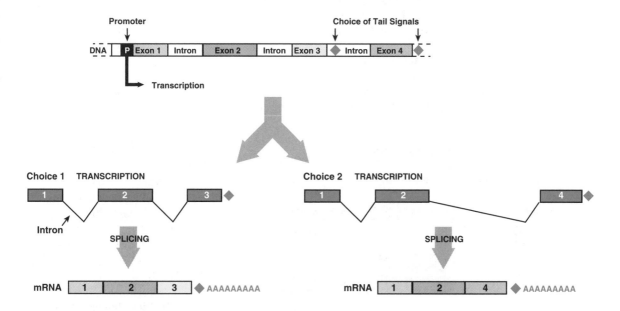

Figure 11.24. Here the primary transcripts are actually the same. They are drawn differently in Figure 11.24 to illustrate the splicing plans. Some cell-type specific factor which recognizes the different possible splice sites must come into play here, but the details are still obscure.

Exon cassette selection occurs in the gene for the skeletal muscle protein troponin T. In the rat this gene has 18 exons. Of these, 11 are always used. Five (exons No. 4 through No. 8) may be used in any combination

11.24 PRINCIPLE OF EXON CASSETTE SELECTION

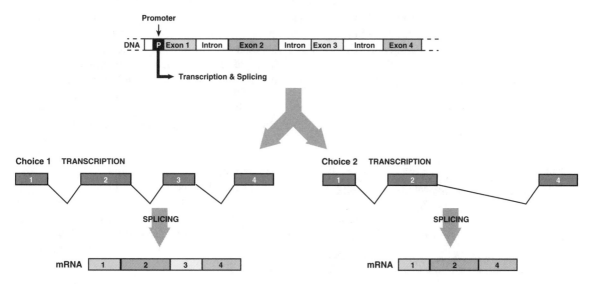

(including "none of the above") and the final two (exons No. 16 and No. 17) are mutually exclusive, and one or the other must be chosen. This gives a mind-boggling 64 possible final messenger RNAs. The result is that muscle tissue contains multiple forms of this, and other!, structural proteins.

trans-splicing making a messenger RNA by joining together segments from two separate original RNA molecules

FOURTH: Trans-Splicing

Until now we have spliced together segments of the same gene. Just as well, you probably think. Alternative splice sites within a single gene are quite confusing enough. Sewing random bits of one gene into the sequence for another would surely cause total confusion. All the same there are some desperate individuals willing to take the risk.

Trypanosomes are parasitic single celled eukaryotes that cause sleeping sickness and other gruesome diseases. They evade immune surveillance by constantly changing the proteins on their cell surfaces by the dirty genetic trick of shuffling gene parts (see Ch. 21). In addition they specialize in trans-splicing of many genes (Fig. 11.25). On the other hand, trypanosomes do not appear to have introns and so do not have normal splicing!

Although it has not (yet!) been found in higher animals, trans-splicing of segments from one RNA molecule into another also occurs in certain primitive worms, the nematodes, and in the chloroplasts of plant cells.

11.25 TRANS - SPLICING

A & B are Primary Transcripts (RNA) from Related Genes

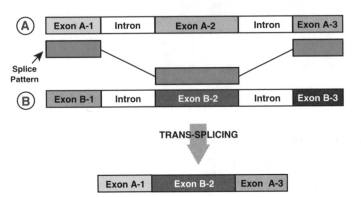

RNA Editing

You thought the weird stuff was over. But no, it gets kinkier. Even more bizarre is RNA editing. Those sneaky enemies, the trypanosomes, are into this too. If you are going to break one rule you may as well break them all. Some of the primary transcripts of trypanosomes are altered by insertion or removal of uridine nucleotides, one at a time, before the final messenger RNA is generated (Fig. 11.26). If the trypanosome did not edit its RNA, the result would be a defective, frame-shifted protein made from an out of phase mRNA.

Transport out of the Nucleus

The nucleus is surrounded by a double membrane. Each nucleus has many pores that allow molecules in or out in a carefully controlled manner. Each nuclear pore is guarded by a platoon of proteins, but the details of just who is allowed in or out are still murky. We do know

11.26 RNA EDITING IN TRYPANOSOMES

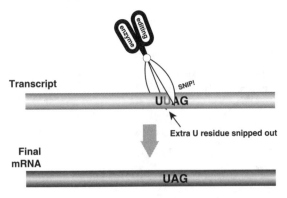

that once messenger RNA has received its cap and tail and had its introns spliced out, it is free to exit the nucleus. The splicing factors (those tricky little snurps) that bind to the RNA, prevent it from leaving until splicing is finished (Fig. 11.27).

Protein Synthesis in Eukaryotes: How the Big Boys Do It

You will be relieved to know that protein synthesis in higher organisms is much the same as in bacteria (see Ch. 7), with few major differences. The ribosomes of eukaryotic cells are a bit bigger than those of prokaryotic cells, and contain several more proteins. Due to this, the ribosomal subunits of higher organisms are referred to as the 40S (small) and 60S (large) subunits. Together they form an 80S ribosome.

In eukaryotic cells, the ribosomes, the sites of protein synthesis, are in the cytoplasm. However, manufacture and processing of mRNA occurs inside the nucleus. Consequently, the messenger RNA molecule must be released from the nucleus before it can be bound to a ribosome and translated into protein. In prokaryotic cells the ribosome can bind to mRNA and may get started making protein before the mRNA has been finished itself (see Ch. 7). Obviously, this cannot occur in eukaryotic cells. The important features of mature eukaryotic mRNA are shown in Figure 11.28.

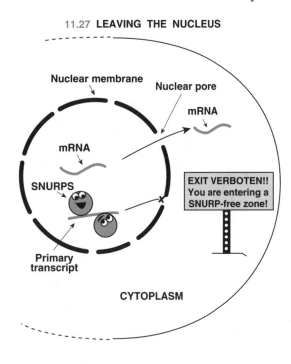

11.27 LEAVING THE NUCLEUS

ribosome   the cell's machinery for making proteins

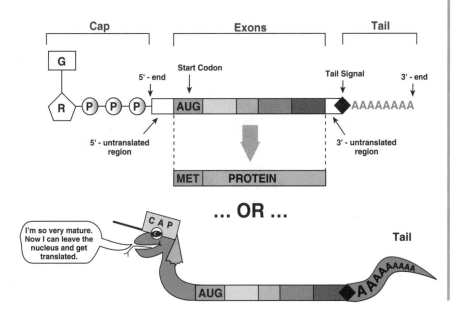

11.28 MATURE EUKARYOTIC mRNA

A whole slew of special proteins, the initiation factors, are needed to
get the eukaryotic ribosome ready for action. The first transfer RNA, the
small (40S) ribosomal subunit, the mRNA plus cap binding protein, and the
large (60S) ribosomal subunit are all bound individually by different initiation factors. The initiation factors assemble the complete ribosome plus the
tRNAs and mRNA in the correct order.

First to tango are the small (40S) ribosomal subunit and the **initiator
tRNA** (Fig. 11.29). Eukaryotic proteins are made starting with the amino
acid methionine, just as in prokaryotes. There is a special, initiator, tRNA,
but, unlike prokaryotes, no formyl-group is used to tag the first methionine.

11.29 FORMATION OF THE 40S EUKARYOTIC PREINITIATION COMPLEX

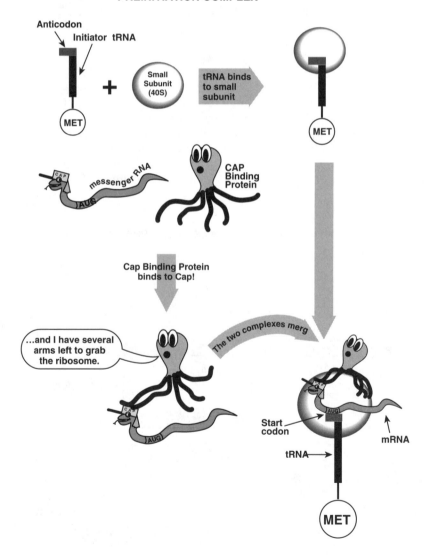

Eukaryotic messenger RNA does not have a ribosome binding site (or Shine-Dalgarno sequence) as in prokaryotes. Instead, it is recognized by the special cap structure at the 5' end. Cap binding protein binds to the cap of mRNA and turns over the mRNA to the small ribosomal subunit. Because there is no Shine-Dalgarno sequence to align the mRNA in eukaryotes, the first AUG codon of the mRNA is recognized as the start site for protein synthesis. Once the initiator tRNA has found the **start codon**, the large (60S) ribosomal subunit binds and protein synthesis gets rolling (Fig. 11.30).

start codon the first codon - AUG - in the coding sequence for a protein

11.30 FORMATION OF THE 80S INITIATION COMPLEX

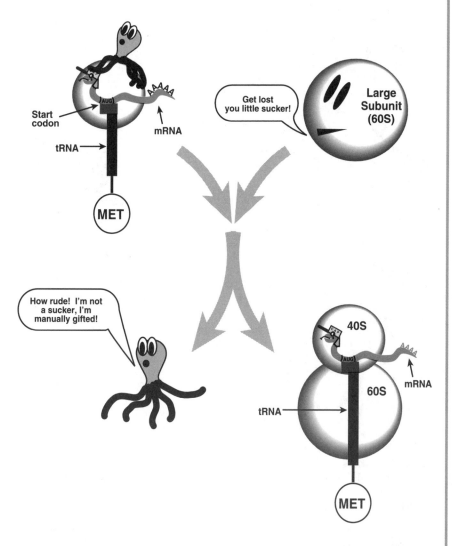

The incoming amino acids are linked into a polypeptide chain, pretty much as in bacteria (see Ch. 7). Eukaryotes only have a single coding sequence on each messenger RNA. Therefore, they make only a single

protein per mRNA. Once the ribosome reaches the stop codon it disassembles. The newly-made protein, the mRNA, and the two ribosomal subunits are all released (Fig. 11.31). This is under the control of a single protein which recognizes the stop codon, eukaryotic **release factor**.

11.31 TERMINATION OF PROTEIN SYNTHESIS

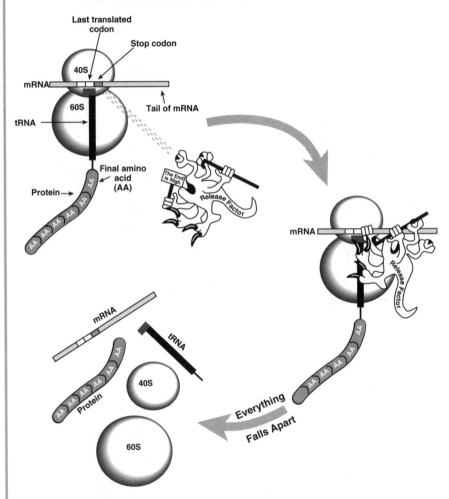

Whew! We're done at last! Now our eukaryotic cells are up and running and happily cranking out lots of proteins. So let's move on and worry about all the things that can go wrong in the next couple of chapters!

Additional Reading

Human Genetics: The Molecular Revolution by McConkey EH. 1993. Jones & Bartlett Publishers, Boston & London

Molecular Biology of the Cell by Alberts B, Bray D, Lewis J, Raff M, Roberts K, & Watson JD. 3rd ed., 1994. Garland Publishing, Inc., New York & London.

Mutations:
Things That Go Bump
In The Night

12

Although few of us resemble the pulsating blobs of green slime which always seem to come crawling out of the sewers after radiation accidents in Sci-Fi movies, we do have something in common with these creatures - we are all mutants!

Nature is not perfect and mistakes can happen. Errors may occur in any of the processes of molecular biology. A mistake in a cell's genetic material is known as a **mutation**. At the molecular level, mutations are alterations in the DNA molecules making up the genes. Because of this, mutations will be passed on from a parent cell to its descendants; they are inherited defects. When humans carry a mutation in their reproductive cells which leads to an observable defect, we talk about inherited disease (see Ch. 13 for more on this). In fact, all of us are mutants - many times over - and we all have quite a substantial number of mistakes in our genes. That includes you!

12.1 WE ARE ALL MUTANTS

Look!! An ugly scary mutant!

mutation an alteration in the genetic information

So why aren't we all dead, dying of some horrible disease or turning into putrid balls of slime which would have to read the rest of this chapter by holding the book in half a dozen tentacles? There are two main reasons why. First, there are many different types of mutations and most of them have only a very minor effect; in fact many appear to cause no noticeable defect at all. Relatively few mutations cause such large changes that they attract our attention.

Secondly (as discussed in Ch. 11), higher organisms have two copies of each gene. This means that if one copy is mutated, there is a back-up copy which can be used. This is just as well. It has been estimated that a typical human carries enough harmful mutations to total approximately eight lethal equivalents. This means that if we were haploid, with only a

12.2 CATS HAVE NINE LIVES

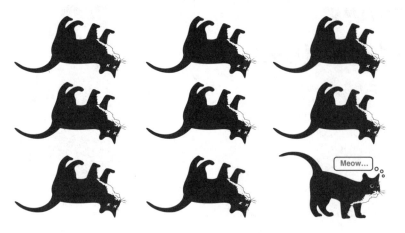

Meow...

single copy of each gene, we would all be dead eight times over. The only survivors would be pussy cats which, as is well known, have nine lives! (Fig. 12.2)

Mutations Alter the DNA

So let's get down to the nitty gritty. A mutation is a change in the base sequence of the DNA. There are many possible changes we can make. These may be illustrated by considering their effect on a well known literary gem:

Wild Type	The cat sat on the mat.
Substitution	The **r**at sat on the mat.
Insertion (single)	The cat s**p**at on the mat.
Insertion (multiple)	The cat**tle** sat on the mat.
Deletion (single)	The c•t sat on the mat.
Deletion (multiple)	The cat ••• •• the mat.
Inversion (small)	The **tac** sat on the mat.
Inversion (large)	**Tam eht no tas tac eht.**

Obviously, such changes alter the meaning of the sentence to varying degrees. Similarly, altering the DNA base sequence has a variety of effects. To understand these, let's recall the **central dogma** of molecular biology (Fig. 12.3):

First, DNA is the genetic material. When a DNA molecule replicates, any changes due to mutations of the original DNA base sequence will be duplicated and passed on to the next generation. In other words, mutations are inherited.

Second, the DNA is used as a template in

central dogma general scheme showing how DNA is replicated and how proteins are made

12.3 CENTRAL **DOG**MA

Replication

DNA

Transcription

RNA

Translation

Protein

transcription to make an RNA copy. Therefore the mutation in the DNA sequence will be passed on to the RNA molecule. Finally, the messenger RNA is translated to give protein. An altered RNA sequence may be translated into an altered protein. Since cells depend on proteins to carry out all their chemical reactions, the final result of a change in the DNA sequence may be a defect in the operation of some vital reaction.

Silent Mutations

A **silent mutation** is not a mutation that stops the cat on the mat from meowing. It is a mutation in the DNA sequence that has no effect on the operation of the cell and is therefore not so much silent as invisible from the outside. In other words, silent mutations do not alter the phenotype.

One obvious way to get a silent mutation is if the base change occurs in the non-coding DNA between genes (Fig. 12.4). Therefore no genes are damaged and no proteins are altered.

Higher organisms possess intervening sequences - **introns** (see Ch. 11) - within many of their genes. Since the intron is cut out and discarded when the messenger RNA is made, an alteration in its sequence will not affect the final protein (Fig. 12.5). Note that not all base changes in an intron are harmless; we must not alter the few important bases at the splice recognition sites or disaster will result (see Ch.11). Nevertheless, most base changes within an intron are also silent mutations.

silent mutation a mutation with no observable effect on cell growth or survival

intron segment of a gene that does not code for protein

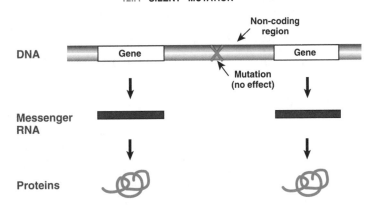

12..4 **SILENT MUTATION**

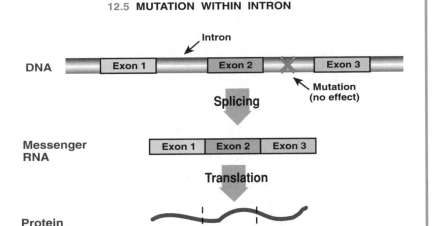

12.5 **MUTATION WITHIN INTRON**

The third main type of silent mutation is within the coding region of a gene and does get passed on to the messenger RNA. How can this be? The key is to remember the genetic code. Each codon, or group of three bases, is translated into a single amino acid in the final protein product. However, because there are 64 different codons, most of the 20 possible amino acids have more than one codon (see Codon Table, Fig. 12.6). So a base change that converts the original codon into another codon that codes for the same amino acid will have no effect on the final structure of the protein.

For example, the amino acid alanine has four codons: GCU, GCC, GCA and GCG. (Note that we are discussing this in RNA language; these are the codons as found on mRNA.) Since they all have GC as the first two bases, any codon of the form GCX (X = any base) will give alanine. So if we start with GCC and mutate the last C to an A, this changes the codon to GCA, but we still get alanine in the resulting protein. Many other amino acids (such as valine, threonine and glycine) also have sets of four codons in which the last base does not matter. This is referred to as **third base redundancy**. If you examine the codon table (Fig. 12.6) you will see that altering the third base may have no effect on the protein that will be made. In other words, about a third of single base changes will be silent, even if they occur within the protein coding region of a gene.

third base redundancy since many amino acids have several codons, the third codon base can often be changed without changing the amino acid for which it codes

12.6 CODON TABLE

1st Base	2nd (middle) Base				3rd Base
	U	**C**	**A**	**G**	
U	UUU Phe UUC Phe UUA Leu UUG Leu	UCU Phe UCC Phe UCA Leu UCG Leu	UAU Tyr UAC Tyr UAA STOP UAG STOP	UGU Cys UGC Cys UGA STOP UGG Trp	U C A G
C	CUU Leu CUC Leu CUA Leu CUG Leu	CCU Pro CCC Pro CCA Pro CCG Pro	CAU His CAC His CAA Gln CAG Gln	CGU Arg CGC Arg CGA Arg CGG Arg	U C A G
A	AUU Ile AUC Ile AUA Ile AUG Mat	ACU Thr ACC Thr ACA Thr ACG Thr	AAU Asn AAC Asn AAA Lys AAG Lys	AGU Ser AGC Ser AGA Arg AGG Arg	U C A G
G	GUU Val GUC Val GUA Val GUG Val	GCU Ala GCC Ala GCA Ala GCG Ala	GAU Asp GAC Asp GAA Glu GAG Glu	GGU Gly GGC Gly GGA Gly GGG Gly	U C A G

Nonsense and Missense Mutations

Now for some bad mutations. When the change in the base sequence alters a codon so one amino acid in a protein is replaced with a different amino acid, this is called a missense mutation.

First a moderately bad mutation: suppose we change the middle base - C - of the codon GCA that codes for alanine to a G. We now have GGA which will give glycine. Glycine and alanine are not identical but they are both relatively small and uncharged amino acids. Replacing alanine with glycine in a protein will probably not radically alter its structure. If we are reasonably lucky, the protein will still work, at least partially. However, if the exchange is made in a critical region of the protein, such as its active site, we may destroy its activity completely. Since the critical regions of most proteins occupy only a small proportion of the total protein sequence, most changes from one amino acid to another amino acid with similar chemical properties will be relatively mild and usually non-lethal (Fig. 12.7). These are known as conservative substitutions.

missense mutation when a sequence change in DNA results in the replacement of one amino acid by another in the encoded protein

active site special site or pocket on a protein where the chemical reaction occurs

conservative substitution replacement of an amino acid in a protein with another that is similar in its chemical properties

12.7 CONSERVATIVE SUBSTITUTION

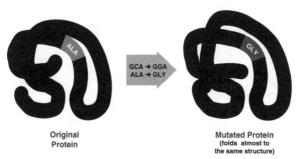

Original Protein

GCA → GGA
ALA → GLY

Mutated Protein
(folds almost to the same structure)

And now for some truly bad mutations: suppose we change the middle base - C - of the codon GCA which codes for alanine to an A. We now have GAA which will give glutamic acid. Glutamic acid is acidic and carries a strong negative charge. It is most definitely not similar to alanine and is therefore referred to as a radical replacement (Fig. 12.8). Unless we are very lucky, replacing alanine by glutamic acid will seriously cripple or even totally incapacitate our protein.

radical replacement replacement of an amino acid in a protein with another that is very different in its properties

12.8 RADICAL REPLACEMENT

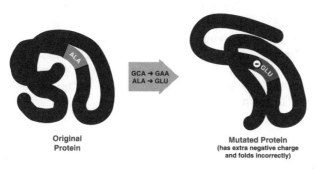

Original Protein

GCA → GAA
ALA → GLU

Mutated Protein
(has extra negative charge and folds incorrectly)

Note that from a political viewpoint, the replacement of a democrat by a republican is NOT a conservative substitution but a radical replacement!

temperature sensitive mutation a mutation with harmless effects at one temperature but noticeable effects at another

permissive temperature the temperature at which a temperature sensitive mutation has no effect or is relatively harmless

restrictive temperature the temperature at which a temperature sensitive mutation is lethal or detrimental

An interesting and sometimes useful type of missense mutation is the **temperature sensitive mutation**. As its name indicates, we get a protein that folds properly at low temperatures but at high temperatures is unstable and unfolds. Consequently, the protein is inactive at high temperatures. If a protein is essential, a missense mutation in it will often be lethal, and it is difficult to study a non-existent organism. However, if we have a temperature sensitive mutant, we can grow it and perform genetic experiments at the lower temperature, the **permissive temperature**, where it is alive. To analyze the damage caused by the mutation, we can shift the temperature up to the **restrictive temperature** at which the protein is inactivated and the organism will eventually die.

Can things get worse? Why even ask. Things can always get worse! Suppose that we start with the codon UCG for serine. Let's change the middle base from C to A. We now have UAG which is one of the three

12.9 TEMPERATURE SENSITIVE MUTANT

Quite a few animals have black tips to their paws and tails, even though they are light colored over the rest of their cute, cuddly, furry bodies. This is due to a temperature sensitive "mutation" in the enzyme responsible for synthesizing melanin, the black pigment in the skin. At the normal temperature of warm-blooded animals, the mutant enzyme is inactive, so melanin is not made over most of the body. However, it is cooler out in the boonies at the tips of the paws, the tail and the nose. Here the enzyme is active, melanin is made, and the tips turn black. What was originally a mutation has become a "normal" form of body coloring for these animals - a useful, or at any rate, a pretty mutation.

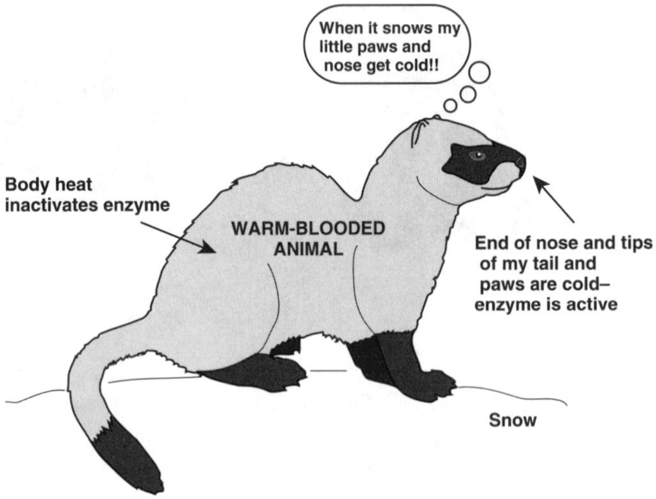

Although real biologists find it difficult to examine non-existent organisms, this does not seem to deter those more venturesome souls who study The Loch Ness Monster or Bigfoot!

nonsense mutation when a sequence change in DNA results in the replacement of the codon for an amino acid with a stop codon, thus producing a shortened protein

STOP codons (UAA, UAG, and UGA = STOP - remember these!). What happens now is that as the ribosome is making our protein it comes to the mutant codon that used to be serine. But this is now a stop codon, so the ribosome, a law abiding citizen, just stops!!! The rest of the protein does not get made. This makes no sense so it's called a **nonsense mutation** (or sometimes a chain termination mutation). Usually we end up with a shortened polypeptide chain that cannot even fold into a properly folded protein (Fig. 12.10). Its fate is sealed. The cell detects and digests unfolded proteins. The result, in practice, is the total absence of this particular protein, which may well have drastic results. Nonsense mutations are often lethal.

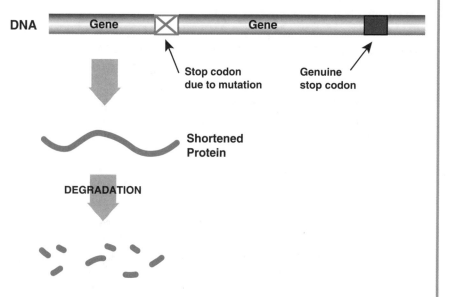

12.10 NONSENSE MUTATION

DNA

Gene | Gene

↙ Stop codon
due to mutation

Genuine
stop codon ↗

**Shortened
Protein**

DEGRADATION

deletion   removal of
one or many nucleotides
from DNA

Deletions and Insertions

So far we have been really rather restrained and only swapped a single base for another. But just as the Aztec priests removed a victim's heart so we can excise one or more bases of the DNA sequence. Mutations in which bases are removed are known as **deletions**.

Obviously, if we delete the DNA sequence for a whole gene, this is pretty serious (Fig. 12.11). If there is no gene, there will be no messenger RNA. If there is no messenger RNA, there will be no protein. If there is no protein, there will be no cell - assuming the protein is essential. Large deletions may remove part of a gene, an entire gene or several genes.

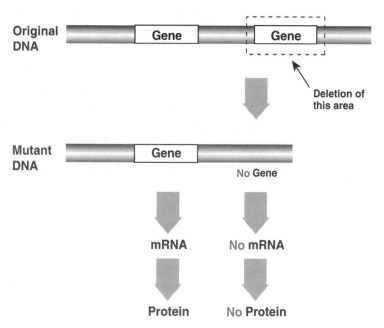

12.11 LARGE DELETION MUTATION

**Original
DNA**

Gene | Gene

Deletion of
this area

**Mutant
DNA**

Gene

No Gene

mRNA | No mRNA

Protein | No Protein

You might think that the more bases we remove the worse the mutation. Not necessarily. Consider the following important piece of RNA message and its translation into protein:

RNA Code: GAG - GCC - GUA - AUC - GAA - UGU - UUG - GCA - AGG - AAA

Protein: Glu - Ala - Val - Ile - Glu - Cys - Leu - Ala - Arg - Lys

Let's delete! First, just one base. Surely, in a DNA molecule with thousands or millions of bases, it will hardly be missed. Fat chance! We'll delete the middle base of the third codon. And here is what happens:

WILD TYPE
GAG - GCC - GUA - AUC - GAA - UGU - UUG - GCA - AGG - AAA

MUTANT
GAG - GCC - G • A - AUC - GAA - UGU - UUG - GCA - AGG - AAA

But remember that bases are read in threes. The • is not actually there, it represents absence of a base. Therefore, when taken three bases at a time, our mutant sequence will be grouped differently. By removing a single base we have changed the **reading frame** (see Ch. 7). The RNA will now be translated as follows:

reading frame one of three possible ways to read off the bases of a gene in groups of three so as to give codons

WILD TYPE
RNA Code: GAG - GCC - GUA - AUC - GAA - UGU - UUG - GCA - AGG - AAA

Protein: Glu - Ala - Val - Ile - Glu - Cys - Leu - Ala - Arg - Lys

MUTANT
RNA Code: GAG - GCC - GAA - UCG - AAU - GUU - UGG - CAA - GGA - AA

Protein: Glu - Ala - Glu - Leu - Asn - Val - Trp - Gln - Gly -----

We have completely changed all of the amino acids after the deletion point. With just a single base deletion, our protein sequence has been completely garbled. Insertion of a single extra base would have much the same effect. Whenever a mutation changes the reading frame, it is known as a **frameshift mutation** and the resulting protein sequence is total drivel.

Deletion or insertion of two bases would also change the reading frame, by two spaces in this case, and would give a similarly garbled protein. However, suppose we delete three bases:

frameshift mutation a mutation which changes the reading frame of the protein encoded by that gene

WILD TYPE
RNA Code: GAG - GCC - GUA - AUC - GAA - UGU - UUG - GCA - AGG - AAA

Protein: Glu - Ala - Val - Ile - Glu - Cys - Leu - Ala - Arg - Lys

MUTANT
RNA Code: GAG - ••• - GUA - AUC - GAA - UGU - UUG - GCA - AGG - AAA

Protein: Glu - ••• - Val - Ile - Glu - Cys - Leu - Ala - Arg - Lys

160

Three bases is a complete codon, so when we translate this sequence to make the protein we delete an amino acid. Although we have deleted an amino acid, we did not get out of step during translation; we have preserved the correct reading frame. Apart from the single amino acid we lost, the rest of the protein is unchanged. Similarly, a three base insertion would add a single amino acid, without affecting the rest of the sequence. If the deleted (or inserted) amino acid is in a relatively less vital region of the protein, we may actually get away with this and make a functional protein (Fig. 12.16).

12.16 IN FRAME DELETION MUTATIONS

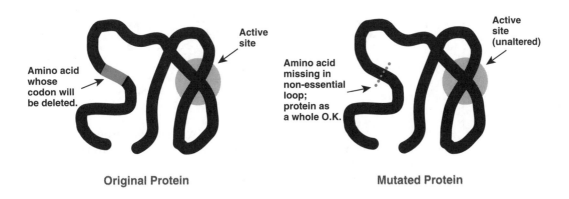

Original Protein

Mutated Protein

We could even get away with adding or deleting more than three bases as long as the number is a multiple of three; in other words, we must add or subtract a whole number of codons in order to avoid the horrible consequences of changing the reading frame.

In much the same way we could add or remove a finger of your hand without killing you; you would just find it hard to get gloves that fit! Compare that to the effects of changing the reading frame, which would be like replacing an arm with a totally different body part (Fig. 12.17).

12.17 CHANGING OR KEEPING THE READING FRAME

In Frame Insertion

Frame Shift

Rearranging DNA: Inversions and Translocations

inversion   when a segment of DNA is removed, flipped and reinserted, facing the opposite direction

An **inversion** is just what its name implies, an inverted segment of the DNA (Fig. 12.18). As you might imagine, reading a stretch of DNA backwards gives a ghastly mess. Inversions are definitely bad news.

12.18 INVERSION MUTATION

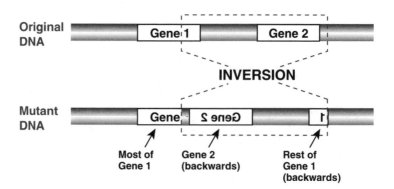

translocation   when a segment of DNA is removed and reinserted in a different place

A **translocation** is when a section of DNA is removed from its original position and moved to another location, either on the same chromosome, or on a completely different chromosome (Fig. 12.19). If an intact gene is merely moved from one place to another, it may still work and little damage may result. On the other hand, if, say, half of a gene is moved and stuck somewhere else in the middle of another gene, the results will be chaotic and severely detrimental.

12.19 TRANSLOCATION MUTATION

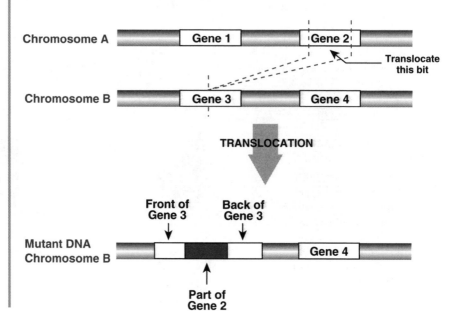

What Causes Mutations?

Mutations may be caused by agents that damage the DNA, and these are often known as **induced mutations**. Agents that mutate DNA are called **mutagens** and are of two main types: toxic chemicals and radiation. Even if there are no dangerous chemicals or radiation around, mutations still occur, though less frequently. These are known as **spontaneous mutations** and they are due to errors in DNA replication. The enzymes of DNA replication are not perfect and sometimes make honest mistakes.

The most common types of toxic chemicals react with DNA and alter the chemical structure of the bases. For example, EMS (ethyl methane sulfonate) is a mutagenic chemical widely used by molecular biologists. It adds a methyl group to bases in DNA and so changes their shape. **Nitrite** is a chemical that replaces amino groups with hydroxyl groups and so converts the base cytosine to uracil (Fig. 12.20).

When the time comes for DNA replication, the DNA polymerase is confused by the altered bases and puts in wrong bases in the new strand of DNA it is making (Fig. 12.21).

induced mutations
mutations caused by chemical damage or radiation

mutagen agent that can cause mutations

spontaneous mutation
mutations which happen without chemical damage or radiation because of mistakes during DNA replication

For all you paranoids out there, **nitrite** is widely used to preserve meat, especially bacon. It is especially popular because in addition to killing bacteria it also binds to hemoglobin in the meat and generates a nice healthy red color that makes the meat look fresh.

12.20 NITRITE CONVERTS CYTOSINE TO URACIL

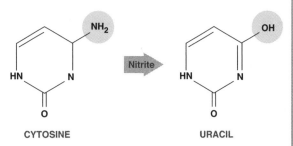

CYTOSINE   URACIL

12.21 ALTERED BASES CONFUSE DNA POYMERASE

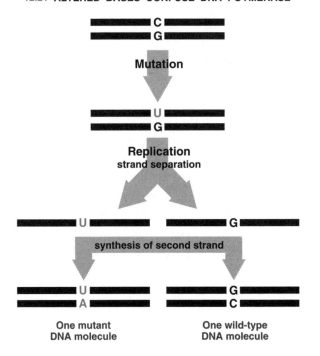

One mutant
DNA molecule

One wild-type
DNA molecule

Another type of chemical mutagen mimics the bases found in natural DNA. For example, the chemical bromouracil resembles thymine in shape. It is converted by the cell to the DNA precursor, bromouridine triphosphate, and DNA polymerase will then insert this by mistake where thymine should have gone. Mimics acting like this are called base analogs. Unfortunately, bromouracil can flip-flop between two alternative shapes like Dr. Jekyll and Mr. Hyde (Fig. 12.22). In its evil Mr. Hyde form, bromouracil resembles cytosine and pairs with guanine. If bromouracil is in its misleading form when DNA polymerase arrives, a G will be put into the new strand opposite the bromouracil instead of A.

A more subtle form of chemical mimicry consists of imitating the structure of a base pair rather than a single base. For example, acridine orange has three rings and is about the size and shape of a base pair. Acridine orange is not actually incorporated into the DNA. Instead it squeezes in between the base pairs in DNA that already exists (Fig. 12.23). This is referred to as intercalation. When it is time for DNA replication, the DNA polymerase thinks the intercalating agent is a base pair and it puts in an extra base when making a new strand. As discussed above, insertion of an extra base will change the reading frame of the protein coded by a gene. Since this will completely destroy the function of the protein, intercalating agents are definitely very bad news.

base analog chemical that resembles a base of a nucleic acid well enough to fool a cell into using it instead

intercalation when a chemical agent inserts itself into DNA between two base pairs

12.22 BROMOURACIL INDUCED MUTATIONS

Br

O

HN

N

H

O

Looks like T
Pairs with A

Br

O

H

HN

N

O

Looks like C
Pairs with G

12.23 ACRIDINE ORANGE IS AN INTERCALATING AGENT

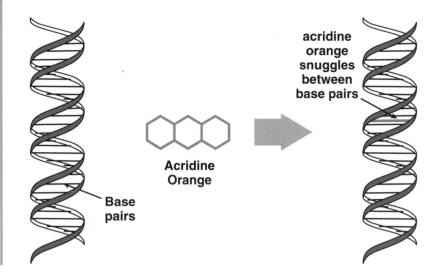

Base
pairs

Acridine
Orange

acridine
orange
snuggles
between
base pairs

A **teratogen** is an agent that causes abnormal development of the embryo leading to "monstrosities," that is to say, gross structural defects (teras means monster in Greek). The most famous example is thalidomide which resulted in the birth of malformed children often missing arms or legs, etc. Teratogens are simply mutagens which have spectacular effects on animals (Fig. 12.24).

Some forms of radiation cause mutations. High frequency electromagnetic radiation - ultraviolet radiation (UV light), X-rays and gamma rays (γ-rays) - directly damage DNA. Ultraviolet radiation makes two neighboring thymine bases react with each other to give thymine dimers (Fig. 12.25). These confuse DNA polymerase which will make mistakes when synthesizing a new strand of DNA.

12.24 TERATOGENS CREATE MONSTROSITIES

teratogen agent that can cause spectacular mutations or "monstrosities"

Ultraviolet radiation is emitted by the sun. Most of it is absorbed by the ozone layer in the upper atmosphere, so it does not reach the surface of the earth (Fig. 12.26). If the ozone layer is destroyed by the chlorinated hydrocarbons used in aerosol sprays and refrigerants, the amount of UV reaching us will increase drastically. But don't worry. Long before you grew two or three extra heads, the increased UV radiation would kill all the plants. There would be nothing even for your single mouth to eat and you would starve in dignity.

12.25 ULTRAVIOLET LIGHT AND THYMINE DIMERS

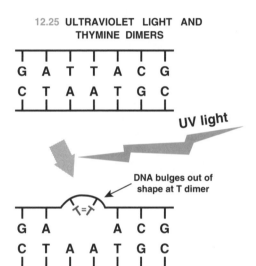

In the early days of molecular biology, X-rays were often used to generate mutations in the laboratory. X-rays tend to produce multiple mutations and often yield rearrangements of the DNA, such as deletions, inversions and translocations. And that is why, when you are given a chest X-ray, your procreative organs are shielded with a lead apron, the geneticist's equivalent of a bullet proof vest.

12.26 ULTRAVIOLET LIGHT IS ABSORBED BY OZONE

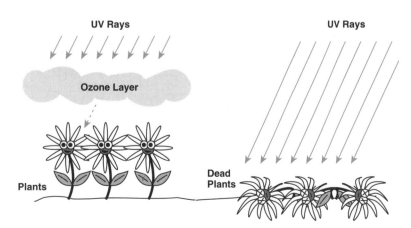

165

In addition to electromagnetic radiation, there are other forms of radiation such as α-particles and β-particles emitted by radioactive materials along with γ-rays. There are also cosmic rays which, as you might guess, come from outer space. Most α-particles are too weak to penetrate skin and it is the β-particles which you need to worry about. Unless you are planning on taking a trip in the space shuttle, forget the cosmic rays. If you stay on the surface of the earth you are more likely to be hit by a falling satellite than be zapped by a cosmic ray.

Mutations Caused by Insertion of Transposons

Insertion of an unrelated stretch of DNA into the middle of a gene will have drastic effects. A variety of DNA sequences are known that can move around from place to place on the chromosome. These are referred to as **transposons,** or jumping genes, and are discussed more fully in Chapter 20. Sometimes, when relocating, they spontaneously insert themselves into the middle of another gene (Fig. 12.27). This disrupts the target gene and completely abolishes its proper function. Although these mutations are insertions, they are really quite distinct in their origin from the smaller insertions described above which are due to chemicals or to mistakes made by DNA polymerase.

12.27 MUTATION BY TRANSPOSON INSERTION

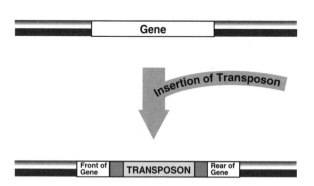

Genetically Engineered Gene Disruption

Mutations that serve to completely inactivate a gene are useful in genetic analysis. So scientists sometimes deliberately insert foreign DNA into genes to disrupt them and then study the results. To do this it is necessary to clone the gene and carry it on some convenient vector such as a bacterial plasmid (see Ch. 9).

For disruption, a deliberately designed segment of DNA is used. Known as a **gene cassette** (Fig. 12.28), it carries a gene for resistance to some antibiotic such as chloramphenicol or kanamycin. This way, the inserted DNA can easily be detected because cells carrying it will become resistant to the antibiotic. At each end, the cassette has several convenient restriction enzyme sites.

The target gene is cut open with one of these restriction enzymes and the cassette is cut from its original location with the same enzyme. The cassette is

12.28 GENE CASSETTE

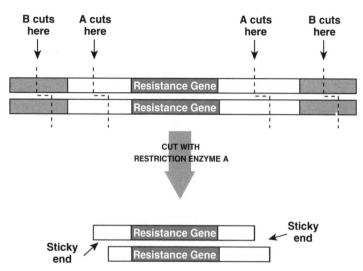

166

then ligated into the middle of the target gene (Fig. 12.29). The plasmid carrying the disrupted gene can now be put back into the organism from which it came (see Chs. 8 and 15).

12.29 GENE DISRUPTION

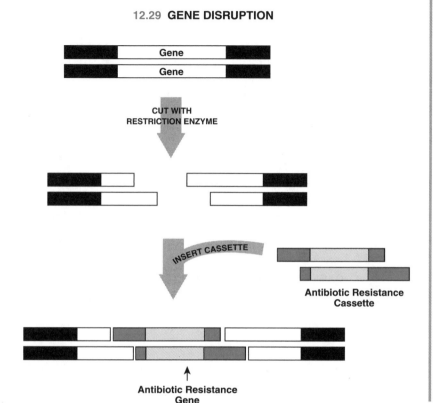

hot spots   areas in the genome usually susceptible to mutation

Mutational Hot Spots

If the same gene is mutated thousands of times, are the mutations all different and are they distributed at random throughout the DNA sequence of that gene? Well, many of them are. However, here and there in the DNA sequence you will find a location where mutations happen many times more often than average (Fig. 12.30). All the mutations occurring at such a site will usually be identical. These are called hot spots.

Most hot spots are due to the presence of occasional methylcytosine bases in the DNA. (See Ch. 9 for modification of DNA by methylation.) These are made from cytosine after DNA synthesis and they pair correctly

12.30 MUTATIONS IN ONE GENE OF *E. COLI* ARE CONCENTRATED IN A "HOT SPOT"

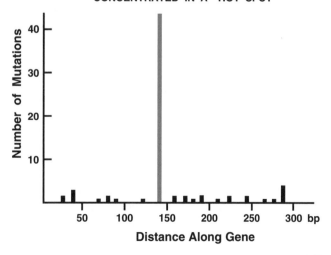

with guanine, just like normal cytosine. However, every now and then one of these methylcytosine bases spontaneously disintegrates to give methyluracil. This pairs with adenine, not with guanine, and so when the DNA is replicated next, an error will creep in.

Reversion and Suppression

Suppose we have a mutant and its DNA gets zapped again. There is a small chance that the second mutation will reverse the effect of the first. This process is called **reversion**. Reversion refers to the observable outward characteristics of our organism: it is a phenotypic term. The likelihood that exactly the one base out of millions that was previously mutated will be the very one to mutate again is extremely low. Those rarities where the original base sequence is exactly restored are **true revertants**.

More often our revertants actually contain a second base change that cancels out the effect of the first one. These are **second-site revertants**. Let's consider two examples. The simplest to understand is if the original mutation was a frameshift mutation due to deletion or insertion of a single base. This alters the reading frame and garbles the protein sequence:

reversion a second mutation that restores original characteristics to a mutant organism

true revertant a revertant in which the original DNA base sequence is exactly restored

second-site revertant a revertant in which a second change in DNA base sequence cancels out the effects of the first

WILD TYPE

GAG - GCC - ATC - GAA - TGT - TTG - GCA - AGG - AAA

Protein: Glu - Ala - Ile - Glu - Cys - Leu - Ala - Arg - Lys

ORIGINAL DELETION MUTANT

DNA: GAG - G•C - ATC - GAA - GTG - TTG - GCA - AGG - AAA

Grouped as: GAG - GCA - TCG - AAT - GTT - TGG - CAA - GGA - AA

Protein: Glu - Ala - Ser - Asn - Val - Trp - Gln - Gly - ------------

But suppose we now insert an extra base a little way farther along the sequence. This second-site insertion will restore the original reading frame:

REVERTANT

DNA: GAG - G•C - AATC - GAA - GTG - TTG - GCA - AGG - AAA

Grouped as: GAG - GCA - ATC - GAA - TGT - TTG - GCA - AGG - AAA

Protein: Glu - Ala - Ile - Glu - Cys - Leu - Ala - Arg - Lys

Although the DNA sequence is not identical to its original state, the protein has been exactly restored. Similarly, an insertion mutation can be corrected by a second-site deletion. The key to success when reverting is to restore activity to the protein, not to get obsessive-compulsive about the DNA sequence.

A less obvious but more frequent case is where the original mutation was a base change. Consider a protein with 100 amino acids whose correct 3-D structure depends on the attraction between a positively charged amino acid at position 25 and a negatively charged one at position 50. Suppose the original mutation changes codon No. 50 from GAA for glutamic acid (negatively charged) to AAA which encodes lysine, a positively charged amino acid (Fig. 12.33). The protein's folding is now disrupted.

We could make a true revertant by replacing AAA with GAA. However, suppose instead we mutate codon No. 25 to give a negatively charged amino acid. We now have a negative charge at position No. 25 and a positive charge at position No. 50. We have now restored the attraction between these two regions and the protein will fold O.K. again (Fig. 12.34). Will the revertant protein work correctly? Sometimes, sometimes not - it depends on other factors, such as whether these alterations damage the active site.

Detecting Mutagenic Chemicals by Reversion

Chemical mutagens can be detected by the Ames test, used routinely by industry and government agencies to screen chemicals for possible mutagenic effects. Food colorings and preservatives, cosmetics such as hair dyes, and many other chemicals are now checked by this test that examines their effect on bacteria in culture.

Mutants of the bacterium *Salmonella typhimurium* carrying mutations in the genes for histidine synthesis are used in the Ames test. Since they can no longer make the amino acid histidine, these mutants cannot grow unless given histidine. When large numbers of these mutant bacteria are placed on growth medium lacking histidine, just a handful of colonies appear. These are revertants, and since reversions are actually mutations back to the

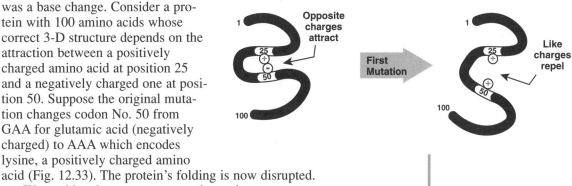

12.33 MISFOLDED MUTANT PROTEIN

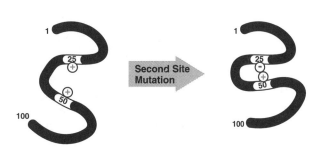

12.34 SECOND SITE REVERSION CURES DEFECT

12.35 THE AMES HAIR DYE TEST

original state, the frequency of reversion is also increased by mutagenic agents. Different types of original mutant, for example, base changes or frameshift mutations, are used to screen for different classes of mutagenic agent. Clever, huh!

DNA Repair

Even if your genes are damaged, all is not lost. Most cells contain a variety of damage control systems and some of these can repair damaged DNA. There are several DNA repair systems, designed to deal with different problems, and they are often rather complicated.

A variety of mutations may result in a base pair that doesn't actually pair properly. In other words, two bases opposite each other do not match and so do not hydrogen bond correctly. This will cause a slight bulge in the DNA double helix that alerts the proteins of the **mismatch repair** system (Fig. 12.36). This repair system cuts out the wrong base and fills in the gap with the right base to make a correctly bonded base pair.

But wait a moment! Which of the two mispaired bases was the wrong one? We need to know which strand came from the mother cell and which was the recently synthesized (and error carrying) daughter strand (Fig. 12.37).

mismatch repair DNA repair system which recognizes wrongly paired bases

12.36 MISMATCH REPAIR SYSTEM

Mismatch repair system homes in on bulge

In bacteria such as *Escherichia coli*, the DNA is tagged to indicate this. Wherever the sequence GATC occurs, it has a methyl group stuck on to the adenine. This modification occurs some time after the DNA has replicated. So the new strand in a recently replicated DNA molecule does not yet have its GATC sequences methylated (Fig. 12.38). This allows the mismatch repair system to tell which is the newest strand. Different organisms use different tagging systems, but the principle remains the same.

12.37 BASE ACCUSATIONS

I'm correct! He should be a G.

I'm correct! She should be a T.

12.38 GATC TAGGING ALLOWS STRAND IDENTIFICATION

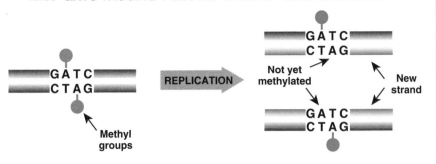

Methyl groups

REPLICATION

Not yet methylated

New strand

G A T C
C T A G

G A T C
C T A G

excision repair (cut and patch repair) cutting out a stretch of damaged DNA and replacing it with new DNA

The most widely distributed system for dealing with mutated DNA is **excision repair** in which a stretch of damaged DNA is cut out and the resulting gap filled in with new DNA. This is often referred to as **"cut and patch" repair.** For example a thymine dimer caused by UV radiation will make the DNA bulge. First a cut is made on one side of the bulge. Then DNA polymerase I manufactures a short replacement strand for the damaged region. As DNA polymerase I moves along, it also nibbles away the old strand. Finally the old strand is snipped off and the new segment is ligated into place (Fig. 12.39). Pol I is better at needlework than your mother; when it has finished patching your genes, there are no visible stitches to show where the new piece is!

Mutations are of vital importance to us all for two main reasons. First, if a mutation occurs in the reproductive cells and can be inherited, it may have major effects on the lives of those who receive it, as described in the next chapter. Second, mutations that are not inherited but arise after birth in cells of the body, may cause cancer, as discussed in Chapter 14.

Additional Reading

Understanding Genetics: A Molecular Approach by Rothwell NV. 1993. Wiley-Liss, New York.

Microbial Genetics by Maloy SR, Cronan JE, & Freifelder D. 2nd edition, 1994. Jones & Bartlett Publishers, Boston & London

Human Genetics: The Molecular Revolution by McConkey EH. 1993. Jones & Bartlett Publishers, Boston & London

12.39 CUT - AND - PATCH STARRING POL-I

5'   3'
3'   5'

INCISION

5'   3'
3'   5'

DISPLACEMENT BY POLYMERASE I

5'   3'
3'   5'

EXCISION

5'   3'
3'   5'

LIGATION

5'   3'
3'   5'

REPAIRED DNA

Inherited Human Disease

13

Genetic defects come in all shapes and sizes. Although we tend to think of inherited conditions like diabetes and muscular dystrophy as diseases, we often refer to cleft palates or color blindness as inherited defects. They are all the result of mutations in our genetic material, the DNA. Not only are some diseases directly caused by mutations but susceptibility to disease is also influenced by genetic constitution. It has been said that all disease, except trauma, has a genetic component.

Mutations in Single Celled Creatures

If you are a single celled organism and your DNA is mutated (Fig. 13.1), the mutation will be passed on to all of your descendants when you divide.

13.1 MUTATION OF A SINGLE CELLED ORGANISM

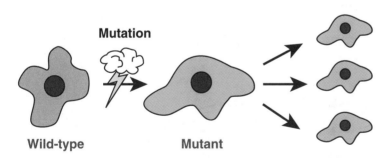

Mutation

Wild-type   **Mutant**

Hereditary Defects in Higher Organisms

But suppose you are a multi-celled creature. In advanced organisms, special cells are set aside for reproductive purposes. These are the **germ line cells**. They will give rise to the eggs and sperm in mature adults. The cells forming the rest of the body are called **somatic cells**. Somatic cells have no long term future. When the body dies, they die as well. Only the germ line cells pass their genes on to the next generation.

germ line cells cells that divide to produce eggs in females and sperm in males

somatic cells cells making up the body and which are usually diploid, that is, they have two sets of genes

You are the descendant of your parents' germ line cells (Fig. 13.2), and not of their big toes or spleens.

13.2 GERM LINE VERSUS SOMATIC CELLS

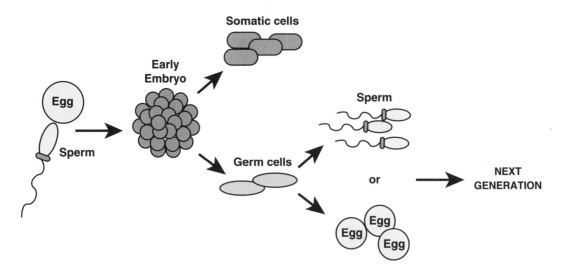

As we said, if the DNA in a cell of your big toe gets mutated, it will not affect your children. On the other hand if the DNA in a sperm or egg cell is mutated, the defect may be passed on to your offspring.

If a somatic cell which is part of an embryo gets mutated early in development, it will pass on the defect to its descendants. Let's suppose this embryonic cell is the precursor of your left eye. Let's also suppose that this defect prevents the manufacture of the brown pigment which makes brown eyes look - well, brown. Your right eye will be brown but your mutant left eye will be blue (Fig. 13.3). Blue eyes are not due to actual blue coloring, they are just lacking the brown pigment. People with eyes that don't match are unusual but not incredibly rare. Such events are called somatic mutations. They are not passed on to children.

13.3 SOMATIC MUTATIONS

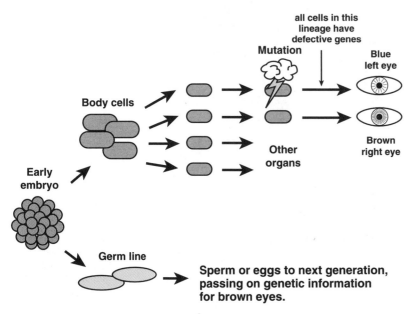

Somatic Mutations and Cancer

Many somatic mutations are relatively harmless, even amusing. Sometimes they may be highly detrimental, as when a cell giving rise to a major organ suffers a serious mutation. However, the most important kinds of somatic mutations are those that damage the regulatory system controlling cell growth and division. This may result in a single cell of your body deciding to start growing and dividing again, long after it is supposed to have settled down as part of some body organ. This is what causes cancer and is dealt with separately in Chapter 14.

Homozygous Recessive Genes and Hereditary Defects

How, you may well wonder, can nice, happy, well-adjusted parents produce the sort of freaks who would write a book like this? Authors, animals and plants are diploid and have two copies of each gene. Therefore, if one copy is damaged by a mutation, you have a second chance. Since almost all mutations are relatively rare, it is unlikely that both copies of a particular gene will carry the same mutation. Furthermore, in most cases, detrimental mutations are **recessive** to the wild-type (Fig. 13.4). That is, the single good copy of the gene is enough for normal function and will cover the effect of the defective copy of the gene. (See Ch. 5 for more on recessive genes.)

13.4 EFFECT OF RECESSIVE GENES

Normal Father with one bad recessive

Normal Mother with one bad recessive

The Authors with a double recessive

recessive gene defective copy of a gene whose properties are not observed because they are masked by a functional copy

homozygous having two identical alleles of the same gene

Nonetheless, as mentioned in Chapter 11, we all have quite a few mutations randomly scattered among our 50,000 genes. Obviously, you will share many of the defects of your close relatives. For example, you share half of your genetic information with your brothers, sisters, father and mother. (Not the same half with each of them, of course!)

Therefore, if a child results from mating between close relatives (for example as in brother/sister or father/daughter), there is a much increased chance of two copies of the same defect being inherited, one copy from each parent (Fig. 13.5). In genetic terminology, the children are **homozygous** for the recessive allele since both copies of the gene are now defective. Consequently, children from the union of close relatives have a much higher incidence of hereditary defects, and the closer the relationship, the higher the chance of a defect resulting. (Hence the high incidence of hemophilia among the royal families of Europe and, possibly, the large, ugly ears observed

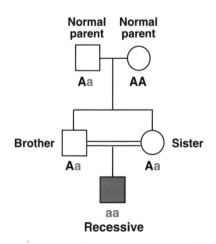

13.5 DOUBLE DEFECTS FROM INCESTUOUS MATING

Normal parent   **Normal parent**

Aa   AA

Brother   **Sister**

Aa   Aa

aa
Recessive

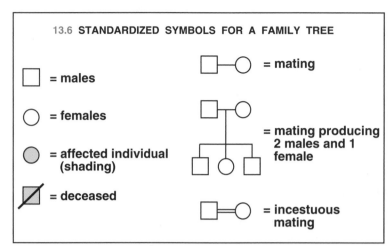

13.6 STANDARDIZED SYMBOLS FOR A FAMILY TREE

☐ = males

○ = females

● = affected individual (shading)

▨ = deceased

☐—○ = mating

☐—○ = mating producing 2 males and 1 female

☐═○ = incestuous mating

in some members of the British monarchy!) This is the underlying reason for the taboos against incest which prevail in most societies, though the degree of forbidden relationships in different cultures varies quite considerably.

Mating between close relatives (brother and sister in the case of Fig. 13.5) makes genetic disease more likely because a rare recessive defect (symbolized by "a") present in one ancestor may be passed down both sides of the family and wind up in two copies in the child.

There is a wide range of genetic defects available for inheritance. Many defects are fatal to the fertilized egg or early embryo so they never see the light of day.

Some of the best known defects due to a single gene are listed in Table 13.1. Unless otherwise noted, they are recessive, and require both copies of the gene to be defective before symptoms appear.

Two heads are better than one!

Maybe so, but I would have preferred to inherit money!

Large genes are bigger targets for random mutation. The genes responsible for cystic fibrosis, muscular dystrophy and phenylketonuria are all abnormally large and all three genetic defects are relatively common.

Hereditary Defects Due to Multiple Genes

Several well known hereditary defects are missing from Table 13.1 because they involve more than one gene. These may be subdivided into two types. Some multi-gene defects are due to the interacting effects of several individual genes, for example, cleft palate, spina bifida, certain cancers and diabetes.

Other multi-gene defects are due to the presence of an extra copy of an entire chromosome. Although most errors involving whole chromosomes are lethal, a few are viable. The best known of these is Down syndrome which causes mental retardation and is due to an extra copy of chromosome No. 21. The overall frequency of Down syndrome is about one in 800, but like most instances of getting a whole extra chromosome, it is due to an accident during cell division in the newly fertilized egg and it is not normally inherited. The relative chances of such a mishap increase with

Down syndrome defective development, including mental retardation, resulting from an extra copy of chromosome No. 21

TABLE 13.1: SOME HUMAN GENETIC DEFECTS

Disease	Frequency
Adenosine deaminase deficiency Enzyme defect causes immune deficiency. First defect approved for human gene therapy.	rare
Cystic fibrosis Defective ion transport with indirect effects on mucus secretion in lungs	1/2,000 (whites) rare in Asians
Duchenne muscular dystrophy Disintegration of muscle tissue. Very large gene is a frequent target for mutations.	1/3,000 males (sex-linked)
Fragile X syndrome Common form of X-linked mental retardation	1/1,500 males 1/3,000 females (sex-linked)
Hemophilia Defect in blood coagulation	1/10,000 males (sex-linked)
Myotonic dystrophy Genetically dominant form of dystrophy	1/10,000
Phenylketonuria Mental retardation due to lack of enzyme. Can be detected in newborn babies by analysis of urine. Special diet prevents symptoms.	1/5,000 (W. Europe) rare elsewhere
Sickle cell anemia Defect in beta-chain of hemoglobin. Heterozygotes are resistant to malaria but homozygotes are sick. First molecular disease to be identified.	1/400 (US blacks) common in Africa

maternal age, but even so, most Down syndrome children are born to young women, simply because more younger women have babies. Extra sex chromosomes are found in about one of every 1,000 people; there are three relatively common possibilities: XXY, XYY and XXX. The XYY individuals are best known for their supposed tendencies toward violent crime, but the others are abnormal too.

Cystic Fibrosis

About one in 2,000 white children suffer from cystic fibrosis. This disease is due to homozygous recessive mutations. In other words you must inherit two defective copies of the gene, one from each parent, to suffer from the disease. Humans with a single defective allele are carriers but

cystic fibrosis   a disease whose major symptom is the accumulation of fibrous tissue in the lungs

carriers   individuals who have a single defective copy of a gene but who show no clinical symptoms

do not show symptoms, as a single wild-type version of the gene is sufficient for normal health.

In marriages between unrelated people (Fig. 13.8), a recessive disease will only occur if an ancestor from each side of the family carries a copy of the defective allele ("a").

The molecular basis of this defect is in controlling the secretion of **chloride** ions across cell membranes. This, in turn, affects a variety of other processes. The most harmful is that the mucus which lines and protects the lungs is abnormally thick. (Lack of chloride ions leads to lack of sufficient water accompanying the mucus.) This not only causes obstructions but allows the growth of harmful microorganisms. Cells lining the airways of the lungs are killed and replaced with fibrous scar tissue, hence the name of the disease. Eventually the patient succumbs to respiratory failure.

The protein encoded by the cystic fibrosis gene is referred to as CFTR (for Cystic Fibrosis TRansporter) and is found in the cell membrane where it acts as a channel for chloride ions (Fig. 13.9). In healthy people, this channel can be opened or shut as needed by the cell.

The general location of the cystic fibrosis gene (to within 2 million base pairs) was found by screening large numbers of genetic markers (RFLPs - see Ch. 18) in members of families with a cystic fibrosis patient. The gene was found on the longer, "q" **arm** of chromosome 7 between bands q21 and q31. After subcloning this region of the chromosome in large chunks, the cystic fibrosis gene was found to occupy 250,000 base pairs and have 24 exons which encoded a protein of 1,480 amino acids (Fig. 13.10). Since 1,480 amino-acids need only 4,440 base pairs to encode them, this means that scarcely 2 percent of the cystic fibrosis gene is actual coding DNA. The rest consists of intervening sequences - introns.

chloride a dissolved inorganic ion found in major levels in body fluids

p arm the shorter of the two arms of a chromosome

q arm the longer of the two arms of a chromosome

13.8 INHERITANCE OF CYSTIC FIBROSIS

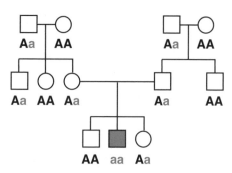

13.9 CHLORIDE CHANNEL

Channel Shut

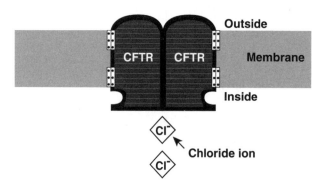

Channel Open

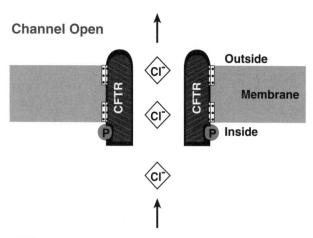

13.10 CYSTIC FIBROSIS GENE - EXONS AND INTRONS

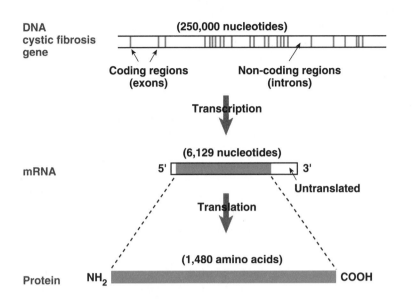

DNA
cystic fibrosis
gene

(250,000 nucleotides)

Coding regions
(exons)

Non-coding regions
(introns)

Transcription

mRNA

(6,129 nucleotides)

5'   3'

Untranslated

Translation

(1,480 amino acids)

Protein   NH₂   COOH

Minding the p's and q's:
To keep track of where a
gene lies on a chromo-
some we must first specify
the p or q arm. The arms
are subdivided for refer-
ence purposes into num-
bered bands.

The cystic fibrosis protein consists of a
series of membrane spanning segments with
a central control module (Fig. 13.11). If the
control module has a phosphate group
attached, the channel is open and when the
phosphate group is removed, it shuts.

About 70 percent of the cystic fibrosis
victims in North America share the same
genetic defect. They all have a small deletion of three bases which code for amino
acid No. 508 (phenylalanine; F in single letter code) of the normal protein (Fig. 13.12).

13.11 THE CYSTIC FIBROSIS PROTEIN

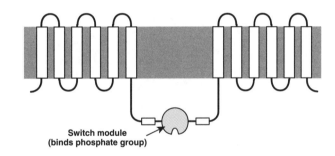

Switch module
(binds phosphate group)

13.12 CYSTIC FIBROSIS PROTEIN – F508 DELETION

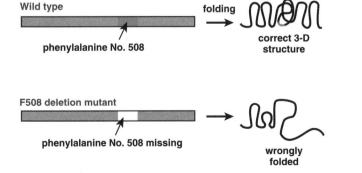

Wild type   folding

phenylalanine No. 508

correct 3-D
structure

F508 deletion mutant

phenylalanine No. 508 missing

wrongly
folded

In Denmark 90 percent of cystic fibrosis cases are due to the F508 deletion, whereas in the Middle East it only accounts for 30 percent. The other cystic fibrosis cases are the result of more than 40 different mutations. This variability makes genetic screening difficult. (See Ch. 12 for possible causes of these mutations).

Muscular Dystrophy

There are several forms of muscular dystrophy. These diseases result in the wasting away of muscle tissue and cause premature death, usually in the late teens or early 20s. There is no known cure. The most common form, Duchenne muscular dystrophy, is sex-linked. In other words, the gene involved is carried by the chromosomes which are responsible for determining sex - the X and Y chromosomes (see Ch. 3; Fig. 3.15 for sex determination).

If you have two X chromosomes you are female, if you have one X and one Y you are male, if you have two Y chromosomes you don't exist. Although X and Y are a chromosome pair, the Y chromosome is shorter than the X chromosome. Consequently, many of the genes are present only on the X chromosome and do not have a corresponding partner on the Y chromosome (Fig. 13.13). Therefore, females have two copies of these genes, while males only have a single copy.

If the single copy of a sex-linked gene present in a male is defective, there is no back-up copy and severe symptoms may result. In contrast, females with just one defective copy will usually have no symptoms, but they will be carriers and half of their male children will get the disease. The result is a pattern of inheritance in which the male members of a family often inherit the disease, but the females are carriers and suffer no symptoms.

Duchenne muscular dystrophy one particular form of muscular dystrophy, a group of degenerative muscle diseases

13.13 SEX - LINKED GENES

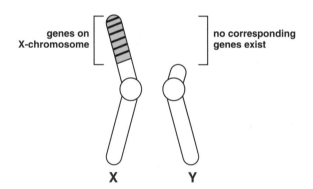

genes on X-chromosome

no corresponding genes exist

X   Y

Figure 13.14 shows a family with several occurrences of an X-linked recessive disease. Males have only one X chromosome and their Y chromosome has no corresponding copy of the gene (symbolized by -). So any male who gets one copy of the defective allele ("a") will get the disease.

About two-thirds of Duchenne muscular dystrophy patients inherit the disease from their mothers and the other third get it as the result of new mutations which arise at a low, but constant, frequency.

13.14 INHERITANCE OF SEX - LINKED DISEASE

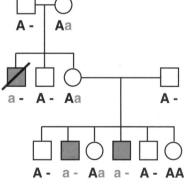

A -   Aa

a -   A -   Aa   A -

A -   a -   Aa   a -   A -   AA

The gene responsible for Duchenne muscular dystrophy is referred to as the *dmd* **gene** and is even more bizarre than the cystic fibrosis gene. It has roughly 75 exons and over 2,000,000 base pairs of DNA of which less than 1 percent are used to encode the protein. Despite this, **dystrophin**, the protein coded for by the *dmd* gene, is gigantic. It has 4,000 amino acids so it is roughly 10 times as large as an average protein. Dystrophin is thought to play a role in attaching the internal muscle fibrils to the membranes of muscle cells.

dmd **gene** gene which, when defective, causes Duchenne muscular dystrophy

dystrophin protein encoded by the *dmd* gene, whose malfunction causes muscular dystrophy

13.15 *dmd* GENE ON CHROMOSOME 15

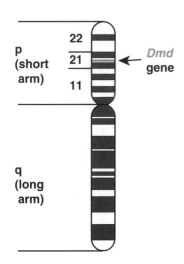

p (short arm)

22
21
11

← *Dmd* gene

q (long arm)

The *dmd* gene is located in the Xp21 band, close to the middle of the short (p) arm of the X-chromosome (Fig. 13.15). (The X chromosome has a shorter, "p" arm, and a longer, "q" arm). In most victims the defect is due to alterations in just one or a few bases of the *dmd* gene. However, in about 10 percent of the victims, the genetic defect responsible for the disease is a deletion of DNA which includes all or part of the *dmd* gene.

Cloning the *dmd* Gene

One Duchenne muscular dystrophy patient had a deletion large enough to locate to the Xp21 region of the X-chromosome using a light microscope.

DNA from this patient was used to clone the DNA which was missing in the deletion. Huh?! How can you clone something that is not there?? Actually you do need a sample of DNA from a healthy person too.

It's done like this: the normal DNA is cut up with a restriction enzyme so it has sticky ends convenient for cloning. Somewhere among these fragments is one carrying the healthy version of the *dmd* gene (Fig. 13.16).

All we need to do now is to get rid of all the DNA that is not part of the *dmd* gene. This is why we need the deletion mutant DNA from the muscular dystrophy patient.

13.16 DNA FOR *dmd* GENE – FRAGMENTATION

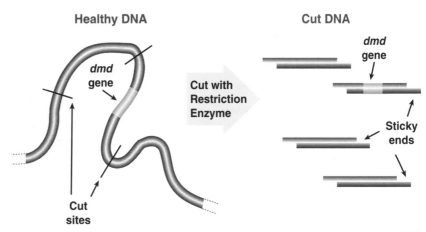

Healthy DNA

dmd gene

Cut sites

Cut with Restriction Enzyme

Cut DNA

dmd gene

Sticky ends

This is also broken into random fragments, but by ultrasonic vibration so it does not have sticky ends (Fig. 13.17).

13.17 SONICATION OF DNA DELETED FOR *dmd* GENE

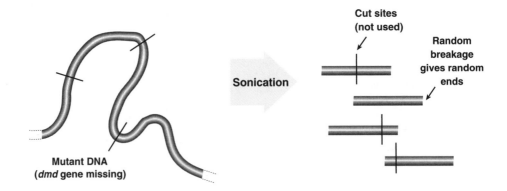

Both lots of DNA are denatured by heating into single strands. Then they are mixed and cooled. The single strands will reassociate. If the deletion mutant DNA is used in a large excess, it will outnumber the fragments of wild-type DNA. All of the wild-type DNA fragments will be paired off with mutant DNA fragments except those carrying the *dmd* gene, which is, of course, completely absent from the mutant DNA sample. The single DNA strands for the *dmd* gene, from the wild-type person, will pair up with each other again to give a segment of dsDNA with two sticky ends (Fig.13.18).

13.18 DNA FOR *dmd* GENE – REANNEALING

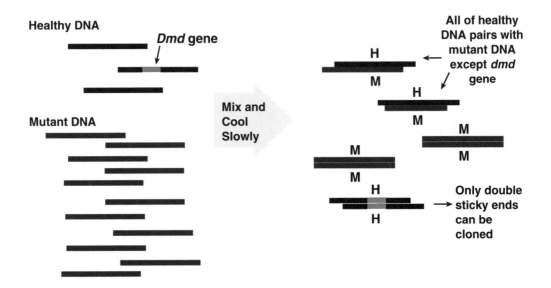

Each pair of mutant/mutant DNA or wild-type/mutant DNA will have at least some random ends resulting from the sonication of the mutant DNA sample. Some of these ends may stick out as single strands, but they do not match any other unpaired ends. So these pieces cannot be cloned. In contrast, the reannealed wild-type/wild-type DNA carrying the

13.19 DNA FOR *dmd* GENE

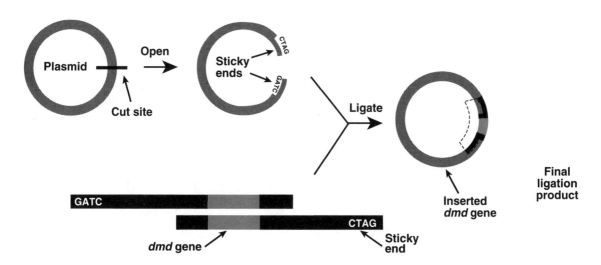

dmd gene has two sticky ends made by the restriction enzyme we original-ly used to cut it up. We can cut open a circle of plasmid DNA using the same enzyme and ligate our wild-type *dmd* gene into it (Fig. 13.19). (See Ch. 9 for cloning procedures). We have now cloned the wild-type *dmd* gene by using the deletion mutant to subtract out all the DNA we didn't want. This general approach is referred to as **subtractive hybridization** and its variants are widely used.

subtractive hybridiza-tion   removal of unwant-ed genes by hybridization, so leaving behind the gene being sought

Genetic Counseling

As explained above, close relatives are more likely to produce malformed children due to harmful recessive genes getting together. However, it is also true that two people from otherwise unrelated families that both have a history of the same heredi-tary disease, would be wise to avoid having children. Today we can do bet-ter than just give general advice. It is possible to test a healthy person from a family with a history of, say, cystic fibrosis, and tell whether or not they have a copy of the defective gene.

13.20 SCREENING FOR PROSPECTIVE PARENTS

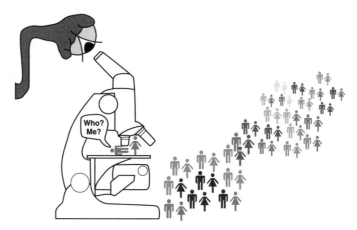

Testing of prospective parents for high risk genetic diseases can now be done for an ever increasing number of genes. The typical approach is to use some sort of gene probe, either a hybridization probe (see Ch. 16) or PCR analysis (see Ch. 17). This will reveal whether a mutant copy of the gene is present, and in whom.

If marital partners both test positive for a recessive defect, then they will have to decide whether or not to take the risk of having children. For a recessive defect, where both parents are carriers, one in four children will get the disease.

It is also possible to examine embryos during early pregnancy by drawing samples of amniotic fluid which contain some cells from the fetus, a procedure called **amniocentesis** (Fig. 13.21). These cells could be genetically screened to see, for example, if the fetus is homozygous for a recessive defect. Embryos doomed to grow up with hereditary defects could then be aborted early if the parents wish to do so.

amniocentesis sampling of cells and fluid from the amniotic cavity

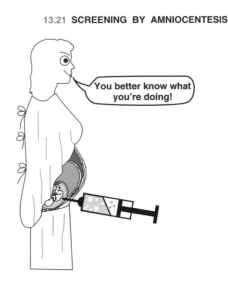

13.21 SCREENING BY AMNIOCENTESIS

You better know what you're doing!

Gene Therapy or Genetic Engineering??

Genetic engineering is generally taken to mean that we alter an organism permanently so changes will be stably inherited. For multicellular organisms this means that changes must be in the DNA of the germ line cells. In contrast, gene therapy (or genetic surgery, as it is sometimes known), is more of a temporary fix. The patient is cured, more or less, by altering the genes in only part of the body. For example, in cystic fibrosis the patient can be cured by introducing the wild-type gene into the lungs. However, these changes are not inherited and the alleles in the germ cells remain defective.

genetic engineering alteration of an organism by deliberately changing its DNA

Genuine human genetic engineering is still in the future. At present, genetic engineering is restricted to lower organisms and has resulted in the creation of **transgenic animals** and **plants** as described in Chapter 15. **Eugenics** is a term used to describe the deliberate improvement of the human race by genetic technology.

transgenic animal or **plant** individual into which genes from another species have been incorporated

Early eugenic proposals were based on picking superior parents by visual inspection or medical screening, and breeding them to get improved offspring in much the same way as it is done for prize pigs and pedigree dogs. (Imagine being denied a marriage license because your nose was not cold or wet enough!!)

eugenics improving a species by genetics

The Napoleonic Wars provided a fascinating example of eugenics by ignorance. Napoleon, who was himself vertically challenged, deliberately recruited tall men into the French Imperial Army. Although Napoleon won many battles, his casualties were enormous and even when victorious he often lost more men than his enemies because of his penchant for attacking in columns many soldiers deep. Cannon balls would often pass right through men in the front rows killing those behind. The combined result of constantly selecting tall men and subjecting them to massive casualties was that the average height of the French nation decreased significantly during this period!

13.22 NAPOLEON SELECTS FOR TALL FRENCHMEN

Since industrialization began, political consciousness of one kind or another has enabled the improvident to produce children who are supported by social programs. On the other hand, overcrowding due to urbanization has greatly aided the spread of infectious disease, especially among the poor. Have we been selecting for or against the less able? No one really knows. Nonetheless it is clear that our actions are changing the human gene pool, whether we like to admit it or not.

Today we possess better techniques not only for screening genetic defects but also for artificial manipulation of genes or whole organisms. The question is, will we use this knowledge or will this issue remain taboo while we continue to make changes to society that alter our genetic heritage in an ignorant and semi-random manner??

13.23 NAPOLEON SELECTS AGAINST TALL FRENCHMEN

Gene Therapy - General Principles

The most obvious use of gene therapy is when we are faced with hereditary disease due to a single gene and which happens only when both copies of the gene are defective, *i.e.*, a recessive condition. The defect can then be cured by introducing a single good copy of the gene. Furthermore, it would obviously simplify matters if the disease mostly affects just one or a few organs.

Assuming that we have identified the genetic defect and have successfully cloned a good copy of the gene involved, how do we actually cure the patient?

First we need to put the good copy of the gene onto a suitable vector. In Chapter 9 we discussed gene cloning and the most commonly used vectors. In the laboratory, most manipulations would be done with genes carried on **bacterial plasmids**. But when we want to put a gene back into a human or other animal we need specialized vectors.

Viruses are Mother Nature's own gene transfer agents (Fig. 13.24). However, when they inject their genes into the target cell, they cause assorted diseases. To be used in gene therapy, viruses first need to be genetically disarmed. Two main groups of viruses have been used in human gene therapy, **retroviruses** and **adenoviruses**.

13.24 VECTORS FOR GENE THERAPY

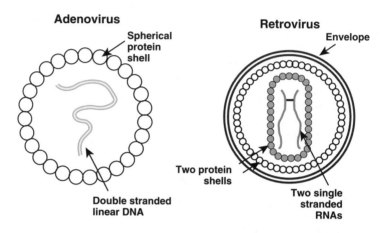

13.25 ADENOVIRUS DNA WITH CYSTIC FIBROSIS GENE

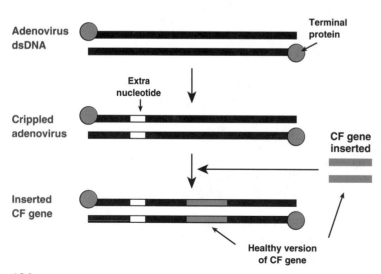

Cystic Fibrosis Gene Therapy by Adenovirus

Because lungs are exposed and relatively easy to get at, cystic fibrosis has been a prime candidate for gene therapy. The healthy version of the cystic fibrosis gene has been cloned and inserted into a crippled adenovirus. This type of virus normally infects the lining of the lungs where it causes colds.

Engineered adenovirus (Fig. 13.25) carrying the cystic fibrosis gene has been tested by spraying it into the noses and lungs, first of rats, and then of human patients (Fig. 13.26). In some instances the healthy cystic fibrosis gene was incorporated and expressed and normal chloride ion movements were also restored. In the near future, most cases of cystic fibrosis will probably be cured by nasal sprays containing genetically engineered viruses carrying the healthy version of this gene.

Note, however, that this sort of gene therapy only cures the symptoms in the lungs, it does not correct the genetic defect in the germ line cells. The defect will still be passed on to the next generation.

Retrovirus Gene Therapy

Retroviruses infect many types of cells in mammals. They need dividing cells for successful infection, and will not infect many tissues where host cell growth and division has come to a standstill. In addition, the genetic material of retroviruses passes through an RNA stage (see Ch. 19). This means that introns must be removed from genes before they are used in a retrovirus.

Despite these extra technical difficulties, a retrovirus has the distinction of carrying the first gene in successful human gene therapy. The *ada* gene on a retrovirus vector is shown in Figure 13.27. Mutations in the *ada* gene cause a deficiency of the enzyme adenosine deaminase which results in severe combined immunodeficiency (SCID). Children with SCID have to be shielded from all contact with other people and are kept inside special sterile plastic bubble chambers. Without immune protection any disease, even a cold, could prove fatal.

13.26 GENE THERAPY BY NASAL SPRAY

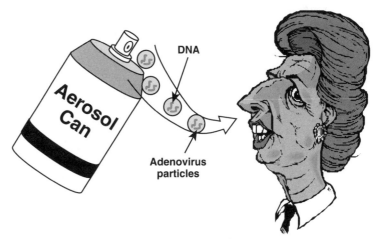

DNA

Adenovirus particles

adenosine deaminase an enzyme that converts adenosine to inosine. Its absence indirectly causes a lack of immune T- and B-cells (see Ch. 21)

severe combined immuno-deficiency (SCID) immune defect due to lack of T- and B-cells. About 20 percent of inherited SCID cases are due to adenosine deaminase deficiency

13.27 RETROVIRUS CARRYING *ada* GENE

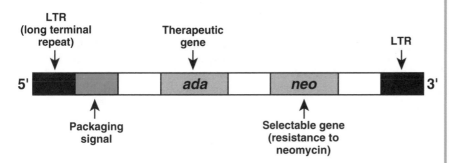

LTR
(long terminal repeat)

Therapeutic gene

LTR

5'   *ada*   *neo*   3'

Packaging signal

Selectable gene (resistance to neomycin)

The affected cells are the peripheral blood lymphocytes that form part of the immune system. Peripheral blood lymphocytes circulate in the blood where they carry out immune surveillance (see Ch. 22). They are produced by the division of bone marrow cells (Fig. 13.28).

13.28 LYMPHOCYTES COME FROM BONE MARROW

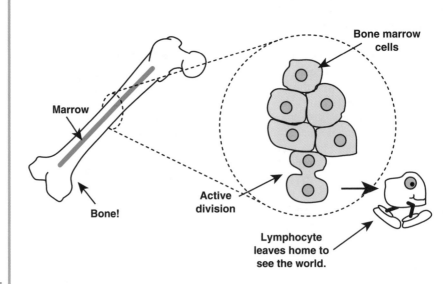

Bone marrow cells

Marrow

Bone!

Active division

Lymphocyte leaves home to see the world.

particle bombardment use of micro-projectiles to insert genes into target animals or plants

13.29 GENE THERAPY VIA CULTURED CELLS

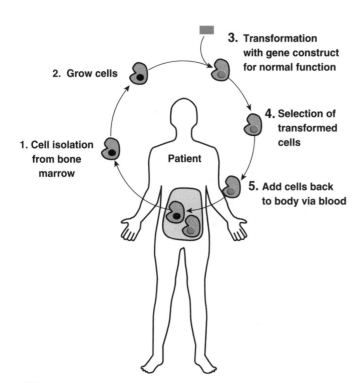

2. Grow cells

3. Transformation with gene construct for normal function

1. Cell isolation from bone marrow

Patient

4. Selection of transformed cells

5. Add cells back to body via blood

Gene therapy involves removing bone marrow cells from the patient and maintaining them in cell culture outside the body. Note that bone marrow cells are rapidly dividing cells and so are suitable for retrovirus infection. While in culture, the bone marrow cells are infected with genetically engineered retrovirus, carrying the *ada* gene, and are then returned to the body (Fig. 13.29).

Since 1991, two children have been successfully treated by this approach, allowing them to leave their plastic bubbles and live normal lives.

Gene Therapy by Particle Bombardment

Gene transfer by **particle bombardment** was originally developed in order to get DNA into plants (see Ch. 15). However, it has also been used successfully to get genes into laboratory animals and is now in the testing stage for humans. Particle bombardment is discussed in Chapter 15.

Aggressive Gene Therapy

So far, we have talked about replacing defective genes in order to treat inherited diseases. However, there is no inherent reason why gene therapy must only be "defensive" and involve replacement of defective genes. We can go on the offensive and provide genes whose products may cure a disease even though the genes we use were not responsible for the problem in the first place.

The best examples of aggressive gene therapy are not actually in curing hereditary defects but in the treatment of cancer. Here the objective is not to replace a gene that is defective or missing but to kill cancer cells. This topic is therefore discussed in the next chapter, which deals with cancer.

Additional Reading

Human Genetics: The Molecular Revolution by McConkey EH. 1993. Jones & Bartlett Publishers, Boston & London.

Human Molecular Genetics by Strachan T & Read AP. 1996. BIOS Scientific Publishers Ltd., Oxford.

Cancer and Aging

14

Somatic Mutations

We have considered mutations and the manner in which they cause inherited diseases. In multicellular organisms a mutation must occur in the germ line cells from which the sperm and eggs are made, in order to be passed on to descendants.

If a mutation occurs in somatic cells - those making up the rest of the body - a variety of possibilities may result (Fig. 14.1). A mutation which occurs early on in embryonic development may be highly detrimental, since each cell of the embryo gives rise to many cells during development. If the single precursor cell that divides, eventually giving rise to a major organ, suffers a serious mutation, the results may be serious or fatal. Most somatic mutations occurring later in development will affect only one or a few cells and will be of little significance.

germ line cells reproductive cells that take part in forming the next generation

somatic cell body cells limited to one individual animal or plant

14.1 SOMATIC MUTATIONS AND DEVELOPMENT

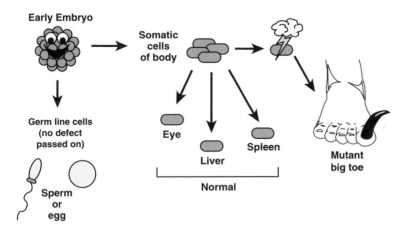

Some Somatic Mutations Cause Cancer

However, some somatic mutations occurring after the organism has reached maturity, are still dangerous. Cancers are the result of somatic mutations that damage the regulatory system controlling cell growth and

cancer disease due to unplanned growth and division of mutant somatic cells

division. This may result in a single cell of the body deciding to start growing and dividing again, long after it is supposed to have settled down to function as part of one of the body's organs (Fig. 14.2).

14.2 CANCER CELL STARTS DIVIDING AGAIN

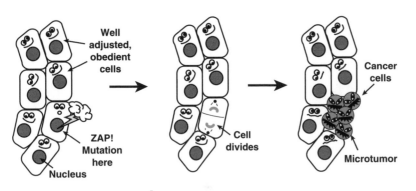

Well adjusted, obedient cells

ZAP! Mutation here

Nucleus

Cell divides

Cancer cells

Microtumor

Cancer occurs in several stages and requires several mutations. First, a cell is mutated and normal control of cell division is lost. Second, the abnormal (mutant) cell divides to form a tumor. If this stays in one place it is known as a benign tumor and can often be cut out by a surgeon resulting in complete recovery. Third, a cancer may gain the ability to invade other tissues and form secondary tumors and is then said to be malignant. Cancers that have spread are much more difficult to cure.

Environmental Factors and Cancer

But isn't cancer caused by smoking cigarettes, by harmful chemicals in the environment and by nuclear radiation? True enough, but these act by causing mutations. Those chemicals and radiation which cause mutations (see Ch. 12) also cause cancer, for the simple reason that cancers are primarily the result of mutations in somatic cells. Cancer-causing agents are often called carcinogens. In practice some cancer-causing mutations are due to environmental factors, and others occur spontaneously due to mistakes made during replication of the cell's DNA.

Most cancers (approximately 80 percent) are derived from epithelial cells, the cells forming the outer covering of tissues. Epithelial cells are the surface cells of the skin and are also found lining the intestines and the lungs. Because the outermost layers are constantly worn away, the underlying layer must keep dividing. Cells from tissues where cell division is rare only occasionally become cancerous. (Nerve and muscle cancers account for only 3 to 4 percent of the total.) In addition, the surface cells are much more likely to suffer exposure to dangerous chemicals and harmful radiation.

Genes That Affect Cancer

It is important to distinguish two general types of genes that affect cancer. Some genes affect your susceptibility to develop cancer. Increased susceptibility to cancer may be inherited just as any other genetic defect (see below). For now, we will stick to worrying about the two classes of genes directly involved in producing cancers as the result of somatic mutations. These are the oncogenes and the tumor-suppressor genes (anti-oncogenes).

To understand the difference between oncogenes and anti-oncogenes let's first remind ourselves that animals and humans are diploid and possess

benign when a tumor stays in one place

malignant when cancer cells from a tumor disperse throughout the body

carcinogen any agent that causes cancer

oncogene mutant gene that promotes cancer

anti-oncogene (tumor-suppressor gene) gene acting to prevent unwanted cell division

two copies of each gene, one carried on each of a pair of homologous chromosomes (see Ch. 11). Mutant oncogenes have a positive effect and promote the development of cancer cells. A single oncogenic mutation in only one of a pair of genes is sufficient to give an effect, *i.e.*, mutations in oncogenes are dominant and the second, wild-type, copy of the gene cannot make up for the defect. In contrast, tumor suppressor genes have a negative effect on cancer development. As their name suggests, they normally suppress division of cancer cells. To allow cancers to grow, both copies of a tumor suppressor gene must be inactivated by mutation. A defective mutation in just one copy of a tumor suppressor gene has no effect, that is, these are recessive mutations.

Oncogenes and Proto-Oncogenes

Oncogenes are cancer-causing genes. Although they were first discovered on cancer causing viruses, they are found in all normal cells, too. Oncogenes exist in three forms: the good, the bad and the ugly.

The "good" form of an oncogene is its original, healthy, unmutated version, *i.e.*, the wild-type allele, sometimes called the proto-oncogene. The wild-type proto-oncogene is necessary for the growth and division of the cell. During development of a multicellular organism, cell division must be closely controlled. Once your spleen is just the right size it should stop growing, not swell up to the size of a prize-winning pumpkin. Clearly, mutations in genes that control cell division are a bad thing. These mutant versions are the "bad" oncogenes (Fig. 14.3).

Third and finally, we have the "ugly" goings on due to retroviruses. These viruses occasionally pick up cellular DNA and carry it off with them. Sometimes they pick up an oncogene and the result is a cancer-causing virus (Fig. 14.4). In fact, all the cancer-causing retroviruses that have been examined closely turn out to be carrying oncogenes from their host cells.

proto-oncogene original, healthy form of gene that may give rise to an oncogene

14.3 PROTO-ONCOGENE TO ONCOGENE BY MUTATION

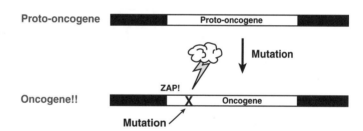

14.4 CANCER - CAUSING RETROVIRUS

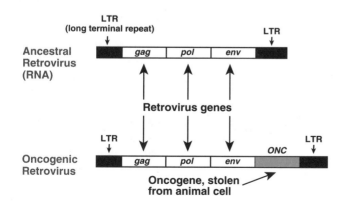

The virus borne version of an oncogene is sometimes written *v-onc* to distinguish it from the cells' version, *c-onc*. Although quite a few cancer viruses are known, most human cancers are not due to viruses, they are due to new mutations of the cellular proto-oncogene to its oncogene form.

Detection of Oncogenes

If DNA is extracted from cells which have gone cancerous and is then inserted into healthy cells it can change them into cancer cells. This is referred to as transformation and if we suspect the presence of an oncogene, we can test this by adding a sample of the suspect DNA to suitable cells in culture.

Normal cultured animal cells usually grow as a thin monolayer on the surface of a culture dish. They are well mannered and do not crawl on top of each other or pile up into heaps (Fig. 14.5). Once the available surface is covered and they are touching neighboring cells on all sides, they stop dividing, a phenomenon known as contact inhibition.

14.5 CULTURED CELLS ARE POLITE

14.6 CANCER CELLS ARE UNINHIBITED

Cancer cells are uninhibited, in fact just plain rude, and they keep on dividing and piling into heaps (Fig. 14.6). These can be seen when DNA containing an oncogene is added to normal cells in culture.

In fact these miniature heaps of cells are really tiny tumors. This can be illustrated by sucking them up into a hypodermic syringe and injecting them into an experimental animal, usually a mouse, which then develops a real tumor (Fig. 14.7).

14.7 CANCER CELLS CAUSE TUMORS IN MICE

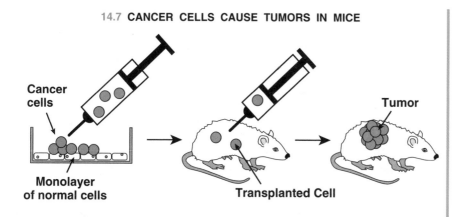

Cancer cells

Monolayer of normal cells

Transplanted Cell

Tumor

Creating an Oncogene

What sort of mutations turn a proto-oncogene into an oncogene? The mutations that create oncogenes result in increased cell division. Furthermore, they are dominant mutations; in other words it is only necessary to mutate one copy of the gene to see an effect. The wild-type duplicate copy of the proto-oncogene, present in all your cells, does not overcome the effect of the oncogene. This is because oncogenic mutations result not from loss of activity but from increased activity of the oncogene. This can be due either to alteration of the sequence of the protein encoded by the proto-oncogene or due to increased production of the protein (Fig. 14.8).

14.8 HOW A PROTO - ONCOGENE GOES BAD

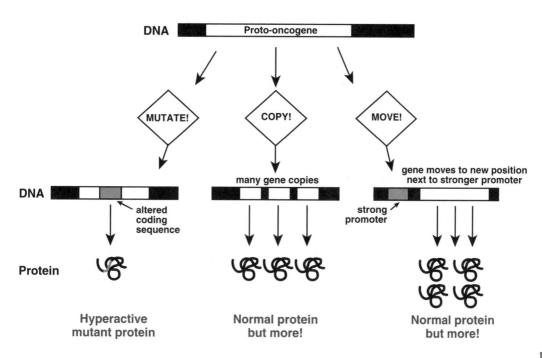

DNA — Proto-oncogene

MUTATE!   COPY!   MOVE!

many gene copies

gene moves to new position next to stronger promoter

DNA

altered coding sequence

strong promoter

Protein

Hyperactive mutant protein

Normal protein but more!

Normal protein but more!

Like any normal gene, a proto-oncogene has a regulatory region plus a structural region that encodes a protein. Some oncogenes are the result of mutations that alter regulation of the proto-oncogene, whereas others are the result of mutations in the structural portion of the proto-oncogene.

The *ras* Oncogene - Hyperactive Protein

The *ras* oncogene is the result of a single base change in the structural region of the gene. This causes an alteration in a single amino acid in the protein encoded by this gene (Fig. 14.9). Most *ras* mutations alter the amino acid at position No. 12, others affect No. 13 or No. 61. Only a few, very specific mutations can create a *ras* oncogene from the proto-oncogene.

14.9 CONSEQUENCES OF CONVERTING PROTO-*RAS* TO *RAS*

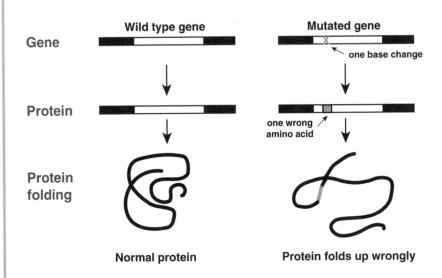

The *ras* protein is involved in transmitting signals concerning cell division in humans, flies and even yeast (Fig. 14.10). These signals are received from outside the cell by receptors, like *ras*, at the cell surface and the job of normal *ras* protein is to pass them on. After receiving a signal, normal *ras* protein binds guanosine triphosphate (GTP) and goes into a signal emitting mode. After emitting a brief pulse of signals, it then splits the GTP into guanosine diphosphate (GDP) plus phosphate, and relapses into standby mode again. The cancer-causing form of the *ras* protein is locked permanently into the signal emitting mode and never splits its GTP (Fig. 14.11). Therefore it constantly floods the cell with signals urging cell division, even when none are being received from outside.

The 3-D structure of the *ras* protein has been worked out by X-ray crystallography (see Ch. 16). Those amino acid residues that are changed by oncogenic mutations all turn out to be directly involved in the binding and splitting of the GTP. The consequence is uncontrolled cell division and the beginnings of a possible cancer.

ras protein   a protein involved in cell proliferation which, when mutated, can cause cancer

14.10 SIGNALS ARE PASSED ON BY *RAS*

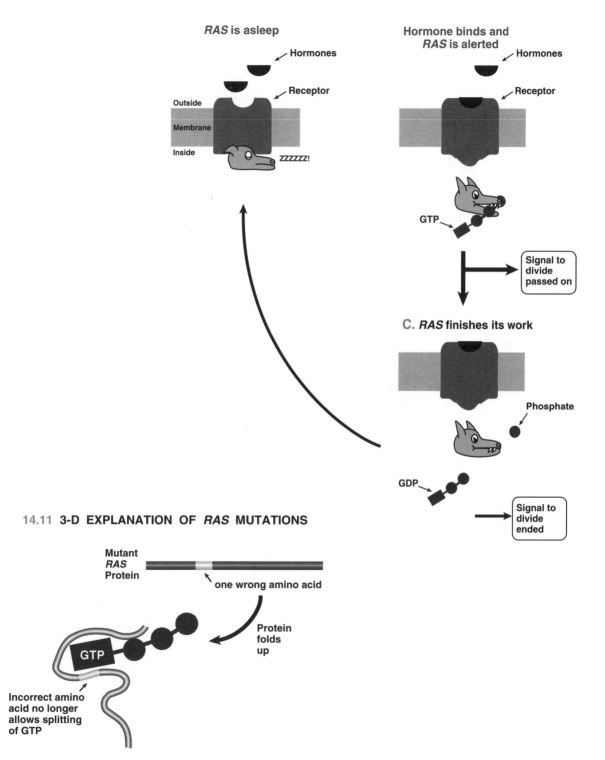

RAS is asleep

Hormones

Receptor

Outside

Membrane

Inside

ZZZZZZ!

Hormone binds and
RAS is alerted

Hormones

Receptor

GTP

Signal to
divide
passed on

C. *RAS* finishes its work

Phosphate

GDP

Signal to
divide
ended

14.11 3-D EXPLANATION OF *RAS* MUTATIONS

Mutant
RAS
Protein

one wrong amino acid

Protein
folds
up

GTP

Incorrect amino
acid no longer
allows splitting
of GTP

The *Myc* Oncogene - Overproduction of Protein

Some oncogenes are due to mutations altering the structure of a protein like *ras*. Other oncogenes are created by changes that affect regulation. Instead of a mutant protein that is stuck in a hyperactive form, many other oncogenes suffer changes that vastly increase the amount of the protein formed, although the protein itself is not changed.

A well known example is the *Myc* oncogene which encodes a protein involved in switching on an array of other genes involved in cell division, *i.e.*, Myc protein is a transcription factor. A Myc protein overdose can occur in two ways. Some *myc*-dependent cancers result from chromosomal changes in which the *myc* gene is duplicated many times. Instead of the normal two copies, 50 to 100 copies may occur as the result of mistaken duplication of the segment of DNA carrying the *myc* gene (Fig. 14.12). The Myc protein will be overproduced by 50- to 100-fold, too.

Alternatively, it is possible to have the standard two copies of the *myc* gene but alter their regulation. In Burkitt's lymphoma, a rare chromosomal translocation swaps segments of two unrelated chromosomes. This splits the *myc* structural gene off from its proper regulatory region and joins it to the regulatory region of another gene (Fig. 14.13). The Myc protein is now produced continuously in substantial amounts instead of being strictly regulated as before.

14.12 THE *myc* ONCOGENE

Before duplication

After duplication

Continued duplication

repeated local duplication of DNA

Normal Role of Proto-Oncogenes

The pathway for activating cell growth and division has several stages. The proto-oncogenes encode the proteins taking part in this scheme (Fig. 14.14). Not surprisingly, mutations that result in hyperactivation of any of the components involved in this can turn proto-oncogenes into oncogenes.

The major components involved are:

1) Growth Factors: These are proteins circulating in the blood and bringing messages from outside to the cell surface.
2) Cell Surface Receptors: These proteins are found in the cell membrane where they receive messages from outside and pass the signal on.
3) Transcription Factors: These proteins bind to and switch on genes in the cell nucleus (see Ch. 11). This results in the synthesis of new proteins, as opposed to the activation of those which are already present (Fig. 14.14).

14.13 TRANSLOCATION OF *myc*

Normal

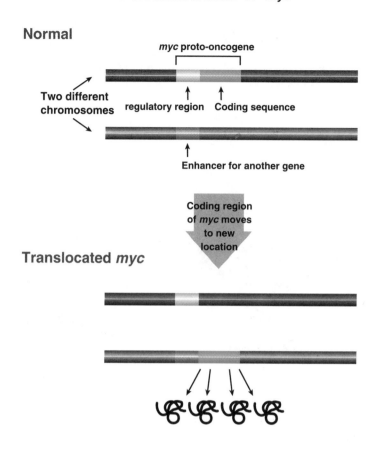

myc proto-oncogene

Two different chromosomes

regulatory region   Coding sequence

Enhancer for another gene

Coding region of *myc* moves to new location

Translocated *myc*

14.14 NORMAL ROLE OF PROTO-ONCOGENES IS MESSAGE TRANSMISSION

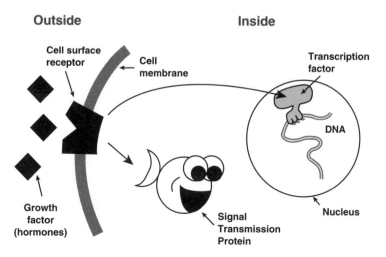

Outside   Inside

Cell surface receptor

Cell membrane

Transcription factor

DNA

Growth factor (hormones)

Signal Transmission Protein

Nucleus

protein kinase an enzyme that switches other enzymes on or off by attaching a phosphate group to them

4) Signal transmission proteins: These pass on the signal from outside the cell to enzymes or genes needed in cell division. **Protein kinases** are signal transmission proteins that act by adding a phosphate group to enzymes that are part of the machinery for growth and division. Some of these enzymes are already hanging around, but in an inactive form. Addition of the phosphate group switches needed enzymes on, and also turns others that are no longer required, off (Fig. 14.15).

14.15 **OPERATION OF PROTEIN KINASES**

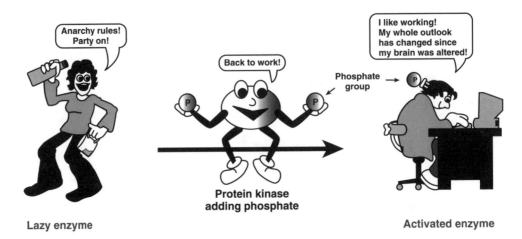

Lazy enzyme

Protein kinase adding phosphate

Activated enzyme

null mutation a mutation that fully inactivates a gene

nullizygous when both copies of a gene are fully inactivated

Tumor Suppressor Genes or Anti-Oncogenes

Proto-oncogenes promote cancer when they are converted to hyperactive variants. It is only necessary to hyperactivate one of a pair of wild-type genes to get an effect, as the oncogene form is dominant. In contrast, a mutation in just one copy of a tumor suppressor gene has no effect as these are recessive mutations. Tumor suppressor genes, or anti-oncogenes as they are also known, normally suppress uncontrolled cell division, so it is necessary to inactivate both copies in order to initiate cancerous growth.

When both copies of a gene have been inactivated by **null mutations,** this is known as the **nullizygous** state.

There are two ways to end up with both copies of a gene inactivated. First, during division of the cells which form the body, two successive somatic mutations may occur (Fig. 14.16).

14.16 **BOTH COPIES OF A TUMOR SUPPRESSOR GENE MUST BE INACTIVATED TO CAUSE CANCER**

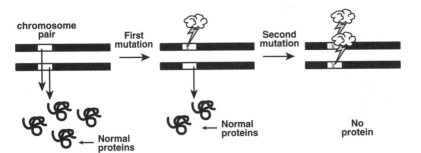

First one copy of the gene is inactivated, and then a second mutation strikes the second copy of the same gene. Although you would have to be pretty unlucky for this to happen, it does occasionally occur.

The second route is more frequent. The first mutation occurs in one copy of the gene in a germ line cell of one of your ancestors. If this defective copy is passed on, when you start life your cells will all have one copy of the gene already inactivated. A somatic mutation, which inactivates the second copy may then occur as your cells divide. In other words you inherit one defective gene and you acquire a second by mutation. Those individuals who inherit a single defective anti-oncogene do not always develop cancer. What they inherit is an increased chance of doing so.

This scheme applies to a dozen or more tumor suppressor genes. Some of them are involved in a wide range of tumors. Other anti-oncogenes are very tissue specific and inactivation of both copies triggers cancer of a particular organ, such as the Wilm's tumor gene responsible for a kidney cancer and the retinoblastoma (Rb) gene responsible for a rare cancer of the retina of the eye (Fig. 14.17).

retinoblastoma (Rb) gene a particular anti-oncogene responsible for retinal cancer if both copies are inactivated

14.17 TWO ROUTES TO RETINOBLASTOMA

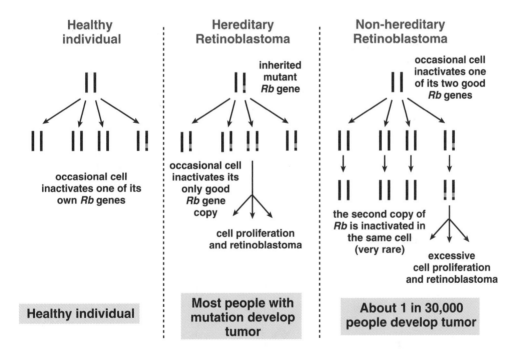

How Do Anti-Oncogenes Work?

They encode proteins whose job is to inhibit growth or prevent cell division. Originally it was thought that chemical signals to start growing all came from outside the cell. In other words, hormones circulating in the blood control growth and development. Some anti-oncogene encoded proteins would then be involved in detecting and acting on these external hormonal signals (Fig. 14.18).

14.18 THEORY OF EXTERNAL SIGNALS FOR CELL DIVISION

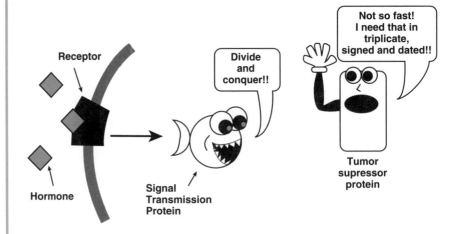

More recently, evidence has accumulated indicating that in addition to hormone control, many somatic cell lines are also pre-programmed. They have a pre-set number of allowed cell divisions. An internal "generation clock" of some sort counts off the number of permitted divisions left and when it reaches zero, growth and division stop. Some anti-oncogenes are part of this system and when they are defective the cell fails to stop dividing.

Many of the proteins encoded by anti-oncogenes are DNA-binding proteins with zinc fingers. These proteins have finger-like bulges of about 30 amino acids each built around central zinc metal atoms (Fig. 14.19). Each zinc finger recognizes three bases of the DNA sequence in the regulatory region of the genes they control.

14.19 ZINC FINGER PROTEINS

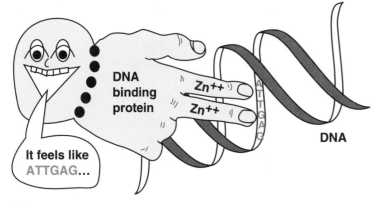

Normal Cell Division: The Cell Cycle

To understand further how anti-oncogenes work, we need to consider the process of normal cell division. The cell cycle has four stages (see Fig. 14.20):

1) G₁ phase - the cell grows
2) S phase - the chromosomes are duplicated
3) G₂ phase - the cell grows and prepares to divide
4) M phase - the cell and its nucleus divide

To move from one stage to another requires the permission

202

of proteins called **cyclins**, one for each major stage. The cyclins act like guards at a security checkpoint. They monitor the environment and also check to make sure that the previous stage of the cell cycle has been finished properly before moving on.

The cyclins have subordinates, the **cyclin dependent kinases**, or **CDK** proteins (Fig. 14.21). When the cyclin for a particular step in the cell cycle decides to go for it, it binds to its CDK protein. This activates the CDK protein which then adds phosphate groups to a whole series of other proteins whose job it is to actually do the work of cell division. These proteins are on standby until they are activated by the phosphate group.

14.20 STAGES OF THE CELL CYCLE

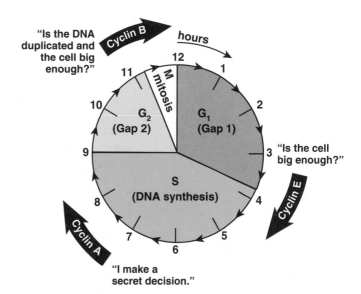

"Is the DNA duplicated and the cell big enough?"

Cyclin B

hours

M mitosis

G₂ (Gap 2)

G₁ (Gap 1)

"Is the cell big enough?"

Cyclin E

S (DNA synthesis)

Cyclin A

"I make a secret decision."

The p53 Anti-Oncogene

Best known of the human anti-oncogenes is the infamous **p53 gene**, found on the short arm of chromosome No. 17. The p53 protein is a DNA-binding protein which, with its buddies p16 and p21, apparently acts as an emergency braking system for the cell cycle.

This gene is involved in a very large number of diverse cancers. The reason why p53 is so often involved is that its behavior differs from that of the standard anti-oncogene. In the case of typical anti-oncogenes, a single mutant allele is recessive and has no effect on cell behavior. In contrast, a single defective p53 allele does show effects, even in the presence of a second, normal copy of the gene.

Why is this? Well, back to biochemistry! The protein encoded by the p53 gene assembles in groups of four, that

14.21 CYCLINS ACT VIA CDK PROTEINS

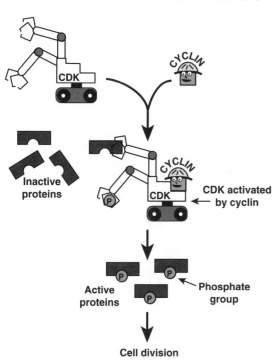

CDK

CYCLIN

Inactive proteins

CYCLIN

CDK

CDK activated by cyclin

Active proteins

Phosphate group

Cell division

cyclin protein that controls the cell cycle

cyclin dependent kinases (CDK) subordinate proteins that transmit orders from a cyclin by adding phosphate groups to the enzymes they control

p53 gene a notorious anti-oncogene, often mutated in cancer cells

is, it forms tetramers (Fig. 14.22). When a cell has one good copy of the p53 gene and one bad copy, it will produce a mixture of good and bad p53 proteins. These will assemble into mixed tetramers and even if a tetramer contains some good proteins, the bad ones will mess up the whole assembly.

14.22 TETRAMERS OF P53 PROTEIN

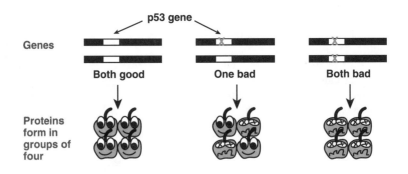

The likelihood of getting a completely good tetramer is the probability of finding four good protein subunits in a row, that is: $1/2 \times 1/2 \times 1/2 \times 1/2 = 1/16$. In other words, only one-sixteenth of the p53 protein assemblies will function correctly even though half of the individual protein subunits are normal. One bad apple spoils the whole basket!

If a cell's DNA is damaged in any way, the p53 protein activates the gene for p21 (Fig. 14.23). The p21 protein then blocks the action of all of the cyclins and freezes the cell wherever it is in the cell cycle until the

p21 protein a protein which blocks cell division by binding to and inhibiting the cyclins

14.23 THE P53 PROTEIN ACTIVATES THE P21 GENE

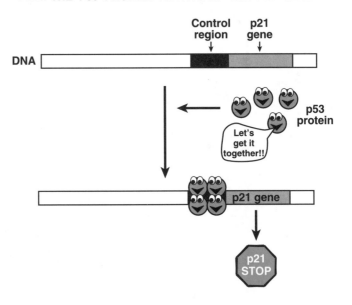

damage can be repaired (Fig. 14.24). The p16 protein acts similarly, but just blocks cyclin E. It stops division at the critical point when the cell is deciding whether to sit tight, grow bigger (but not divide) or to grow and divide.

The p53 protein is not necessary for normal cell division. Mice with both copies of the p53 gene knocked out grow normally. Their only problem is they all die of cancer after three or four months. The role of p53 then, is to shut down cell division in emergencies, when, for example, the DNA is damaged by ultraviolet radiation. Well over half of all human cancers are defective in p53.

Formation of a Tumor

The actual generation of a real tumor requires several steps. In practice more than one somatic mutation is necessary for the production of most cancers (Fig. 14.25). For example, many colon cancers carry the following defects:

1) Inactivation of the APC anti-oncogene
2) Activation of the *ras* oncogene
3) Inactivation of the DCC anti-oncogene
4) Mutational loss of the p53 gene

Thus, cancer development goes through a series of four or more mutational stages before a full blown tumor results. Even then, the cancer cells will all stay in one place, and if the tumor is cut out by surgery, all may still be well. However, cancers do not stay put forever. Eventually they bud off cancer cells that travel around the body, settle down in other tissues and grow into secondary tumors. This cancer colonialism is called metastasis and once things reach this stage it is virtually impossible to find and remove all of the cancers.

14.24 BLOCKAGE OF CYCLINS BY P21 PROTEIN

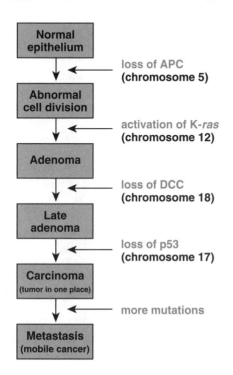

14.25 DEVELOPMENT OF COLON CARCINOMA

gene knock out technically, knocking out a gene means disrupting it by inserting foreign DNA (see Ch. 15)

Apc and Dcc anti-oncogenes two particular anti-oncogenes often involved in colon cancer

metastasis spreading of cancer cells from their original site to form new secondary cancers

Other mutations that aren't fully understood are necessary for cancer cells to start traveling.

These include mutations which result in:

1) Loss of adhesion to neighboring cells in the home tumor, and,
2) Ability to bind to and penetrate the membranes surrounding other tissues of the body (Fig. 14.26).

14.26 TRAVELING CANCER CELLS MUST BIND, DIGEST AND BREAK THROUGH A CONFINING BASAL LAMINA

Binding to Laminin

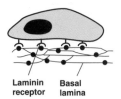

Laminin receptor   Basal lamina

Digestion of Basal Lamina by Collagenase

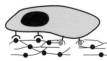

Movement through Basal Lamina

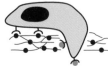

Around 10^{16} cell divisions happen in a human body during a typical lifetime. That's 10,000,000,000,000,000 to those of you with a liberal arts orientation! Even in the absence of artificial mutagens, an average gene will mutate about once in a million generations. So somewhere, in cells that are part of you, each of your genes has mutated about 10^{10} times. Clearly, if a single mutation in a single gene were enough to cause cancer, we would all be walking masses of tumors. Calculations based on the frequency of cancer indicate that from four to six separate mutations are required in its development, depending upon the type of cancer.

mutator gene   a gene that will cause an increased rate of mutation if it is defective

mismatch repair   repair system that corrects mispaired bases

Inherited Susceptibility to Cancer

It is thought that 5 to 10 percent of cancers may be largely due to inherited defects. Many of the genes involved in this are poorly understood, but we should briefly mention three general categories.

First, as we have already seen using the *Rb* gene as an example, it is possible to inherit one dud copy of an anti-oncogene. This means that every one of your somatic cells starts life with one faulty copy and only a single somatic mutation is needed to completely inactivate the pair of anti-oncogenes. What if you inherit two bad copies of an anti-oncogene? If you had, you would not be reading this now. Artificially engineered mice that are double negative for such genes generally die before birth. So homozygous recessives for anti-oncogenes are generally lethal when inherited.

To be politically correct, we must not forget breast cancer. Inheriting a single defective copy of the *BRCA1* (yes, it stands for BReast CAncer) gene predisposes women to cancer of both breast and ovary. About 0.5 percent of U.S. women carry mutations in *BRCA1*. As with other tumor-suppressor genes, the second copy must mutate during division of somatic cells for a cancer to arise.

Secondly, mutations in certain special genes affect the rate at which mutations occur during cell division. Such genes are known as **mutator genes**. As you might imagine, these include the genes involved in DNA synthesis, such as the genes encoding DNA polymerase (see Ch. 5).

Some mutator genes are more subtle and are involved in DNA repair (see Ch. 12). These have been analyzed in detail in bacteria, but in humans the details are vague. However, it appears that certain inherited forms of colon cancer are due to defects in a mutator gene, probably those involved in **mismatch repair** (see Ch. 12). This, in turn, increases the rate of mutation in all other genes including the tumor-suppressor genes. In humans, mutator gene defects are recessive, like tumor-suppressor genes themselves.

Third, there are indirect effects due to genetic differences between races or populations. For example, some skin cancers are caused by mutations due to ultraviolet radiation from the sun. White people, especially those exposed to high levels of sunshine in the tropics or Australia, develop skin cancer much more often than Blacks. The reason is obvious, the more black pigment the less UV radiation penetrating to your DNA.

Cancer Causing Viruses

Although **retroviruses** that cause cancer in chickens and mice were important in the discovery of oncogenes, very few human cancers are due to retroviruses.

The most famous human retrovirus is **AIDS** (see Ch. 19). Although some AIDS patients die of cancers, the AIDS virus does not cause cancer directly. What it does is knock out the immune system that would otherwise kill off most cancer cells before they get too far out of control.

The first cancer-causing retrovirus to be discovered was **Rous Sarcoma Virus, RSV**

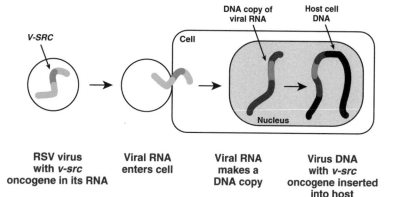

| RSV virus with *v-src* oncogene in its RNA | Viral RNA enters cell | Viral RNA makes a DNA copy | Virus DNA with *v-src* oncogene inserted into host chromosome |

(Fig. 14.27). Long ago, the ancestor of this virus picked up a copy of the chicken *src* oncogene. The nucleic acid inside a retrovirus particle is RNA. When RSV infects chickens it converts the RNA into a DNA copy and incorporates the *src* oncogene into the host cell DNA and this causes muscle tumors. Ever wonder how they got those fast food chicken nuggets??!!

In fact only about 15 percent of human cancers are due to viruses and most of these are due to DNA viruses (**papillomaviruses** and **herpesviruses**). The DNA tumor viruses act by blocking the cell's tumor suppressor genes. First the DNA tumor virus integrates its DNA into the host cell's chromosome. Second, it makes a virus protein which binds to the cell's *Rb* and p53 proteins (Fig. 14.28 next page). This activates cell division and may lead to a tumor. **Simian Virus 40 (SV40)** that causes cancer in monkeys has been studied most. Human papillomaviruses act in a similar manner but usually produce only benign growths - warts. Only rarely do they cause dangerous tumors.

Cancer Revisionism: Diagnosis or Cure?

Since the "War on Cancer" was started, most of the major advances have been in the diagnosis and understanding of cancer, rather than its cure. Because of the massive vested interests in the cancer research industry, cancer statistics are often presented in an overly optimistic way.

Let's suppose that you began to develop a serious cancer which is eventually detected in the year 2000, that you survive for five more years and they bury you in the year 2005. Now let's consider what happens when the ability to diagnose cancer improves. Your cancer is detected in 1995, but you still kick off in 2005. In the first case you "survived for only five years with cancer," whereas in the second case you "survived for 10 years with cancer." Wow!!! A 100 percent increase in survival - right??? Wrong!! You did not live any longer at all; it was merely that you were diagnosed earlier. Statisticians call this "lead time bias." In fact, you just get to worry about cancer for five years longer.

retrovirus type of virus which has its genes as RNA in the virus particle but converts this to a DNA copy inside the host cell

AIDS (acquired immunodeficiency syndrome) the disease caused by the HIV retrovirus

Rous sarcoma virus (RSV) a retrovirus that causes cancer in chickens

Src oncogene one particular oncogene which encodes an enzyme acting as a protein kinase

papillomavirus family of DNA-containing viruses that sometimes cause tumors

herpesvirus family of DNA-containing viruses causing a variety of diseases and sometimes tumors

simian virus 40 (SV40) a DNA-containing monkey tumor virus of the papovavirus family

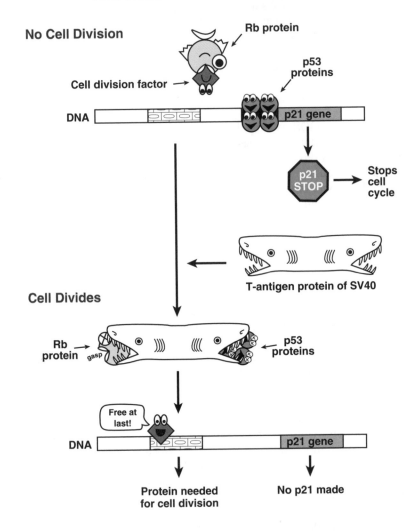

No Cell Division

Rb protein

Cell division factor →

p53 proteins

DNA

p21 gene

p21 STOP → Stops cell cycle

T-antigen protein of SV40

Cell Divides

Rb protein *gasp*

p53 proteins

Free at last!

DNA

p21 gene

Protein needed for cell division

No p21 made

By using this bogus measure of survival time, we can make it look as if great advances have been made in curing cancer. In fact, (using data from 1950 to 1982) great improvements in genuine life expectancy have been made only in one or two relatively rare forms of cancer.

The basic problem is that although most cancers are not inherited via the germ line, cancer is nonetheless a genetic disease and perhaps what we need is a genetic approach to its cure.

Aggressive Gene Therapy for Cancer: Today and Tomorrow

In the case of hereditary disease we may take the approach of replacing the defective component by genetic means. In the case of an already developed cancer we give way to the primal urge to destroy the cancer cells. For one thing, replacing defective genes is too late when faced with a full-blown tumor.

Two plans of attack may be used, direct and indirect. In the direct plan of attack, a gene that helps kill cancer tissue is used. For example, the TNF gene encodes Tumor Necrosis Factor (Fig. 14.29). This is normally produced by white blood cells known as tumor-infiltrating lymphocytes or TILs. These are blood cells with ninja training who normally sneak into tumors and release TNF, which is quite often effective at snuffing out small cancers. But, alas, even a ninja can be overwhelmed by superior numbers.

To attack a large cancer that is out of control, we can try to hype up the TNF system. So first we clone the TNF gene. Then we suck out some white blood cells from the patient and keep them alive in culture. Next, we introduce extra copies of the TNF gene - or perhaps an improved TNF gene with enhanced activity - into the white cells. Then we inject them back into the patient (Fig. 14.30).

14.29 TUMOR NECROSIS FACTOR

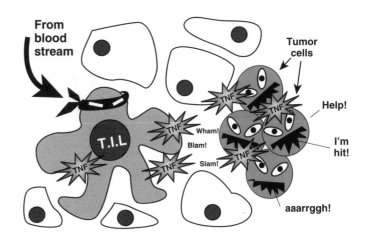

tumor necrosis factor (TNF) short protein that kills cancer cells

tumor-infiltrating lymphocyte (TIL) white blood cell which secretes TNF

14.30 *EX - VIVO* GENE THERAPY

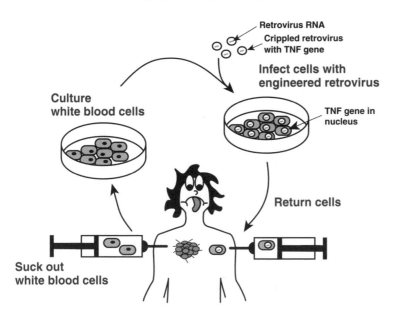

The indirect approach is to rely on the body's natural defenses. Our immune systems are really quite effective at killing cancers, provided they find them in time. To be successful, a cancer has to somehow evade

14.31 INDIRECT GENE THERAPY FOR CANCER

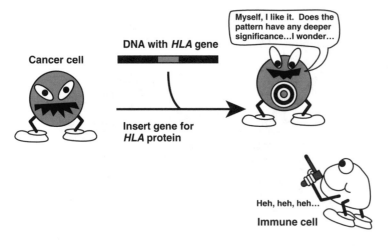

Cancer cell

DNA with *HLA* gene

Myself, I like it. Does the pattern have any deeper significance...I wonder...

Insert gene for *HLA* protein

Heh, heh, heh...

Immune cell

HLA human leukocyte antigens are proteins found on the cell surface that allow cells to be recognized by the immune system

the body's immune surveillance. So in this approach we insert into the tumor cells a gene that betrays the tumor as an alien intruder to be eliminated. Then we sit back and let the immune system take care of the tumor (Fig. 14.31).

An example of this approach is to use an HLA gene. HLA genes form a family of genes whose encoded proteins are situated on the cell surface. There they act in cell recognition (see Ch. 21). Different individuals have different combinations of HLA genes, which act as a sort of cell level fingerprint so that cells of the body recognize each other as "self." What must be done for a cancer patient is to insert (but only into the tumor cells) HLA genes which are not originally present in that particular individual. The tumor then appears to be foreign, and the immune system will mount an assault.

Getting Genes in by Lipofection

Lipofection is sort of the opposite of liposuction. During liposuction, large greasy globs of fat are sucked out of your ugly body in the hope of aesthetic improvement. (Fat chance!!)

In lipofection, hollow microscopic spheres of fat - liposomes - are filled with DNA. These tiny fat globules will merge with the membranes surrounding most normal animal cells and whatever was inside the liposome will end up inside the cell. Although lipofection does work, it is rather non-specific as the liposomes tend to glop on to any cells. Nonetheless, "armed" liposomes can be injected directly into tumor tissue.

Liposomes can, of course, be filled with other goodies, not just DNA. In fact, they are probably of more use in delivering proteins, something not feasible when using viruses as genetic engineering vectors. For example, killer proteins can be packaged inside liposomes and injected into tumor tissue (Fig. 14.33). Cancer cells take them up and the dangerous proteins are then released inside the cancer cells which are (we hope!) killed.

14.32 LIPOSUCTION VERSUS LIPOFECTION

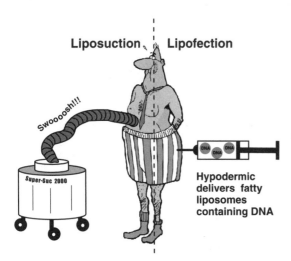

Liposuction   Lipofection

Swoooosh!!!

Super-Suc 2000

DNA   DNA   DNA

Hypodermic delivers fatty liposomes containing DNA

Immortality and the Philosophers' Stone: Aging

Why threescore years and ten? Why aging in the same chapter as cancer? Surely you have guessed! Over the years we all accumulate ever increasing numbers of mutations in the cells of our bodies. Cells that get several mutations in growth regulating genes will give rise to microtumors. Sooner or later one of these will sneak past immune surveillance and you will get cancer. Ultimately, your only way of avoiding cancer is to die of something else first!

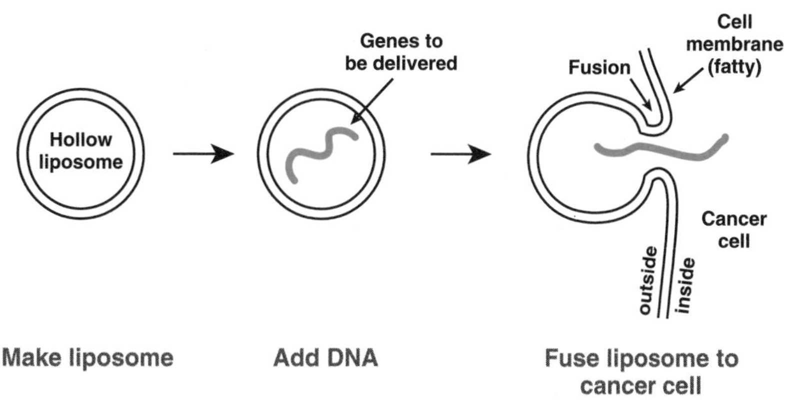

Make liposome   **Add DNA**   **Fuse liposome to cancer cell**

But perhaps not all is lost. One mechanism for avoiding cancer is to repair mutations in the DNA before the mutated genes even get a chance to express themselves. There are a variety of DNA repair systems, which are often quite complex in their operation (see Ch. 12 for examples). However, the ability to repair DNA declines with the passing years. If this could be somehow remedied, perhaps the number of mutations could be reduced and tumor formation could be put off a few years longer.

"The End is Nigh" (Again): Telomeres

Another link between aging and cancer concerns telomeres. These are the structures at the ends of eukaryotic chromosomes (see Ch. 11). Human telomeres consist of several hundred repeats of the sequence TTAGGG and a few of these are lost each time the chromosomes divide. This limits the number of times that a chromosome, and hence the cell containing it, can divide to about 50 generations.

Further division requires an enzyme, telomerase to restore the repeated sequences of the telomere. As you grow older, telomerase fades away in most somatic cells. Of course this does not apply to those specialized tissues that continue actively dividing, nor to the germ-line cells! Consequently, old people have shorter telomeres in most of their cells than do younger people.

For a cell to become cancerous it must regain the ability to divide and this means that the gene encoding telomerase must be reactivated by one or other of the mutations responsible for causing cancer. Can we bring the division of cancer cells to an end by inhibiting telomerase?

Telomerase carries around with it a short piece of RNA that reminds it of the correct sequence of DNA to add to the chromosome tips. Let's make a piece of artificial RNA complementary in sequence to the telomerase memory RNA. This is known as antisense RNA and should bind to and block the telomerase RNA (Fig. 14.34, next page). In fact such antisense RNA does indeed inhibit the division of cancer cells in culture. Whether telomerase inhibitors will prove useful in treating cancer is for the future to decide.

telomere   special end structure found on a linear eukaryotic chromosome

telomerase   enzyme that adds DNA to the end, or telomere, of a chromosome

antisense RNA   an RNA molecule which is complementary in sequence to, and so will base pair with, a target RNA molecule with some functional role in the cell

Antisense - also a language widely used by politicians to communicate with voters

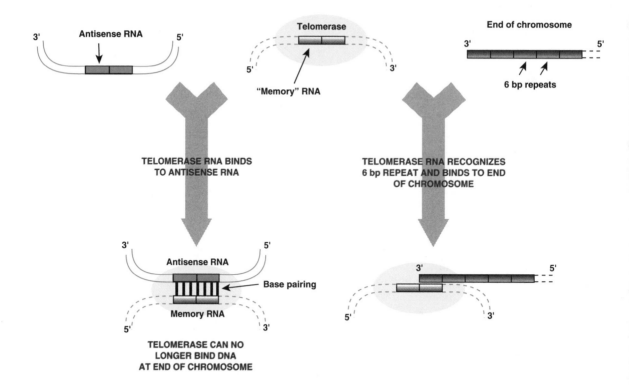

**TELOMERASE RNA BINDS
TO ANTISENSE RNA**

**TELOMERASE RNA RECOGNIZES
6 bp REPEAT AND BINDS TO END
OF CHROMOSOME**

**TELOMERASE CAN NO
LONGER BIND DNA
AT END OF CHROMOSOME**

Caenorhabditis elegans
a species of nematode or
roundworm, widely used
in studying animal devel-
opment because of its
relative simplicity

Daf-2 gene   gene
involved in life length
determination in nematode
worms

Older and Wiser Worms

Obviously, some people do chug on to 100, or occasionally even older, without sprouting tumors, so there is clearly room for improvement for most of us. Although not much has been done yet for people, science has managed to lengthen the life of certain primitive worms. We have to start somewhere!

A tiny round worm, *Caenorhabditis elegans*, which normally lives about 25 days, gets to live more than twice as long when it carries a mutation in its *daf-2* gene (Fig. 14.35). What's more, the worms stay active and wriggling healthily till just a day or two before they finally peg out.

14.35 LIVE LONGER AND PROSPER

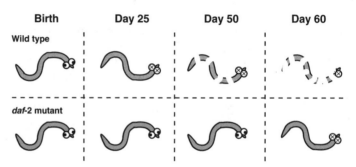

Although other worm genes also involved in age control have been identified, we still have no idea how they work.

Presumably there is some sort of analogous genetic system in higher animals. Although we live longer than our medieval ancestors, this is due to better food and improved hygiene; our longevity is due more to avoiding premature termination rather than to an enhanced life span. However, as far as biotechnology is concerned, nothing much has been done to extend the life of higher animals apart from the Energizer Bunny, which just keeps going, and going, and going.

Programmed Cell Death

When a frog changes from a tadpole to an adult it loses its tail. Just at the correct time for this to happen, the cells in the tail that are no longer needed, kill themselves (Fig. 14.36). Such pre-programmed cell death is known as apoptosis. It sure saves on retirement pay! Apoptosis is Greek for "dropping off" and if you want to fool other people into thinking you are educated, you must not pronounce the second "p" - it is silent like the "p" in pterodactyl.

apoptosis programmed suicide of unwanted cells for the good of the whole animal

14.36 **PTADPOLES PTURN INTO PFROGS BY APOPTOSIS; PFROGS PTURN INTO PRINCES BY KISSING**

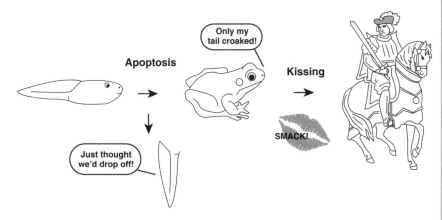

When cells die as a result of injury, they normally swell and burst. This is quite a different matter and is known as necrosis. It causes inflammation that attracts cells of the immune system to the trouble spot.

Even for those of us who lost our tails several generations back, apoptosis is still important. For example, it prevents humans from being born with the webbed fingers found in early fetuses. Less romantically, throughout adult life, it phases out old, exhausted body cells so they can be replaced by new ones. It is beneficial to the organism to have specific cell types die in specific locations at specific times. When it is time to go, it happens fast. The cell's DNA forms insoluble clumps and the cell disintegrates, leaving a cluster of membrane fragments (Fig. 14.37). It is all over in half an hour. After this public-spirited sacrifice, their mortal remains are eaten by surrounding cells. Ritual suicide followed by cannibalism!

necrosis unplanned death of cells as the result of injury

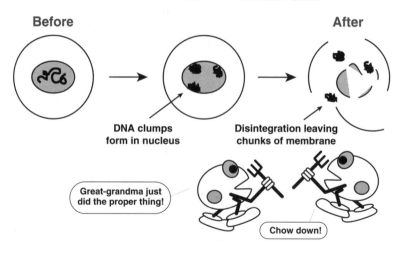

14.37 APOPTOSIS OF A SINGLE CELL

Before

After

DNA clumps
form in nucleus

Disintegration leaving
chunks of membrane

Great-grandma just
did the proper thing!

Chow down!

Death Is Only A Phone Call Away

Even more amazing is that many animal cells need a constant signal to stay alive. Once they no longer receive permission to stay alive, they commit suicide. A cell's fate hangs in the balance, depending on the function of just a single gene. This gene, *ced-9*, has been identified in our little wriggly friend, *Caenorhabditis elegans*. This worm is used because it only produces a total of 1,090 cells, instead of the millions upon millions found in humans. Of these, 131 commit suicide during development so that an adult worm has exactly 959 cells.

When *ced-9* is defective, cells that should stay alive commit suicide and the worms die prematurely. If *ced-9* is mutated to a kinder and gentler form which gives permission to stay alive to every cell all the time, cells which should have sacrificed themselves to the public good, stay alive. The supervision of cell death involves a group of *ced* genes, a molecular death squad. The *ced-9* gene gives the orders and two other genes, *ced-3* and *ced-4*, are the molecular hit men. If *ced-3* or *ced-4* are defective it doesn't matter what orders *ced-9* gives.

Humans are more complex than worms and rather than just a single gene that decides whether our cells live or die, we may have a sort of molecular jury system which takes account of a variety of factors.

In humans and other higher animals the *bcl-2* oncogene may resemble the worm's *ced-9* gene in some

Ced genes genes which control apoptosis (*i.e.,* planned cell death) in nematode worms

bcl-2 gene a gene that probably controls apoptosis in humans and may contribute to cancer when mutated

14.38 GENETIC DEATH SQUAD

Apoptosis now!!

CED 9

CED 3

CED 4

ways. When it is hyperactivated, cells no longer die and a cancer may result. When it is non-functional, premature cell death results. Engineered mutant mice with both copies of the *bcl-2* gene defective are normal at birth. But within a week or two, massive abnormal apoptosis destroys the thymus and spleen and causes major kidney damage (Fig. 14.39). Only about a quarter of the mice survive this disaster. And a few weeks later they all turn completely gray!!! (Mice grow a new coat of hair during puberty, and it is this new hair that comes out gray.)

14.39 **APOPTOSIS TURNS MICE GRAY**

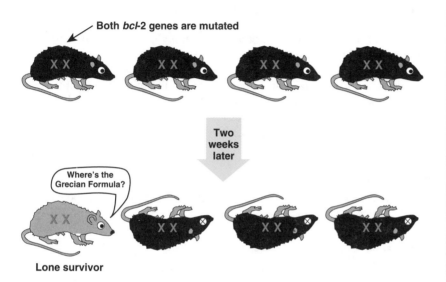

Both *bcl*-2 genes are mutated

Two weeks later

Where's the Grecian Formula?

Lone survivor

Mitochondria are Maternal, Not Eternal

Although most body cells have no long-term future and lose the ability to divide, this still does not explain why they should grow old and eventually degenerate. Let's be honest, we do not yet understand the aging process. However, some interesting snippets of information are now being revealed by molecular biology. Let's finish by brooding over the declining energy aging animals and people display.

The cells of animals contain hundreds of mitochondria that generate energy (see Ch. 11). These structures are roughly the same shape and size as bacteria. Although they do not contain all the genes for their own manufacture, mitochondria do possess circular DNA molecules that carry a few of their own genes (Fig. 14.40). In particular, about 75 percent of human mitochondrial DNA encodes proteins involved in energy generation.

mitochondrion organelle found in eukaryotic cells that produces energy

14. 40 **HUMAN MITOCHONDRIAL DNA**

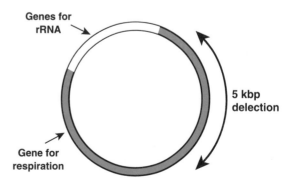

Genes for rRNA

5 kbp deletion

Gene for respiration

Just as mutations accumulate over the years in the chromosomal DNA of somatic cells, so the same happens to mitochondrial DNA. These mutations do not cause cancer since mitochondria are not in charge of their own multiplication. However, many of them do damage genes needed for energy production.

In addition to single base change mutations, mitochondrial DNA often suffers a particular deletion of 5,000 base pairs (Fig. 14.40). This removes several whole genes involved in energy production. Such mutations accumulate gradually over the years and the cells of older people contain increasing numbers of mitochondria which have taken early retirement. This leads to a gradual decrease in the energy generated by heart cells, brain cells, etc., and so we slowly get less active and alert.

Some people age faster than others. It is thought this may often be due to their starting life with some of their mitochondria already mutated. Egg cells are large enough to contain many mitochondria, whereas sperm are relatively tiny with specialized mitochondria located in the tail to provide energy for swimming. In practice, mitochondria are inherited only from your mother. Other factors aside, your life expectancy depends mostly on your mother not your father. So make sure you pick a healthy and long-lived female to give birth to you!

Additional Reading

Human Genetics: The Molecular Revolution by McConkey EH. 1993. Jones & Bartlett Publishers, Boston & London

Molecular Biology of the Cell by Alberts B, Bray D, Lewis J, Raff M, Roberts K, & Watson JD. 3rd ed., 1994. Garland Publishing, Inc., New York & London.

Molecular Biology for Oncologists edited by Yarnold JR, Stratton M, & McMillan TJ. 2nd ed., 1996. Chapman & Hall, London.

Down on the Farm: Transgenic Plants and Animals

15

Better Homes and Gardens

For thousands of years mankind has improved crop plants and domestic animals by selective breeding, mostly at a trial and error level. Obviously, the more we know about genetics the faster and more effectively we can improve our crops and livestock.

During the 1960s the yield per acre of many major crop plants doubled as a result of applying genetics (NOT genetic engineering) to breeding programs. This was the Green Revolution and its major global effect was that Western Europe, which used to be the world's biggest importer of food, became self-sufficient. In contrast, agriculture in the ex-colonies of Australia, Canada, New Zealand and the United States, lost a major part of its export market. A substantial part of the recent decline in the American balance of trade is due to the genetic improvement of crops world-wide which has lowered demand, at least by those countries that can afford to import food.

Better Clones and Genomes

Today it is possible to improve plants, animals and even humans by genetic engineering. Let's start at the bottom of the heap, with the vegetables. Most modern emphasis in plant improvement is not so much on increasing overall yield as on such things as increasing resistance to plant diseases and predators, reducing fertilizer requirements, and, for the very clever, altering the products the plant makes.

Weed killers cost the world's farmers more than $10,000,000,000 each year. Yet despite this, around 10 percent of their crops are lost due to weeds. One problem is that many of the chemicals available are "equal opportunity" herbicides and do not discriminate. So they kill any plant they touch, not just weeds.

One clever approach is to make the crop plants resistant by genetic engineering and then spray everybody (Fig. 15.2). The weeds are snuffed out but the good guys survive.

15.2 ENGINEERED PLANTS SURVIVE WEED KILLER

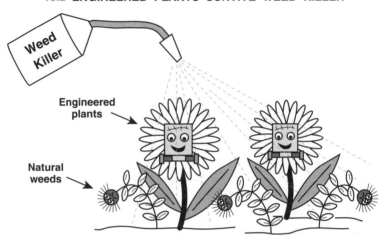

The herbicide glyphosate kills plants and bacteria by blocking the pathway for synthesizing the essential aromatic amino acids, phenylalanine, tyrosine and tryptophan. As shown in Figure 15.3, glyphosate inhibits one particular enzyme in this pathway, EPSPS (5-enolpyruvoyl shikimate-3-phosphate synthase). Glyphosate is sold by Monsanto under the trade name of "Round-Up" and is environmentally sensitive as it quickly breaks down to non-toxic compounds in the soil.

15.3 ACTION OF GLYPHOSATE

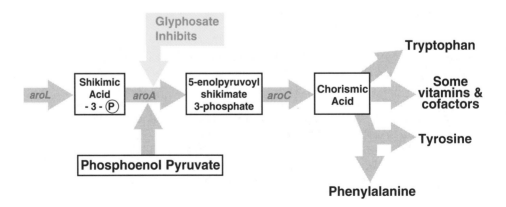

To make crops herbicide resistant, first we select mutant bacteria which are resistant to glyphosate. These have a mutation in the *aroA* gene that codes for the EPSPS protein making it resistant to the effects of glyphosate. Since we can grow and screen zillions of bacteria in a day or two, this is much quicker and cheaper than using plants directly.

aroA gene   the gene which codes for EPSPS

This glyphosate resistant bacterial *aroA* gene has been cloned (see Ch. 9 for cloning) and inserted successfully into plants such as tomatoes and potatoes (Fig. 15.4). Now we can slaughter the weeds knowing that our crops are safe even if they get sprayed too.

15.4 USE GLYPHOSATE - RESISTANT GENE FROM BACTERIA

Mutation in *aroA* gene

a) mutate
b) select with glyphosate

Resistant Mutant

Cut out DNA with *aroA* (mutated)

Insert into plant

Although weeds are a nuisance, even worse enemies of plants, and correspondingly more expensive to farmers, are:

1) Insects and roundworms
2) Fungal diseases (molds, blights, rusts and rots)
3) Viral diseases of plants

It is possible to engineer plants for resistance to all of these, but let's just consider the insects. Instead of spraying the insects ourselves, we need to equip the plants with chemical weapons to defend themselves. Again we go to our little friends the bacteria for help. Several bacteria produce toxic proteins that kill insects yet are harmless to higher animals (Fig. 15.5). The best known of these is *Bacillus thuringiensis* whose toxin has not only been cloned but has been improved by genetic engineering.

Bacillus thuringiensis type of bacterium which makes toxins for killing insects

15.5 INSECT TOXIN FROM *BACILLUS THURINGIENSIS*

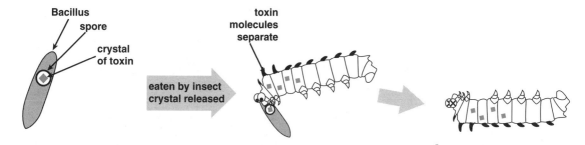

Bacillus
spore
crystal of toxin

toxin molecules separate

eaten by insect crystal released

When the cloned toxin gene was inserted into tomato plants, it gave some protection against assorted caterpillars. However, the plants only made low levels of the toxin. This is because the toxin gene is from a bacterium and is adjusted to work well in bacteria not plants. So we need to make improvements (Fig. 15.6).

First, the original toxin is a big protein that has 1,156 amino acids. However only the front 650 or so are actually needed to kill insects.

So the gene was shortened by cutting the DNA with a restriction enzyme. Now the gene gives rise to a shorter protein requiring less energy and material to make, so the yield increases.

Second, the toxin gene was placed under the control of a promoter that expresses the toxin gene at a high level all the time. Small plant viruses have only a few genes but express them at high levels when they invade plant cells. So in a blaze of poetic justice, a strong promoter was borrowed from the cauliflower mosaic virus (CMV). This was placed in front of the insect toxin gene and gave a 10-fold increase in production.

promoter DNA sequence in front of a gene which binds RNA polymerase

cauliflower mosaic virus a small virus with circular DNA which attacks cauliflower and related plants

15.6 IMPROVED INSECT TOXIN GENE

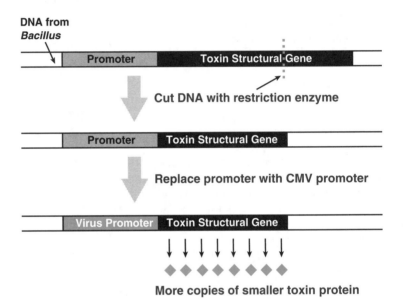

More copies of smaller toxin protein

When genes from one organism are expressed in a very different host cell we will also face a third and more subtle problem. This is codon usage. As explained in Chapter 7, the genetic code is redundant in the sense that several different codons can encode the same amino acid.

So even though the protein we want to make has a fixed amino acid sequence, we still have considerable choice in which codons to use. Different organisms favor different codons for the same amino acid. For example, the amino acid lysine can be coded by AAA or AAG (see Ch. 7). In *E. coli* AAA is used 75 percent of the time and AAG only 25 percent. In contrast, *Rhodobacter* does the exact opposite and uses AAG 75 percent of the time even though both *E. coli* and *Rhodobacter* are bacteria.

But why should codon usage effect how much protein is made? The reason is that codons are read by transfer RNA (see Ch. 7) and when a cell uses a particular codon only rarely, it has lower levels of the tRNA needed for the rare codon. So if a gene with lots of AAA codons is put into a cell that almost never uses AAA for lysine, the tRNA is in such short supply that protein synthesis slows down (Fig. 15.7).

Rhodobacter type of aquatic bacterium that lives by photosynthesis

15.7 **RARE CODONS SLOW PROTEIN SYNTHESIS**

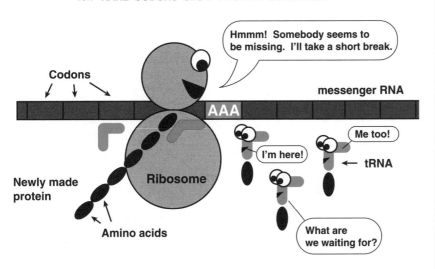

The insect toxin gene has been altered by changing many of the bases in the third position of redundant codons. Almost 20 percent of its bases were altered so as to make it more plant-like in its codon usage. This did not alter the amino acids encoded and, therefore the toxin protein itself was not affected by this procedure. However, the rate at which the protein was assembled inside plant cells greatly increased, and the result was another 10-fold increase in toxin production. Plants with "vegetized" toxin genes driven by a virus promoter can really kick caterpillar butt!!

15.8 **THE CATERPILLAR'S NIGHTMARE**

Getting Genes into Plants

One advantage of plants is that they can often be regenerated from just a single cell. So plant cells are grown in tissue culture and then altered genetically. After engineering them in culture, we grow them back into real plants (Fig. 15.9). This is not something that works with animal cells.

15.9 ENGINEERED PLANT CELLS CAN GROW INTO NEW PLANTS

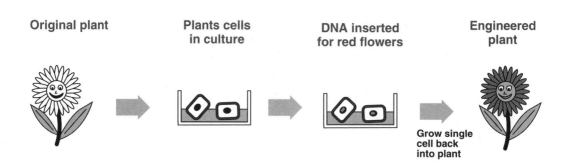

Original plant   **Plants cells in culture**   **DNA inserted for red flowers**   **Engineered plant**

Grow single cell back into plant

Ti-plasmid (tumor-inducing plasmid) plasmid that carries the genes necessary for causing tumors in plants

Agrobacterium type of bacterium which infects plants and lives inside the tumors it causes

acetosyringone one of several related chemicals released by wounded plants

We have already mentioned some useful genes to improve our plants, but how exactly do we get them into the veggies? We'll discuss two methods. The first is to use a plasmid which causes plant tumors and replace the tumor causing genes with our plant improvement genes. The second is to use a gene gun and cowboy style, just blast the DNA in. These are discussed separately.

Plants suffer from tumors though these are quite different in nature from the cancers of animals (Fig. 15.10). The most common cause is a **Ti-plasmid (tumor-inducing plasmid)**. These are carried by soil bacteria of the *Agrobacterium* group. *Agrobacterium* is attracted by chemicals such as **acetosyringone** which are released by wounded plants. It then sneaks in via the wound and transfers Ti-plasmid DNA into the plant cells. The result is a "crown gall" tumor that provides a nice home for the *Agrobacterium* at the expense of the plant.

15.10 CROWN GALL TUMOR OF PLANTS

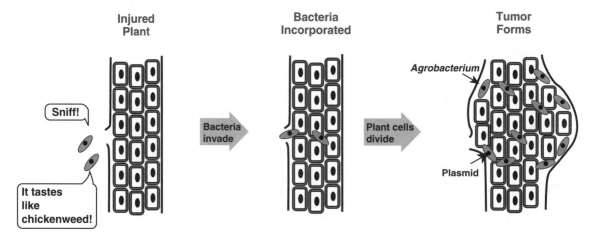

Injured Plant   **Bacteria Incorporated**   **Tumor Forms**

Sniff!

It tastes like chickenweed!

Bacteria invade

Plant cells divide

Agrobacterium

Plasmid

The Ti-plasmid consists of several regions (Fig. 15.11), but only one segment, the **T-DNA** (tumor-DNA), is actually transferred into the plant cell. The **virulence genes** on the plasmid are in charge of transferring the T-DNA but do not themselves enter the plant cells.

T-DNA the DNA segment from a Ti-plasmid which actually inserts into the chromosomes of a plant cell

virulence gene gene involved in aiding infection or causing symptoms of disease

15.11 STRUCTURE OF Ti - PLASMID

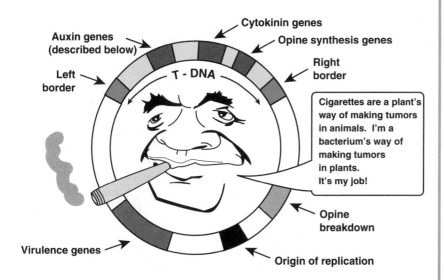

Acetosyringone also induces the virulence genes and this results in DNA transfer from the bacteria into the plant cells. The mechanism resembles bacterial conjugation (see Ch. 8). The plasmid replicates by a rolling circle mechanism and the single stranded T-DNA region enters the plant cell. The T-DNA then integrates into a chromosome in the plant cell nucleus (Fig. 15.12).

Once inserted, the genes in the T-DNA are switched on. The proteins they code for manufacture two plant hormones, **auxin** and **cytokinin**. Auxin makes plant cells grow bigger and cytokinin makes them divide. If this happens rapidly and is uncontrolled, the result is a tumor.

The T-DNA also carries a gene that subverts the plant cell into manufacturing **opines**. These are unusual nutrient molecules that are made at the

auxin plant hormone that typically promotes cell growth

cytokinin plant hormone that typically promotes cell division

opines special nutrient molecules that can only be used by bacteria possessing a Ti-plasmid

15.12 T - DNA IN THE PLANT CELL NUCLEUS

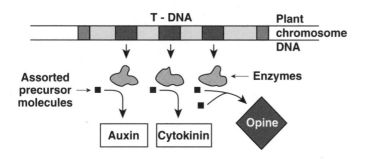

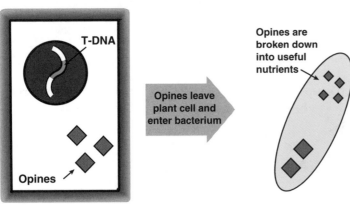

Plant Cell   **Agrobacterium**

Opines leave plant cell and enter bacterium

Opines are broken down into useful nutrients

T-DNA

Opines

expense of the plant cell. They can only be used by bacteria that carry special genes for opine breakdown (Fig. 15.13).

These genes are found on the part of the Ti-plasmid that does not enter the plant cell. So the *Agrobacterium* can grow by using the opines but the plant cannot use them. By this sneaky maneuver *Agrobacterium* gets to steal the plant's nutrients. Other bacteria which may be hanging around hoping for a free lunch cannot use the opines as they do not have the opine breakdown genes either.

So how do we use Ti-plasmids to improve plants? First we must disarm the Ti-plasmid by cutting out the genes for plant hormone synthesis in the T-DNA, as we want to make a better plant not a bigger tumor! We also get rid of the opine synthesis genes as we want a healthy plant not one constantly losing nutrients to bacteria. Then we insert the useful genes, such as insect toxin genes, into the T-DNA region of the Ti-plasmid (Fig. 15.14). Now when the T-DNA enters the plant cell and integrates into the chromosome, it will simply bring in the insect toxin genes instead of causing a tumor.

15.14   ENGINEERED Ti PLASMID

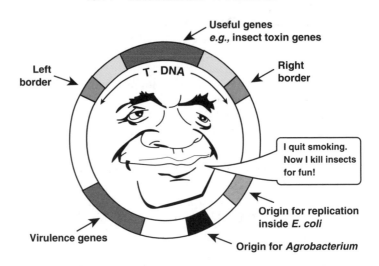

Left border

Right border

Useful genes *e.g.,* insect toxin genes

T - DNA

I quit smoking. Now I kill insects for fun!

Origin for replication inside *E. coli*

Origin for *Agrobacterium*

Virulence genes

In practice, we make a couple of other alterations, too. Ti-plasmids are often very large, with lots of extra genes which are not involved in moving T-DNA. Since smaller plasmids are much easier to engineer, we get rid of the unnecessary sequences. In addition, as we actually do most of our genetic engineering in *E. coli*, we insert an origin of replication that allows the plasmid to replicate in *E. coli*. Once we have finished our genetic engineering, we move the plasmid back into *Agrobacterium* for transfer to the plant cells.

Ti-plasmids work with most broad-leaved plants, including most fruits and vegetables, tobacco, roses, etc. However, they do not work with narrow-leaved plants such as grasses and cereals, nor with conifers. For these we normally use microprojectile bombardment. Since this approach is also used on animals, we will deal with it on page 228.

Transgenic Animals

Woollier sheep and smarter sheep dogs have both been improved through many generations of selective breeding. Today, making transgenic animals is all the rage. Most experiments are done with mice but trials are now underway with larger farm animals. The overall scheme is as follows:

1) We clone a useful gene - the "transgene" - that we intend to put into some animal (see Ch. 9).
2) The transgene is injected into the fertilized egg cells (Fig. 15.15). Just after fertilization, the egg contains its original female nucleus plus the male nucleus from the sperm that "scored." These two "pro-nuclei" will soon fuse together. Before this happens, the DNA is injected into the male pronucleus which is larger and so is a better target for microinjection.
3) The engineered eggs are implanted into the womb of a female animal, the "foster mother."
4) Some of the babies will have the transgene stably integrated into their chromosomes. In others the process fails and the transgene is lost.

transgenic animal animal containing a foreign gene, the transgene, inserted into its chromosomes

transgene a foreign gene that has been moved into a new host organism by genetic engineering

15.15 MICROINJECTION TO MAKE TRANSGENIC MICE

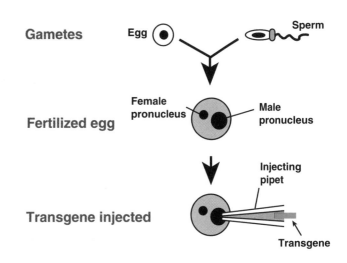

Those mice that received the transgene and maintain it stably are mated together to form a new line of mice carrying the transgene. They are therefore referred to as "founder mice" (Fig. 15.16). Note that the founder mice have only a single copy of the transgene on one chromosome, *i.e.*, they are heterozygous for the transgene. When two such founder mice are bred together, 25 percent of the progeny will get two copies of the transgene and will be homozygous, 25 percent will get zero copies and the remaining 50 percent will get one copy (see Ch. 3 for inheritance). Homozygous transgenic animals are usually more useful since if these are further interbred, all of their descendants will get two copies of the transgene.

founder mice original engineered mice which receive a single copy of a transgene and are bred together to found a line of transgenic animals

heterozygous having two different copies, or alleles, of the same gene

homozygous having two identical copies, or alleles, of the same gene

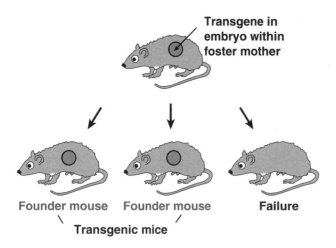

Transgene in embryo within foster mother

Founder mouse   Founder mouse   Failure

\ **Transgenic mice** /

Growth Hormone in Transgenic Mice

Growth hormone, also called somatotropin, is a good example of a protein hormone. In 1982 the somatotropin gene of rats was cloned and put into fertilized mouse eggs. The eggs were then inserted into foster mother mice who gave birth to the genetically engineered mice. These transgenic mice were larger (about twice normal size), although not as large as rats. This was the first case where a gene transferred from one animal to another was not only stably inherited, but also functioned more or less normally.

To express the rat somatotropin gene, it was put under the control of the regulatory region (*i.e.*, the promoter) from an unrelated mouse gene, metallothionein (Fig. 15.17).

growth hormone (somatotropin) the hormone made by the pituitary gland whose major effect is on skeletal growth in young animals

15.17 **CLONED SOMATOTROPIN GENE**

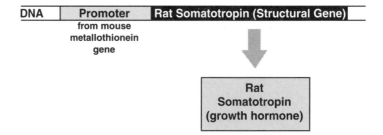

DNA	Promoter from mouse metallothionein gene	Rat Somatotropin (Structural Gene)

Rat
Somatotropin
(growth hormone)

Instead of being made in the pituitary gland, the normal site for growth hormone, the rat somatotropin was mostly manufactured in the liver of the transgenic mice. The liver is the main site for synthesis of metallothionein. Human somatotropin has also been tried in mice. It also gives bigger mice, but, so far, they cannot talk like Mickey and Minnie.

15.18 **BIGGER MICE**

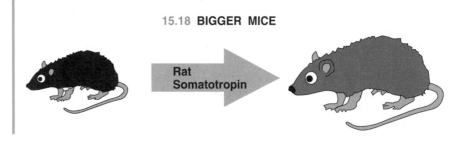

Rat
Somatotropin

Since then, the somatotropin gene from cows has been cloned and expressed by bacteria. The bacteria produce large amounts of the hormone, known as **rBST** (recombinant Bovine SomatoTropin), which is then injected into cows (Fig. 15.19). Boosting an adult cow's own somatotropin levels by injection results in increased milk production, rather than giant cows. Milk from treated cows is now sold in several states. Despite the usual hysterical opposition from the anti-technology crowd, it has been shown to be udderly safe!

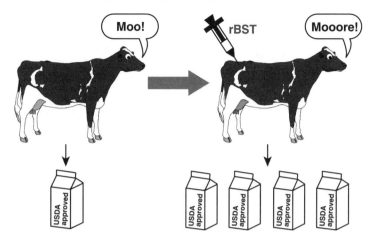

15.19 **IMPROVED MILK PRODUCTION BY COWS**

Size does not depend on growth hormone alone. African pygmies rarely grow taller than 4 feet, 10 inches, yet they have normal levels of human somatotropin. It appears that pygmies are short on growth hormone receptors. These provide the docking site for somatotropin circulating in the bloodstream and are necessary for the hormone to work on its target tissues.

Dwarfism, among non-pygmies, may be due to defects in production of somatotropin or to a shortage of receptors. Snow White would be happy to know that **recombinant human somatotropin (rHST)** is now used to treat the hormone deficient type of dwarf. Receptor deficient dwarves remain short-changed for now.

recombinant bovine somatotropin (rBST) genetically engineered version of cow growth hormone

recombinant human somatotropin (rHST) genetically engineered version of human growth hormone

recombinant tissue plasminogen activator (rTPA) genetically engineered version of tissue plasminogen activator, an enzyme which helps dissolve blood clots

Pharming Using TransCows

At present most products of recombinant DNA technology, human insulin for example, are made by bacteria, such as *E. coli,* growing in culture. Such high-tech products are expensive and require a highly-trained work force. However, dairy cows produce 10,000 quarts of milk each per year and an industry to collect and process this liquid already exists (Fig. 15.20). Moreover, milkmaids don't have to be paid as much as PhDs. On the other hand, bacteria don't suffer from "mad coli disease!"

What we must do is to place our cloned gene under the control of a regulatory region that will only allow expression of the gene in the mammary gland. That way, the gene product will come out in the milk. Experimentally, on a smaller scale, transgenic goats have been made to produce **recombinant tissue plasminogen activator (rTPA)**, which is used for dissolving blood clots. Of course, the ultimate question is, when do we make transgenic humans?

15.20 **TRANSGENIC COWS FOR BIOTECH. PRODUCTS**

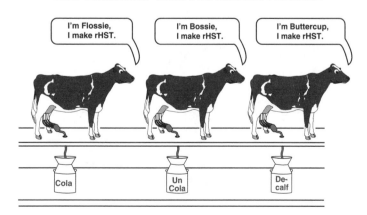

Knockout Mice for Medical Research

In a more serious vein, transgenic animals, mostly mice, have been of great value in the genetic analysis of inherited diseases and cancer. Here we are interested, not so much in adding a cloned transgene, as in discovering the function of genes already present. The approach is obvious. We inactivate, or "knock out," the gene of interest and then ask what defect this causes.

First, we must obtain the DNA for the gene of interest by cloning it (see Ch. 9). Then we disrupt the gene by inserting a gene cassette into the coding sequence of the DNA. The presence of this intruding chunk of DNA prevents the disrupted gene from making the proper protein product and so abolishes its function. In addition, most gene cassettes include an antibiotic resistance gene so allowing us to easily detect their presence. We now have an inactive copy of the gene which we insert back into the animal. The incoming DNA, carrying the disrupted gene, will sometimes replace the original, functional, copy of the gene by homologous recombination (see Ch. 3). By following the procedure outlined above for transgene insertion, we can obtain mice with a single disrupted gene or, by breeding them together, mice with both copies disrupted. Such mice are known as "knockout mice" and will completely lack gene function (Fig. 15.21). If the gene in question is essential, double knockout mice may not survive, or may live only a short time.

15.21 PRODUCTION OF KNOCKOUT MOUSE

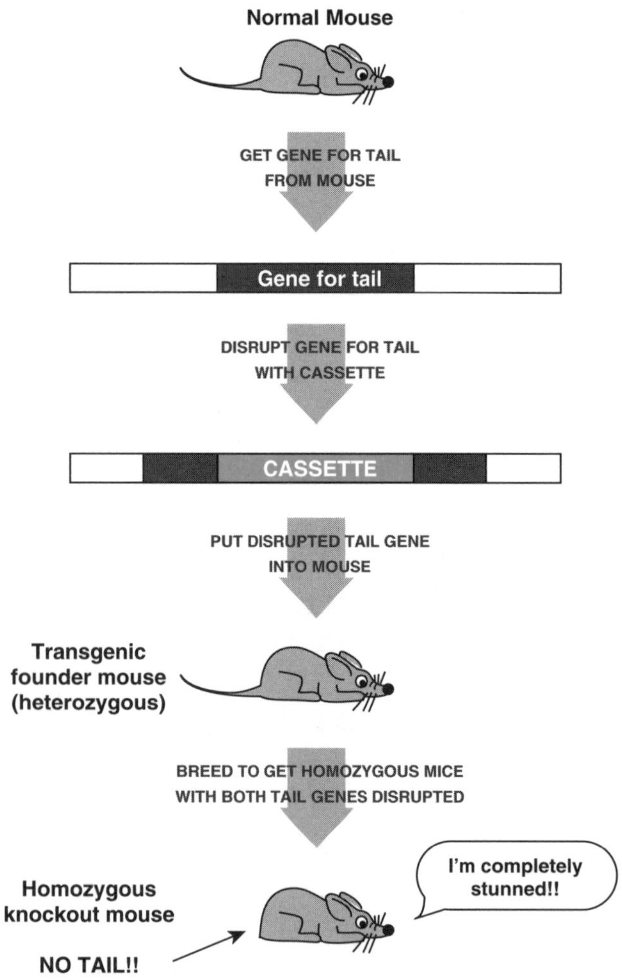

Normal Mouse

GET GENE FOR TAIL
FROM MOUSE

Gene for tail

DISRUPT GENE FOR TAIL
WITH CASSETTE

CASSETTE

PUT DISRUPTED TAIL GENE
INTO MOUSE

Transgenic founder mouse (heterozygous)

BREED TO GET HOMOZYGOUS MICE
WITH BOTH TAIL GENES DISRUPTED

Homozygous knockout mouse

I'm completely stunned!!

NO TAIL!!

Particle Bombardment Technology

One major problem with vegetables, apart from convincing children to eat them, is just how to get DNA into them. If we wish to make improved plants by genetic engineering then we must somehow get the engineered DNA into the plant. The snag is that unlike animals which are soft and squishy, plant cells are surrounded by tough cell walls made of cellulose and lignin. Consequently, standard transformation procedures are usually impossible.

One increasingly popular solution is to blast the DNA through the plant cell walls with a particle gun (Fig. 15.22). Philosophically speaking, the DNA particle gun is based on the traditional American virtue of shooting first and asking questions later.

15.22 PRINCIPLE OF THE DNA PARTICLE GUN

transformation the uptake of pure DNA by bacteria, plant or animal cells (Caution: this word is also used to describe conversion of a normal cell to a cancer cell)

particle gun or gene gun device used to fire particles carrying DNA into cells

Uranium slugs are used by the military to penetrate tank armor. Similarly, heavy metal particles are used to carry the DNA during particle bombardment (Fig. 15.23). Tungsten particles were originally used. In practice, tungsten is fine from a physical point of view, but is moderately toxic to some plants. Nowadays, gold is more often used as it is chemically inert and therefore nontoxic, which is why gold is used to plug teeth.

15.23 METAL PARTICLES CARRY DNA THROUGH CELL WALL

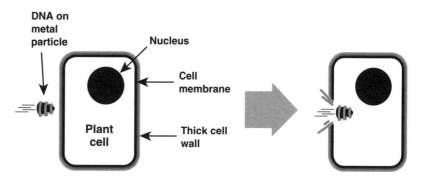

The actual DNA particle guns vary from weapons a cowboy would recognize to sophisticated laboratory set-ups using compressed gases or electrical discharges to project the particles. In the more "traditional" approach, a modified pistol is loaded with blanks. The DNA coated microprojectiles are about 1 to 4 microns in diameter and are carried in a macroprojectile which is a sort of fake bullet with a cavity at the front end. The macroprojectile is propelled down the barrel and crashes into a stop plate. The stop plate has a small central hole which allows the microprojectiles to continue on their way, much in the same way that motorists who didn't buckle up are projected through the windshield when their cars come to a violent stop. This projects DNA into a circle of target cells 9 or 10 cm in diameter. Most of the transformed cells are clustered around the center of the target area, surrounding a central zone of death where the cells are destroyed by heavier concentrations of projectiles.

Sadly, most plants are now bombarded with airguns or electric discharges rather than cowboy pistols. One version of an electric discharge gun is shown in Figure 15.24. The initial projectile is a droplet of water which is charged up and then fired down the tube by an electrical discharge. The water droplet smacks into the carrier sheet that holds the DNA microprojectiles. The force of the impact throws the carrier sheet against a fixed screen with small holes which allow the DNA microprojectiles through on their way towards the target.

15.24 **DNA PARTICLE GUNS**

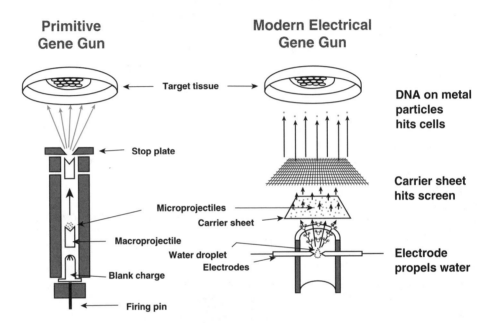

reporter gene   a gene that is easy to detect and which is inserted for diagnostic purposes

npt gene   gene that confers resistance to the antibiotic neomycin

How do we know whether any DNA got into the target cells? Reporter genes (see Ch. 16) whose effects are easy to detect and measure accurately, are used to test the procedure. One widely used reporter gene is *npt* which encodes neomycin phosphotransferase (Fig. 15.25). This enzyme inactivates the antibiotic neomycin by sticking on a phosphate group.

15.25 **ANTIBIOTIC SELECTION OF CELLS GETTING DNA**

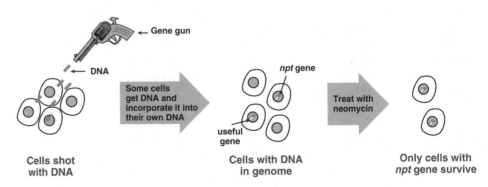

Cells successfully incorporating DNA carrying the *npt* gene are no longer killed by neomycin. This means that you can directly select for trans-formed cells because treatment with neomycin will kill any cells that did not get the DNA.

A kinder and gentler diagnostic method is to include on the DNA a gene that codes for **luciferase**. This enzyme emits light when provided with a chemical known as luciferin (Fig. 15.26). Luciferase is found naturally in assorted luminous creatures, from pretty fireflies to the ghostly glowing and hellish denizens of the ocean deeps. (Sad to say, luminous deep-sea fish and giant squid don't actually make their own luciferase; it is provided by kindly bacteria living inside them.) If DNA carrying the *luc* gene is successfully incorporated into the target cell, light will be emitted when luciferin is added. Although high level expression of luciferase can be seen with the naked eye, usually the amount of light is small and must be detected with a sensitive electronic apparatus such as a scintillation counter.

luciferase the enzyme which generates light

15.26 LUCIFERASE REPORTER GENE

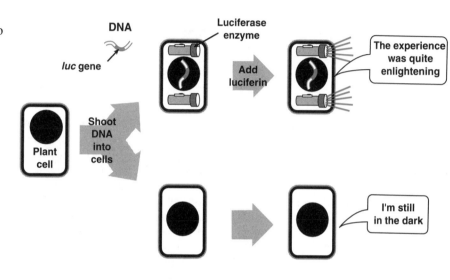

The *luc* gene from fireflies has been successfully inserted into tobacco plants. When the plants were watered with a solution of luciferin, they pro-duced enough light to be visible in the dark. (In the near future perhaps we can look forward to the production of a cigarette that lights itself!)

luc gene gene from fireflies which encodes luciferase

Evergreen trees (pine, spruce, etc.) are virtually impossible to engineer by standard methods. However, particle bombardment of white spruce has allowed creation of transgenic trees that are resistant to that nefarious caterpillar, the **spruce budworm**. Actually, the DNA was shot into spruce embryo tissue in culture, not blasted into whole trees!! Resistance to the caterpillars is provided by bacterial genes for protein toxins which kill insect larvae, as described above.

spruce budworm nasty caterpillar which munches on spruce and related trees

Although particle bombardment was originally developed for plants because of their tough cell walls, it has been applied to animals as well. Although cultures of individual cells have been used as targets, far more amusing, and also more practical, is the use of genuine chunks of real organs. DNA particle bombardment is by far the most efficient way to get

DNA into chunks of tissue (Fig. 15.27). Other, more "standard" methods such as lipofection, chemical transformation or electroporation (see Ch. 8), work well with single cells but aren't very effective with organ slices.

15.27 GENE GUN FOR ANIMAL TISSUES

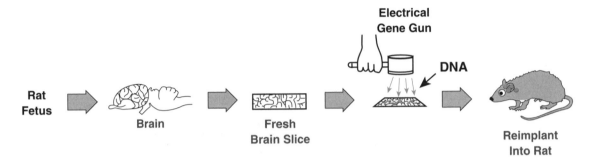

The general idea is that we open up an animal, slice out part of an organ, maintain it in culture outside the body, insert DNA into the organ slice and then re-implant it into the original animal.

For example, clumps of fetal rat brain tissue have been bombarded with DNA and then re-implanted into rats. Not only have reporter genes been tested successfully in rat brains, but the gene for **tyrosine hydroxylase** was inserted and expressed successfully. Inserting the gene for tyrosine hydroxylase may be used in future gene therapy to cure **Parkinson's disease**.

Just for the insensitive among you, we should mention the use of mammary gland organoids. Using a gun to blast genes into sliced off rat breasts may seem fairly crude, even by the cultural standards of a nation which believes the right to bear arms is more important than the right to bare breasts! Nonetheless, it may well lead to genetic engineering-based methods to combat breast cancer by shooting in genes that kill cancer cells.

tyrosine hydroxylase enzyme involved in the synthesis of the dopamine family of neurotransmitters, substances that carry signals between brain cells

Parkinson's disease degenerative disease of certain brain cells which make the neurotransmitter dopamine and which leads to muscle tremors and weakness

Why bother to cut the organ out and re-implant it? Why, indeed?! In recent experiments, rats have been opened to expose their internal organs and then had DNA blasted directly into their livers, spleens, etc. After surgery, the rats are sewn up, given antibiotics to prevent infection, and provided with small vases of flowers and miniature bedside TVs (just kidding for the last two! - but then you sometimes wonder, don't you?).

Most of the rats recover and quite often express the foreign genes in the targeted organ (Fig. 15.28). Gold particles with DNA carrying the gene for

15.28 RODENT RECOVERY ROOM

human growth hormone have been blasted nearly half a millimeter into the livers of living mice. Human growth hormone was produced by the mice and detected in their blood. In some cases foreign genes have been expressed for over a year, quite a large proportion of the life span for small animals like rats and mice.

Although organ targeted gene therapy does not affect the germ line and will not prevent inherited defects from being passed on, it is quite a good remedy for an actual patient suffering from a defect mostly confined to one particular organ. So the day may not be far off when your doctor may say to you, "Sorry, you have colon cancer - *butt* - don't worry!" And then he'll just whip out his DNA Special and say, "Bend over, this will hurt you more than it hurts me!!"

Well, the fun's over for a while. The next couple of chapters are pretty heavy and describe many of the most common procedures of genetic engineering. If you surmount this obstacle, you will eventually reach our cheerful discussions of crime and disease from a genetic viewpoint.

Additional Reading

Molecular Biotechnology: Principles and Applications of Recombinant DNA by Glick BR & Pasternak JJ. 1994. American Society for Microbiology, Washington, DC.

Just Do It!
Techniques of Molecular Biology

16

In this chapter we deal with the principles behind some of the techniques widely used in molecular biology. More detailed applications of these methods to particular topics are cross-referenced. Cloning techniques, PCR and DNA sequencing are biggies and have chapters of their own.

Gel Electrophoresis

Perhaps the most widely used physical method in all of molecular biology is gel electrophoresis. The idea behind electrophoresis is that positive charges attract negative charges and vice versa (Fig. 16.1). Conversely, two charges of the same sign repel each other.

electrophoresis movement of charged molecules toward an electrode of the opposite charge. Used to separate nucleic acids

gel electrophoresis electrophoresis of charged molecules through a gel meshwork in order to sort them by size

16.1 OPPOSITES ATTRACT

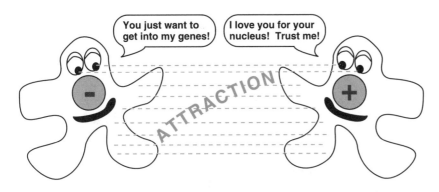

To perform electrophoresis we can dissolve molecules which carry electrical charges in water and stick into the solution two electrodes, one positive and the other negative. When we switch on the current, the negatively charged molecules are attracted to the positive electrode and move through the solution until they reach it. The positively charged molecules move in the opposite direction (Fig. 16.2). Molecules carrying charges are known as ions.

ion any molecule carrying an electrical charge

16.2 ELECTROPHORESIS IN SOLUTION

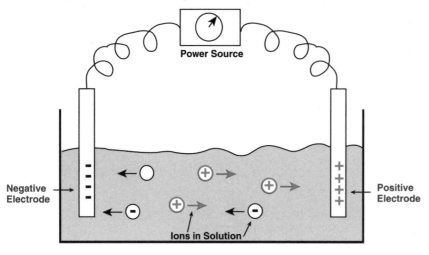

The greater the charge, the faster a molecule will swim under the influence of an electrical attraction. On the other hand, the larger the molecule, the more force needed to get it moving. Molecules of nucleic acid have exactly one negative charge for each nucleotide so these two factors cancel out. Consequently all molecules of DNA or RNA will move at the same speed towards the positive electrode as long as they are free in solution.

If we want to separate our DNA or RNA on the basis of size, we must bring in an extra handicap to slow down the larger molecules - the gel. The gel is a meshwork of cross-linked polymer chains, usually of agarose for nucleic acids. The molecules of DNA are slowed as they try to wriggle through the gaps in the gel meshwork (Fig. 16.3). The larger they are, the harder it is to squeeze through the holes. The result is that a mixture of DNA molecules separates according to size, the smaller molecules moving through the gel much faster than the larger ones.

gel a semi-solid made by a polymer that forms a cross-linked meshwork in water

agarose polysaccharide from seaweed used for making gels

16.3 GEL ELECTROPHORESIS

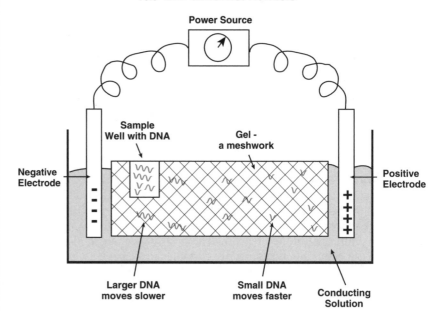

After running the gel, we need some way of visualizing the DNA. The DNA may be radioactively labeled and detected by autoradiography which is explained on page 243. Alternately, the gel can be stained with ethidium bromide which binds tightly to DNA or RNA. The gel must then be examined under UV light where ethidium bromide bound to DNA fluoresces bright orange. In either case, we will see a dark or colored band where the DNA has ended up.

Nucleic acids all come with a built in negative charge, however proteins are not so convenient. Some of the amino acids from which proteins are built have a positive charge and some have a negative charge while most are neutral. So, depending on its overall amino acid composition, a protein may be positive, negative or neutral.

To avoid these complications proteins are boiled in a solution of the detergent sodium dodecyl sulfate (SDS). Boiling destroys the folded 3-D structure of the protein, *i.e.*, the protein is denatured. The SDS molecule has a hydrophobic tail with a negative charge at the end. The tail wraps around the backbone of the protein and the negative charge dangles in the water. The protein is unrolled and covered from head to toe with SDS molecules which give it an overall negative charge (Fig. 16.4).

sodium dodecyl sulfate (SDS) a detergent used to unfold proteins

denature to destroy the 3-D folding of a protein or other polymeric molecule

hydrophobic water hating

16.4 SDS GIVES PROTEINS A NEGATIVE CHARGE

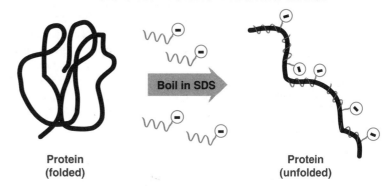

Protein (folded) → Boil in SDS → Protein (unfolded)

What's more, the number of negative charges bound is proportional to the length of the protein. So now we can separate proteins according to their sizes by running them through a gel (Fig. 16.5). Because proteins are a lot smaller on average than DNA or RNA, we normally use a gel made of the artificial polymer, polyacrylamide, which gives smaller gaps in its meshwork than agarose.

polyacrylamide artificial polymer used to make gels for separating proteins by electrophoresis

16.5 GEL ELECTROPHORESIS OF PROTEINS

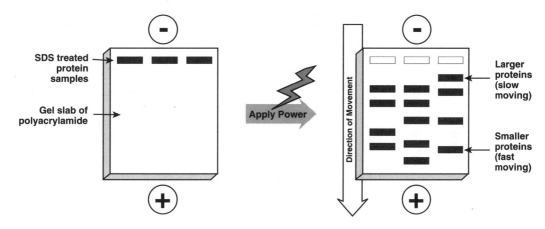

SDS treated protein samples

Gel slab of polyacrylamide

Apply Power

Direction of Movement

Larger proteins (slow moving)

Smaller proteins (fast moving)

Once we have run our gel, we must stain it to see the protein bands. The two favorite choices are Coomassie Blue, a blue dye that binds strongly to proteins, or silver compounds. Silver atoms bind very tightly to proteins and yield black or purple complexes. Silver staining is more sensitive and, yes, more expensive!

Gel Retardation and Footprinting

When we are trying to work out how a gene is controlled, we need to know which regulatory proteins affect the gene in question. In practice, this means measuring whether or not the regulatory protein binds to the regulatory region in front of the gene. To test this, we need purified DNA carrying the gene and its regulatory region, and we also need a supply of purified regulatory protein. The procedure is known either as a bandshift assay or as gel retardation (Fig. 16.6).

First, the DNA carrying the gene is cut with a convenient restriction enzyme to get a series of fragments. After cutting our DNA we split it into two samples. To one of these we add the protein we are testing. Both samples are then run side by side on an agarose gel. If the protein binds to one of the DNA fragments, the complex formed will be larger and run slower than the original DNA, *i.e.*, that fragment will be retarded. To visualize the DNA fragments after running the gel, the DNA should be "labeled" (usually by making it radioactive or fluorescent) before starting the experiment.

16.6 GEL RETARDATION

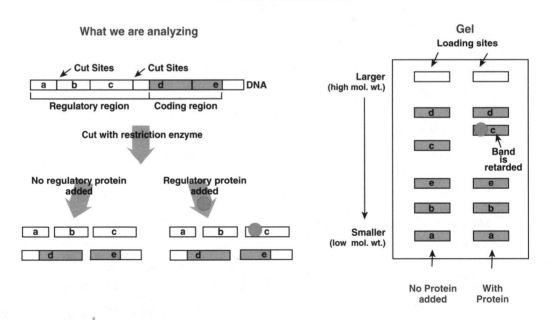

An average protein has a molecular weight of about 40,000. A chunk of DNA of 1,000 base pairs has a molecular weight of about 700,000. If a typical protein is bound to a length of DNA much bigger than this, the relative change in size, and therefore in mobility, would be 5 percent or less. Such a small change would impossible to observe. Consequently, for gel retardation analysis, we choose a restriction enzyme which gives us segments of DNA in the range of 250 to 1,000 base pairs.

Gel retardation tells us which chunk of DNA binds a protein. To locate the binding site more precisely we perform a footprint analysis.

In footprinting we take the fragment of DNA that binds the protein and label it at one end with radioactivity or fluorescence. We split our sample into two portions and allow the protein to bind to one batch. Then we treat both samples of the DNA with a small amount of a reagent which breaks DNA strands. Deoxyribonuclease I (DNAse I) is often used because it is not as smart as restriction enzymes and cuts DNA between any two nucleotides without any specificity. The DNA will be attacked and degraded except in the region covered, and thus protected, by the protein (Fig. 16.7).

16.7 PROTECTION OF DNA BY PROTEIN

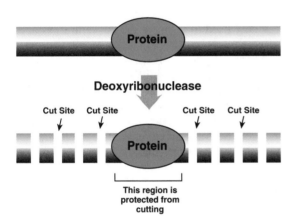

The unprotected DNA will give rise to a series of fragments of all possible lengths. We use just a little DNAse, only enough to cut each molecule of DNA once on average, in a random position. The sample of protected DNA will have certain fragments missing. When the two samples are run on a gel side by side we see a "footprint" (Fig. 16.8).

In practice, the footprint is usually run side by side with a sequencing reaction (see Ch. 22) which allows matching the footprint with the DNA sequence.

Measuring DNA and RNA with Ultraviolet Light

The bases found in DNA and RNA absorb ultraviolet light. If a beam of UV light is shone through a solution containing nucleic acids, the

footprint when a protein binds to DNA at a specific site it can protect the DNA from being cut in this region. The result may be visualized as a missing group of bands when the DNA is run on a gel

deoxyribonuclease I (DNAse I) an enzyme that degrades DNA by cutting between individual nucleotides in a non-specific manner

ultraviolet light invisible radiation of higher energy than visible light

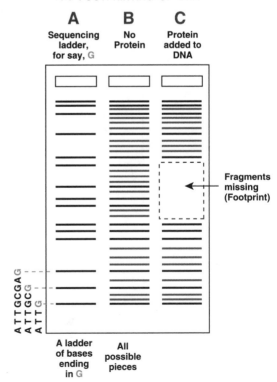

16.8 FOOTPRINTING OF DNA

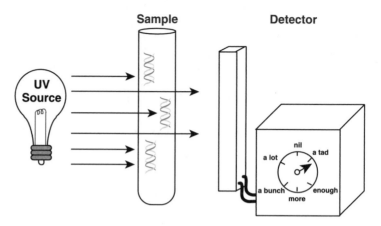

proportion of the UV absorbed depends on the amount of DNA or RNA (Fig. 16.9). This approach is widely used to measure the concentration of a solution of DNA or RNA.

The UV light is actually absorbed by the aromatic rings of the bases (A, G, T, C & U). In a solution of unlinked nucleotides, the bases are more spread out. In a DNA double helix, the bases are stacked on top of each other and get in each other's way so that relatively less UV light is absorbed (Fig. 16.10). In RNA that is single stranded (or in single stranded DNA), the situation is intermediate. So for a given amount of nucleotides, DNA absorbs less than RNA, which absorbs less than free nucleotides.

16.10 UV ABSORBANCE OF DNA AND FREE NUCLEOTIDES

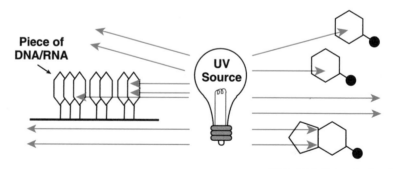

Free bases spread out and absorb more

Radioactive Labeling

One matter we need to tackle is the question of how to detect and measure DNA or RNA when, for example, it binds to a filter or runs on a gel. We also need to distinguish the DNA by its source. Since DNA itself is colorless, tasteless and odorless we must somehow label the DNA.

Originally this was done by making the DNA to be traced radioactive. The DNA we are following around would be "hot" (radioactive) and any other DNA involved in our experiment would be non-radioactive or "cold" (Fig. 16.11).

radioactive emitting radiation due to unstable atoms that break down releasing alpha-, beta-, or gamma-rays

"hot" slang for radioactive

"cold" slang for non-radioactive

16.11 DETECTION BY RADIOACTIVITY

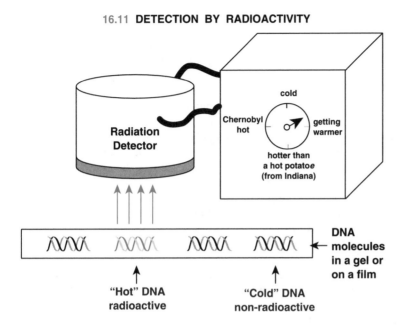

Radioisotopes are the radioactive forms of an element. In molecular biology, two are especially important: the radioactive isotopes of phosphorus, ^{32}P, and sulfur, ^{35}S. Nucleic acids consist of nucleotides linked together by phosphate groups, each containing a central phosphorus atom. If ^{32}P is inserted at this position we have radioactive DNA or RNA (Fig. 16.12). The half life of ^{32}P is 14 days, which means that half of our radioactive phosphorus atoms will have disintegrated during this time period, so we need to get on with our experiment fast!

16.12 POSITIONING OF ^{32}P IN DNA

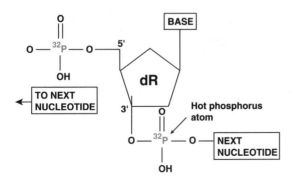

16.13 PHOSPHATE VERSUS PHOSPHOROTHIOATE

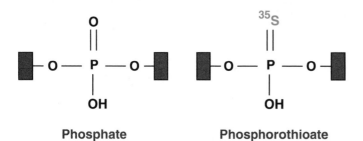

Phosphate

Phosphorothioate

The sulfur isotope, ^{35}S, is also widely used. Since sulfur is not a normal component of DNA or RNA we make what is called a phosphorothioate derivative. A normal phosphate group has four oxygen atoms around the central phosphorus. If we swap one of these for a sulfur we get a phosphorothioate (Fig. 16.13).

In order to introduce ^{35}S into DNA or RNA, these modified phosphate groups are used, with the sulfur atom radioactive, to link together the nucleotides. Despite this, ^{35}S is usually preferable to ^{35}P in most molecular biology applications. There are two reasons: first the half life of ^{35}S is 88 days so it doesn't all disappear so fast; second, the radiation emitted by ^{35}S is of lower energy than for ^{35}P. What this means in practice is that the radiation doesn't travel so far, so the radioactive bands are more precisely located and not so fuzzy. In short it is more accurate.

The two methods most widely used in molecular biology to measure radioactivity are **scintillation counting** and **autoradiography**. If our sample is in liquid or on a strip of filter paper we use scintillation counting. If our sample is flat, a gel or a blot for example, and we need to know the location of radioactive bands or spots, we use autoradiography.

scintillation counting detection and counting of individual microscopic pulses of light

autoradiography allowing radioactive materials to take pictures of themselves by laying them flat on photographic film

scintillation counter machine that detects and counts pulses of light

Scintillation Counting

Many of the commonly used radioactive materials emit high energy electrons known as beta-particles. If these are absorbed by special chemicals called scintillants, they result in the emission of a flash of light.

The light pulses are detected by a photocell (Fig. 16.14). A **scintillation counter** is simply a very sensitive machine for recording faint light pulses. Radioactive samples to be measured are added to a vial containing scintillant fluid and loaded into the counter.

Scintillation counters can also be used to measure light generated by chemical reactions. In this case, the light is emitted directly so no scintillant fluid is needed and the luminescent sample is merely inserted directly. The detection of light emitted by luciferase or lumiphos in genetic analysis is described on page 262.

16.14 SCINTILLATION COUNTER

Sample

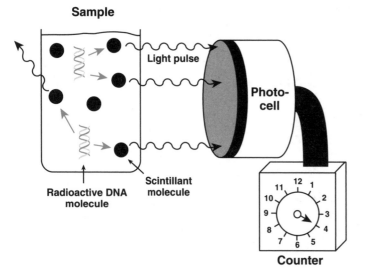

Light pulse

Photo-cell

Scintillant molecule

Radioactive DNA molecule

Counter

Autoradiography

The DNA or RNA we want to detect may be in the form of bands on a gel after separation by electrophoresis, or of spots on a filter from a blotting experiment. If so, then we use autoradiography, which means that the radiation emitted by the radioactive DNA is allowed to take its own picture on a sheet of photographic film (Fig. 16.15).

16.15 AUTORADIOGRAPHY

Gel ⟶ *Autoradiograph*

Gel

gel with
radioactive
and invisible,
bands of DNA

film

gel

lay film on gel and keep in
dark, then develop film

Film

film shows
position
of bands

fluorescence when a molecule absorbs light of one wavelength and then emits light of another, longer wavelength of lower energy

In practice, the gel or filter is dried. Next, a sheet of photographic film is laid on top of the gel or filter and left for several hours or sometimes even for days. The film will be darkened where the radioactive DNA bands or spots are found. Exposing the film must be carried out in a dark room to avoid visible light.

Non-Radioactive Detection

Most classic work in biochemistry and molecular biology has been done with radioactive tracers and probes. Although the low levels of radioactivity used in laboratory analyzes are of little hazard, the massive burden of ever-increasing government regulation (Fig. 16.16) has made it relatively cheaper and quicker to use other detection methods.

Fluorescence happens when a molecule absorbs light of one wavelength and emits light of lower energy at a longer

16.16 THE BURDEN OF GOVERNMENT REGULATION

No, it wasn't the radiation. He was crushed by a pile of government manuals.

wavelength (Fig. 16.17). Detection of fluorescence requires both a beam of light to excite the dye and a photodetector to detect the fluorescent emission.

Fluorescent dyes can be attached to DNA molecules and modern automated methods for DNA sequencing make use of such fluorescent tagging (see Ch. 22). In this case, the dyes are zapped by a laser and then emit light of a lower wavelength than the laser.

Another instrument that uses fluorescence is the molecular biologist's version of the fax machine, known as FACS. This stands for Fluorescence Activated Cell Sorter and its job, originally, was sorting cells labeled with a fluorescent tag from those that were untagged. The new generation of more sensitive FACS machines are capable of sorting chromosomes labeled by hybridization with fluorescent tagged DNA probes (Fig. 16.18).

16.17 PRINCIPLE OF DETECTION BY FLUORESCENCE

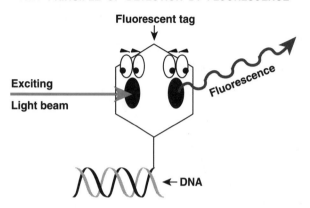

FACS or Fluorescence Activated Cell Sorter machine that sorts particles, such as cells or chromosomes, according to their fluorescence

16.18 FACS FOR SORTING CHROMOSOMES

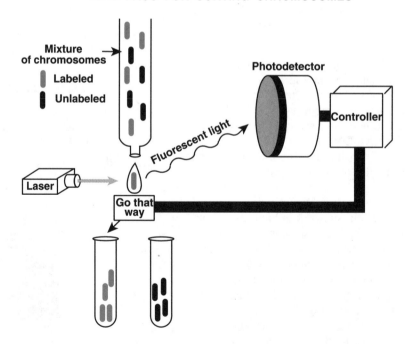

biotin   a vitamin

digoxigenin   steroid molecule from the foxglove plant. Commonly used as a linker for attaching a fluorescent tag to DNA

Biotin (a vitamin) and digoxigenin (a steroid from foxglove plants) are two tags widely used for labeling DNA. They are not colored or fluorescent themselves and are detected in a two stage process.

244

To label the DNA we use biotin or digoxigenin attached to uracil (Fig. 16.19). Uracil is normally found in RNA where it replaces the thymine found in DNA. Nonetheless, although DNA polymerase will not use uridine triphosphate (UTP), it mistakes <u>deoxy</u>UTP for thymidine triphosphate. This allows the incorporation of the tagged uracil. The biotin or digoxigenin tags stick out from the DNA without disrupting its structure.

Why biotin? Biotin is a vitamin required both by animals and many bacteria. Chickens lay nice, nutritious eggs that would be a paradise for bacteria looking for somewhere convenient to grow and divide. One of the defense mechanisms the chicken deploys is a protein known as avidin found in egg whites (Fig. 16.20). This binds biotin so avidly that invading bacteria become vitamin deficient.

Molecular biologists use avidin to bind the biotin tag. Attached to the backside of the avidin is another molecule that provides the actual detection system. Digoxigenin is used similarly. In this case an **antibody** that recognizes and binds to the digoxigenin is used. (Antibodies are proteins made by the immune system which specifically recognize foreign molecules; see Ch. 22.) As before, the detection system is tacked to the back of the antibody.

So what is the detection system? Here we have a couple of choices. An enzyme that generates a colored product may be attached to the avidin

antibody protein of the immune system that recognizes and binds to foreign molecules

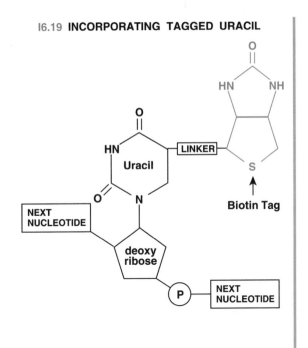

I6.19 **INCORPORATING TAGGED URACIL**

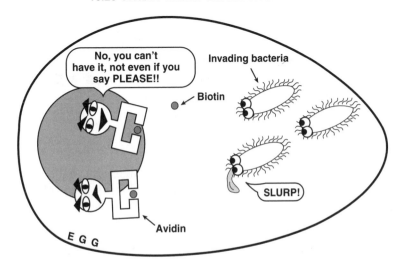

16.20 **AVIDIN BINDS BIOTIN AVIDLY**

245

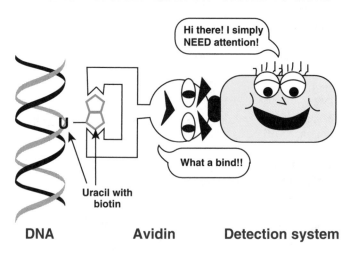

DNA   Avidin   Detection system

alkaline phosphatase
an enzyme that chops
phosphate groups from a
wide range of molecules

X-phos substance split
by alkaline phosphatase so
yielding a blue dye

or antibody (Fig. 16.21). One good example is **alkaline phosphatase**, an enzyme that snips phosphate groups from a wide range of molecules. We provide the alkaline phosphatase with a molecule, **"X-phos,"** consisting of a dye precursor bound to a phosphate group. The enzyme splits this, releasing the dye-precursor which is converted to a blue dye by oxygen in the air (Fig. 16.22).

16.22 DETECTION BY DYE RELEASE

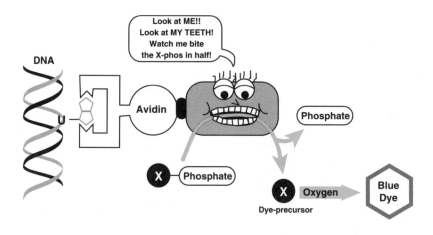

chemiluminescence
emission of light as a side
product of a chemical
reaction

Alternatively, an enzyme which produces light by a chemical reaction may be used instead. This is known as **chemiluminescence**. In this case the gel or filter carrying the DNA to be analyzed must be covered with photographic film or scanned by an instrument capable of detecting light emissions. We can use alkaline phosphatase again, but this time we provide it with a molecule, "lumi-phos," consisting of a light-emitting group

246

bound to the phosphate group (Fig. 16.23). When the phosphate is split off, the unstable luminescent group emits light.

16.23 **DETECTION BY CHEMILUMINESCENCE**

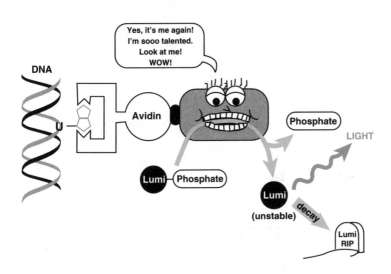

Restriction Fragment Length Polymorphisms (RFLPs)

As their name suggests, RFLPs are found as a result of chopping up DNA with restriction enzymes (see Ch. 9). Each restriction enzyme recognizes a specific sequence of four, six or eight bases. If we change even a single base within this sequence, we will prevent the enzyme from cutting the DNA (Fig. 16.24).

restriction enzyme an enzyme that binds to DNA at a specific base sequence and then cuts the DNA

16.24 **ALTERING A SITE STOPS CUTTING**

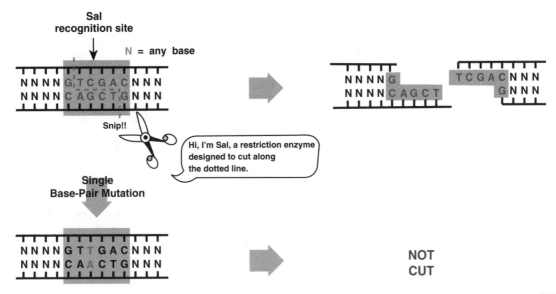

If we examine the same region of a chromosome from two related, though not identical, organisms, we will find that the DNA sequence is similar but not quite the same. Obviously, the amount of difference depends on how closely related the organisms are. Consequently, restriction sites that are present in one version of a sequence may be missing in its relatives. If we cut up two related, but different, DNA molecules with the same restriction enzyme, we may get segments of different lengths and when these are separated on a gel we will see bands of different sizes (Fig. 16.25). Such a difference between two organisms is a restriction fragment length polymorphism (RFLP).

restriction fragment length polymorphism (RFLP) differences in lengths of fragments made by cutting the DNA with restriction enzymes

16.25 **RESTRICTION FRAGMENT LENGTH POLYMORPHISMS**

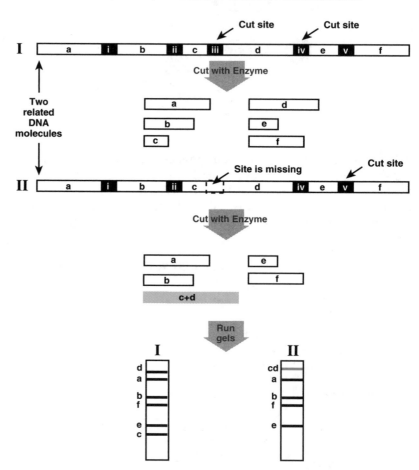

RFLPs may be used to identify organisms or analyze relationships even though we do not know the function of the altered gene. In fact, since we are examining the DNA directly, the alteration may be in non-coding DNA or an intervening sequence. It does not need to be in the coding region of a gene. RFLPs are widely used in forensic analysis, described in Chapter 18.

Hybridization of DNA and RNA

As you know, DNA consists of two strands of nucleotides twisted around each other to form a double helix. The two strands are held together by hydrogen bonding between the bases in one strand that pair with the bases in the other strand. (We hope you remember Ch. 4!)

If we heat a solution of DNA, the input of heat energy makes the molecules vibrate and the hydrogen bonds start coming apart. If the temperature is high enough, the DNA comes completely apart into two separate strands (Fig. 16.26). This is known as melting or more technically, denaturation.

hydrogen bonding bonding resulting from the attraction of a positive hydrogen atom to both of two other atoms with negative charges

melting when used of DNA, refers to its separation into two strands as a result of heating

denaturation the loss of its correct 3-D structure by any biological polymer; in the case of DNA this amounts to separation into two strands

melting temperature the temperature at which the two strands of a DNA molecule are half-way unpaired

16.26 HEATING SEPARATES DNA INTO SINGLE STRANDS

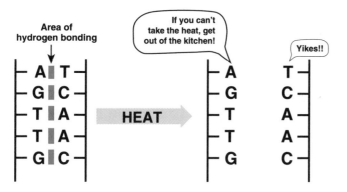

Area of hydrogen bonding

If you can't take the heat, get out of the kitchen!

Yikes!!

HEAT

Let's gradually raise the temperature and watch the DNA as it melts. Since the GC base pair has three hydrogen bonds compared to the two holding AT together, GC base pairs are stronger than AT base pairs. Therefore, as the temperature rises, AT pairs come apart first (Fig. 16.27). The regions of DNA with lots of GC base pairs hold on longer.

The melting temperature of a DNA molecule is defined as the temperature at the half-way point on the melting curve. (We use the half-way point because it is more accurate than trying to guess where exactly

16.27 A T REGIONS OF DNA SEPARATE FIRST

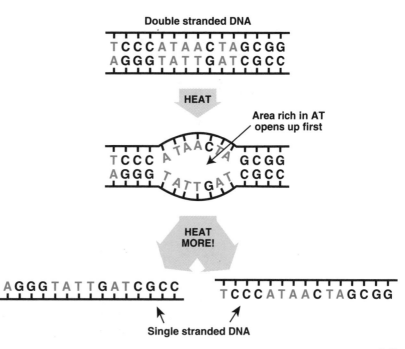

Double stranded DNA

TCCCATAACTAGCGG
AGGGTATTGATCGCC

HEAT

Area rich in AT opens up first

TCCC ATAACTA GCGG
AGGG TATTGAT CGCC

HEAT MORE!

AGGGTATTGATCGCC   TCCCATAACTAGCGG

Single stranded DNA

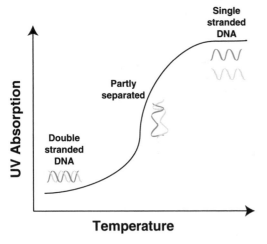

16.28 **MELTING CURVE FOR DNA**

(Figure axes: UV Absorption (vertical) vs Temperature (horizontal); labels: Double stranded DNA, Partly separated, Single stranded DNA)

melting is complete.) Melting is followed by measuring the UV absorption since disordered DNA absorbs more UV light (Fig. 16.28). Overall, the higher the proportion of GC base pairs, the higher the melting temperature of a DNA molecule.

Now let's take our single strands and cool them. The single DNA strands will recognize their partners by base pairing and the double stranded DNA will re-form. This is referred to as **annealing**. For proper annealing, we must cool the DNA slowly to allow the single strands time to find the correct partners.

Suppose we start with two completely different DNA molecules. We mix them, melt them and reanneal the single strands. Each single strand will recognize and pair with its original complementary strand (Fig. 16.29).

On the other hand we could use two closely related DNA molecules. Although the sequences may not match perfectly, nonetheless, if they are similar enough some base pairing will occur. We will then get some **hybrid DNA** molecules.

annealing the rejoining of separated single strands of DNA to form a double helix

hybrid DNA artificial double stranded molecule of DNA formed by two single strands from two different sources

16.29 **ANNEALING OF SINGLE STRANDED DNA**

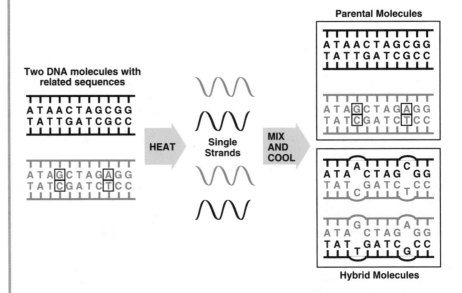

The formation of hybrid DNA molecules has a wide variety of uses. For example, we can test how closely two DNA molecules are related. To do this a sample of the first DNA molecule is melted by heating. We then attach the single strands to a suitable filter (Fig. 16.30).

16.30 BINDING OF DNA TO FILTER

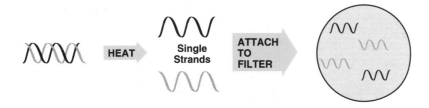

We then take the second DNA sample, melt it also, and pour the solution through the filter (Fig. 16.31). Some of the single strands of DNA molecule No. 2 will base pair with the single strands of DNA molecule No. 1 and will stick to the filter. The more closely related our two molecules are, the more hybrid molecules will be formed and the higher the proportion of molecule No. 2 which will be bound by the filter.

16.31 FILTER BINDING ASSAY FOR DNA

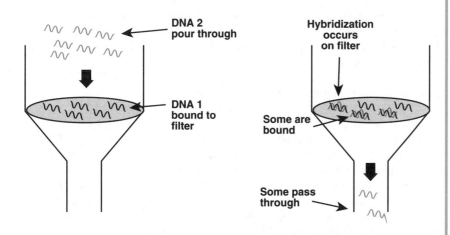

For example, we might bind the DNA for a human gene, say for hemoglobin, to the filter. We could then test the binding of DNA for the same gene but taken from lots of different animals. We might expect chimpanzee DNA to bind strongly, shark DNA to bind weakly and mouse DNA to be intermediate.

Another use for hybridization is in finding genes for cloning. Suppose we already have the human hemoglobin gene and want to isolate the corresponding gorilla gene. First we bind the human DNA to the filter as before. Then we get a sample of gorilla DNA and cut it into short segments with a suitable restriction enzyme. We heat the gorilla DNA to melt it into single strands and pour it through the filter. The DNA fragment that carries the gorilla gene for hemoglobin will bind to the human hemoglobin

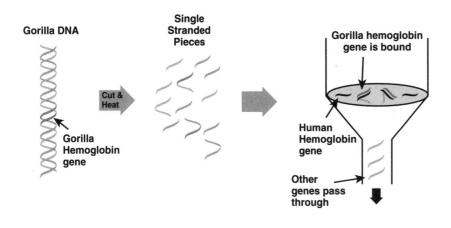

Gorilla DNA

Gorilla Hemoglobin gene

Cut & Heat

Single Stranded Pieces

Gorilla hemoglobin gene is bound

Human Hemoglobin gene

Other genes pass through

gene and remain stuck to the filter. Other, unrelated genes will pass through (Fig. 16.32).

This approach allows us to isolate new genes as long as we already have a known gene closely related enough in sequence to hybridize well.

A whole slew of methods based on hybridization are used for analysis in molecular biology. The basic idea in each case is that we have a known DNA sequence which acts as a "probe." Generally, the probe molecule is labeled, by radioactivity or fluorescence, for ease of detection. The probe is used to search for identical or similar sequences in our experimental sample of target molecules. Both the probe and the target DNA must be treated so as to get single stranded DNA molecules that can hybridize to each other by base pairing. This is done either by heating or by alkaline denaturation.

Southern, Northern and Western Blotting

Suppose we have a large DNA molecule, such as a plasmid, or even a chromosome, and we wish to locate a particular gene whose sequence is known or is similar to a known sequence. First we cut up the target DNA with a restriction enzyme and separate the fragments on a gel. Then we transfer the DNA fragments to a nylon membrane. Finally we dip the membrane in a solution of labeled DNA probe molecules (Fig. 16.33). The probe binds only to those fragments with sequences similar enough to base pair. Usually, autoradiography is used to locate the bound probe. Hybridization of DNA to DNA is known as Southern blotting.

Southern blotting hybridization technique in which DNA binds to DNA

16.33 SOUTHERN BLOTTING

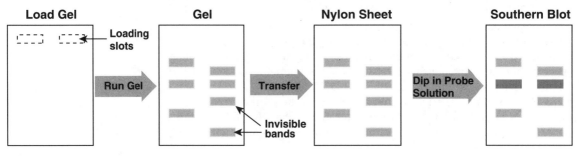

Load Gel

Loading slots

Run Gel

Gel

Invisible bands

Transfer

Nylon Sheet

Dip in Probe Solution

Southern Blot

Although Southern blotting was actually named after its inventor, Edward Southern, it set a geographical trend (Fig. 16.34). When hybridization uses RNA as the target molecule it is called Northern blotting. For example, we could use a DNA probe to locate the messenger RNA molecule that corresponds to the same gene. The mixture of RNA is run on the gel and transferred to the filter. The filter is then probed just as above.

16.34 TABLE OF BLOTTING SYSTEMS

	Material on Nylon Screen	Probe
Southern	DNA	DNA
Northern	RNA	DNA
Western	PROTEIN	ANTIBODY
South-Western	PROTEIN	dsDNA

Northern blotting hybridization technique in which a DNA probe binds to an RNA target molecule

Western blotting detection technique in which a probe, usually an antibody, binds to a protein target molecule

South-Western blotting detection technique in which a DNA probe binds to a protein target molecule

Western blotting does not even involve nucleic acid hybridization. Proteins are separated on a gel, transferred to a membrane and detected by antibodies (see Ch. 21) or other methods. In South-Western blotting, proteins suspected of binding to DNA are stuck to the membrane for testing. The probe is a DNA fragment, in this case double stranded, since DNA binding proteins bind to the DNA double helix.

Zoo Blotting

Zoo blotting is not so much a separate method as a neat trick using Southern blotting. One problem encountered when cloning human genes is that most DNA in higher animals is non-coding DNA (see Ch. 11). The coding sequences are what we want, yet they are only a small fraction of the total DNA. So how do we identify a coding region?

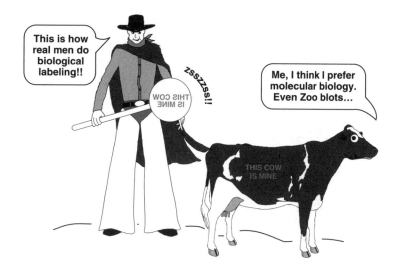

16.35 "MID - WESTERN" BLOTTING

zoo blotting comparative Southern blotting using DNA target molecules from several different animals to test whether the probe DNA is from a coding region

During evolution, the base sequence of non-coding DNA mutates and changes rapidly, whereas coding sequences change much more slowly and can still be recognized after millions of years of divergence between two species (see Ch. 23).

So we extract samples of DNA from a human, monkey, mouse, hamster, cow, chicken, democrat, etc. The "zoo" of DNA samples are each cut up with a suitable restriction enzyme and the fragments are run on a gel and transferred to a nylon membrane. They are probed using the DNA we suspect of being human coding DNA. If the DNA really is from a coding sequence it will probably hybridize with some fragment of DNA from most other closely related animals (Fig. 16.36). If the DNA is non-coding DNA, it will probably only hybridize to the human DNA.

16.36 **ZOO BLOTTING**

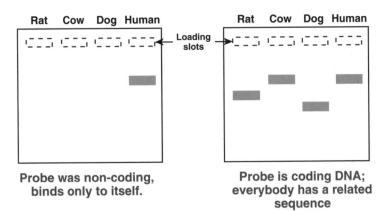

Probe was non-coding, binds only to itself.

Probe is coding DNA; everybody has a related sequence

Chromosome Walking

As a result of genetic investigation, we may know the approximate chromosomal location of a gene responsible for, say, some inherited disease. Using this information to clone the gene is referred to as positional cloning. One of the simplest versions of this is chromosome walking, a method based on hybridization.

16.37 **CHROMOSOME WALKING**

The chromosome which we wish to walk along is cut into manageable fragments with a suitable restriction enzyme. We need a cloned fragment of DNA to get started, and we use this as our first probe. We then test each fragment for hybridization to our starting probe. Obviously, it must overlap at least one or two neighboring fragments. We then take the newly-isolated clones and use them as probes in a second cycle of hybridization, and so on. For simplicity we have shown a walk in one direction only in Figure 16.38, though obviously in real life we could cruise along the chromosome both ways from our central starting point.

With each cycle of hybridization, we identify segments of DNA that overlap the one we started with. Gradually we work outwards from our starting position and eventually we could end up mapping and cloning the whole chromosome, bit by bit.

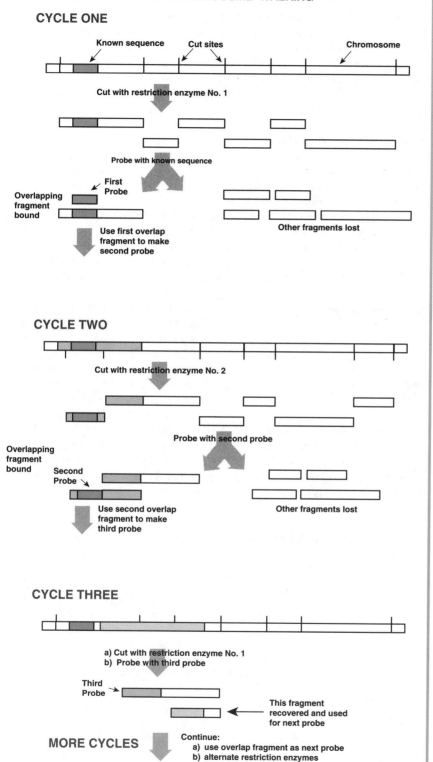

16.38 **CHROMOSOME WALKING**

CYCLE ONE

Known sequence   Cut sites   Chromosome

Cut with restriction enzyme No. 1

Probe with known sequence

Overlapping fragment bound

First Probe

Use first overlap fragment to make second probe

Other fragments lost

CYCLE TWO

Cut with restriction enzyme No. 2

Probe with second probe

Overlapping fragment bound

Second Probe

Use second overlap fragment to make third probe

Other fragments lost

CYCLE THREE

a) Cut with restriction enzyme No. 1
b) Probe with third probe

Third Probe

This fragment recovered and used for next probe

MORE CYCLES

Continue:
a) use overlap fragment as next probe
b) alternate restriction enzymes

In practice this approach is used when we already have one cloned segment of DNA and we think that neighboring genes might be of interest. For example, inheritance of a particular RFLP may accompany a genetic defect responsible for a hereditary disease. Since RFLPs are visualized as bands of DNA on a gel, we can cut a slice from the gel and extract the DNA. This gives us a piece of DNA to get started. We then need to cruise along the chromosome until we find the actual gene causing the inherited disease (see Ch. 13 for examples).

Subtractive Hybridization

subtractive hybridization removal of unwanted genes by hybridization, leaving behind the gene being sought

Subtractive hybridization is a technique used to isolate a gene that is not there. Hmmmmm! you say, is that possible? Yes, we already used this approach to clone the Duchenne muscular dystrophy gene in Chapter 13 (see Fig. 13.16 to Fig. 13.19).

Suppose that a hereditary defect is due to the deletion of the DNA for a particular gene. So we have a chromosome with a certain sequence of DNA missing. Since we want to find the missing DNA, we obtain the corresponding chromosome from a healthy individual. Our situation is as shown in Figure 16.39.

We proceed as usual by cutting our two chromosome samples into fragments of convenient size (Fig. 16.40). Suppose we were to hybridize these two batches of DNA. We would get hybrid molecules for all regions of the DNA except the region of the deletion which is only present in the healthy chromosome.

16.39 SUBTRACTIVE HYBRIDIZATION

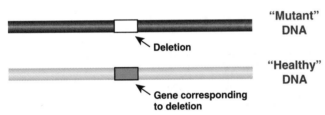

"Mutant" DNA

Deletion

"Healthy" DNA

Gene corresponding to deletion

If a large surplus of mutant DNA is used, all fragments of the healthy chromosome will be hybridized to mutant fragments except the region corresponding to the deletion, which will be left over. The single strands of this lone fragment will have to pair with each other. Thus, we have subtracted out all the segments of DNA that we do not want.

16.40 SUBTRACTIVE HYBRIDIZATION II

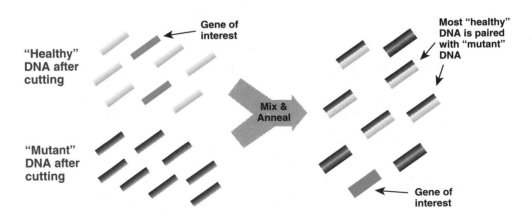

Gene of interest

"Healthy" DNA after cutting

"Mutant" DNA after cutting

Mix & Anneal

Most "healthy" DNA is paired with "mutant" DNA

Gene of interest

In practice we need some means of getting hold of this left over "deletion fragment." One way is to chop up the two batches of DNA with different enzymes (see Ch. 13 for a variant of this approach). This results in hybrid molecules having non-matching ends, whereas the self-paired fragment will have matching sticky ends and can be cloned.

Subtractive hybridization can be used for a variety of purposes. Suppose we want to isolate the messenger RNA only for those genes which are switched on at high temperature, or any other environmental stimulus that floats your boat. We grow batches of cells under normal conditions and under the test condition.

Next we extract the mRNA from both batches. How do we get mRNA? Simple - hybridization again!! Since mRNA ends in a poly-A tail, we make a piece of poly-T and attach it to glass beads which we put in a column. We pour the cell extract through the column and the poly-A tails hybridize to the poly-T and the mRNA is trapped! (Fig. 16.41). For the truly trendy, the poly-T can be attached to tiny magnetic beads that can be pulled out of the mixture using a magnet.

sticky ends   ends of a DNA molecule with short single stranded overhangs

poly-A tail   string of adenine residues on the 3' end of messenger RNA from eukaryotes

16.41 TRAPPING mRNA BY ITS TAIL

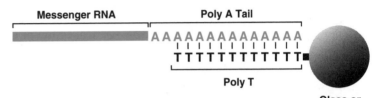

We want to use the sample of mRNA from the normal culture to subtract out the "normal" mRNA from the high temperature sample. This will leave behind only the mRNA specific for high temperature. But RNA molecules are single stranded and so will not hybridize to each other. Not to worry! Taking the normal mRNA and using reverse transcriptase, we make a double stranded DNA corresponding to the mRNA (Fig. 16.42). This complementary DNA (cDNA) actually has one strand identical in sequence to the original mRNA and the other strand is complementary.

reverse transcriptase enzyme that starts with RNA and makes a DNA copy of the genetic information

complementary DNA (cDNA)   the DNA sequence complementary to an RNA sequence, usually to mRNA

16.42 MAKING cDNA FROM mRNA

Messenger RNA

Reverse Transcriptase

Same sequence as mRNA

cDNA (double stranded)

Complementary to mRNA

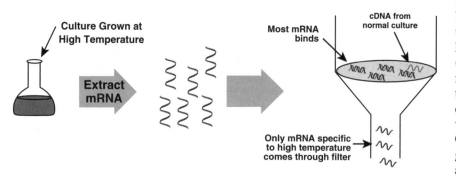

The cDNA from the normal culture can be used to hybridize to the mRNA from the hot culture (Fig. 16.43). All of the mRNA corresponding to genes expressed during normal growth will be removed. Only the mRNA from genes expressed only at high temperature will be left. (If what we really wanted were genes rather than mRNA we could also convert the mRNA from the heat-induced genes to their cDNA analogs.)

A cDNA Library is a Collection of DNA, Sans Introns

Clone libraries are collections of cloned genes which are big enough to contain at least one copy of every gene from an organism. They are carried on an appropriate plasmid or virus vector. Eukaryotic genes have intervening sequences of non-coding DNA (introns). Hence it is usually more convenient to use the cDNA version of the sequence. Since messenger RNA has the introns removed during processing, the cDNA versions of genes are made by reverse transcription of messenger RNA. A cDNA library is a collection of genes in their cDNA form (lacking introns). The cDNA version of a eukaryotic gene can often be successfully expressed in bacteria, whereas the original version cannot. This is because bacteria cannot process the RNA to remove the introns.

"FISH" - Fluorescence *In Situ* Hybridization

Fluorescence *In Situ* Hybridization, better known as FISH, is used to detect the presence of a gene, or the corresponding messenger RNA, in a living cell (Fig. 16.44). The gene investigated must be cloned and is used as the probe. As the name indicates, the DNA probe is labeled with a fluorescent dye whose localization will eventually be observed under a fluorescence microscope, that provides the beam of light needed to excite the fluorescent tag.

We could treat a thin section of tissue with a DNA probe for a gene we know comes from the animal we sliced up. The probe will hybridize to the DNA in the nucleus of all the cells since they all have the same genes.

cDNA library a collection of cloned genes present as their cDNA versions and carried on an appropriate plasmid or virus vector. The cDNA is generated by the reverse transcription of messenger RNA

FISH (Fluorescence *In Situ* Hybridization) using a fluorescent tagged probe to see a molecule of DNA or RNA in its natural location

16.44 **FISH – THE PRINCIPLE**

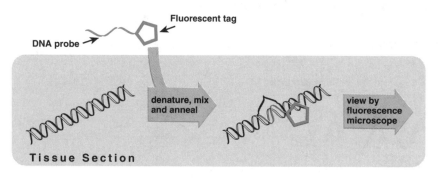

This tells us that the genes are in the nucleus, which we knew anyway, no fun and less use. When we go FISHing the following variations are much more exciting:

1) If we used a virus gene as a probe we could see which cells contain virus genes, and whether the virus genes are in the cytoplasm or have penetrated the nucleus (Fig. 16.45).

2) We could make a chromosome smear and bind the probe to this. The place where the probe binds tells us which chromosome carries our gene and, if we have good enough equipment, we can even localize the gene to a region on the chromosome (Fig. 16.46).

16.45 **FISHing FOR VIRUS GENES**

16.46 **FISHing LOCATES GENE ON CHROMOSOME**

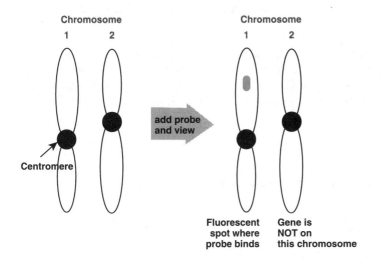

3) If we use a DNA probe, we can detect mRNA within our target tissue since one of the two strands of the DNA will bind to the RNA. Cells actively transcribing the gene of interest will have lots of the corresponding mRNA and will bind probe and light up (Fig. 16.47). What a bright idea! The greater the gene expression the brighter the cell will fluoresce. Actually, since hybridization needs single strands, we would have to treat the target tissue with heat or alkali to denature the DNA it contains. If we omit the denaturation step, only RNA will be detected, as it is naturally single stranded.

259

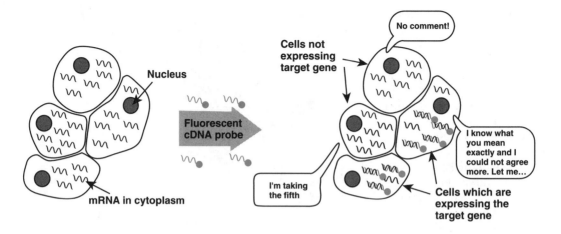

Reporter Genes

reporter gene a gene that is easy to detect and which is inserted for diagnostic purposes

antibiotic resistance gene gene conferring resistance to an antibiotic

Reporter genes are molecular spies that are supposed to report what is happening where they are stationed. Suppose we are trying to put a DNA molecule, such as a cloning plasmid, into a new host cell. How do we know if the DNA has made it in? How can we tell which cells have the plasmid?

The crudest type of reporter gene is an **antibiotic resistance gene** (Fig. 16.48). We can include the gene for resistance to our favorite antibiotic on our plasmid. After attempting to get the plasmid into the target cells, they are treated with the antibiotic. Those getting the plasmid become antibiotic resistant; those not getting the antibiotic resistance gene are killed.

16.48 ANTIBIOTIC RESISTANCE AS REPORTER GENE

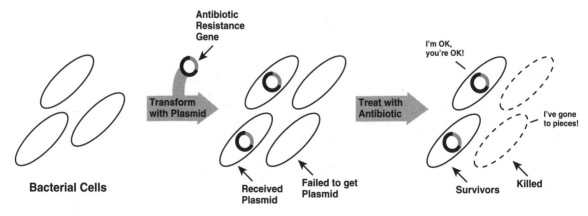

The most widely used reporter gene is the *lacZ* gene encoding β-(beta)-galactosidase (Fig. 16.49). This enzyme normally splits lactose, a compound sugar found in milk, into the simpler sugars glucose and galactose.

lacZ gene   the gene that encodes β-galactosidase

β (beta)-galactosidase enzyme that splits lactose and related compounds

16.49 β – GALACTOSIDASE SPLITS LACTOSE

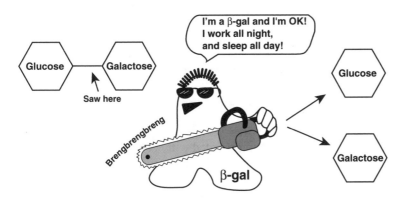

If we give β-galactosidase artificial compounds of galactose instead, it will split these also (Fig. 16.50). The two most common are **ONPG** and **X-gal**. ONPG (o-nitrophenyl galactoside) is split into o-nitrophenol and galactose. The o-nitrophenol is yellow so it is easy to measure.

ONPG (o-nitrophenyl galactoside) substance split by β-galactosidase and yielding yellow nitrophenol

X-gal substance split by β-galactosidase and yielding a blue dye

16.50 β - GALACTOSIDASE SPLITS ONPG OR X - GAL

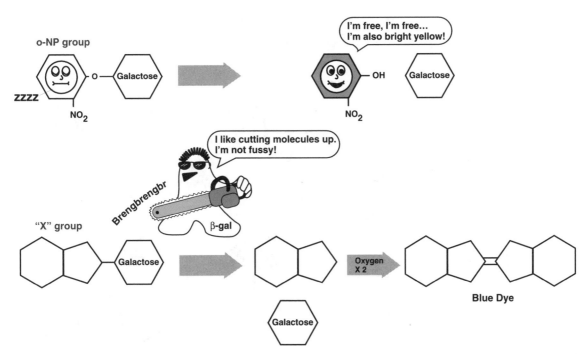

X-gal is split into galactose and the precursor to an indigo type dye. Oxygen in the air converts the precursor to an insoluble blue dye that precipitates out at the location of the *lacZ* gene.

Another example is the *phoA* gene that encodes alkaline phosphatase. This enzyme is even less fussy than β-galactosidase and chops phosphate groups off from a wide range of molecules. Like β-galactosidase, we can provide alkaline phosphatase with a variety of artificial substrates:

1) o-Nitrophenyl phosphate is split releasing yellow o-nitrophenol.
2) X-phos consists of an indigo dye-precursor joined to phosphate. The enzyme splits this, and exposure to air converts the dye-precursor to the blue dye.
3) Lumi-phos is split releasing an unstable chemical fragment that emits light.

A more sophisticated reporter gene is for luciferase (Fig. 16.51). This enzyme emits light when provided with a chemical known as luciferin. Luciferase is found naturally in assorted luminous creatures from bacteria to deep sea squid. The *lux* genes from bacteria and the *luc* genes from fireflies produce different brands of luciferase but both work well as reporter genes. If DNA carrying a gene for luciferase is incorporated into a target cell, it will emit light when luciferin is added. Although high level expression of luciferase can be seen with the naked eye, usually the amount of light is small and must be detected with a sensitive electronic apparatus such as a scintillation counter.

16.51 LUCIFERASE REPORTER GENE

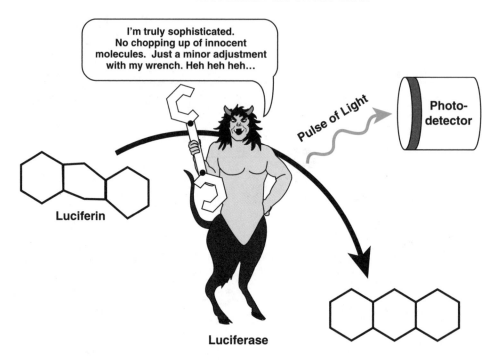

I'm truly sophisticated. No chopping up of innocent molecules. Just a minor adjustment with my wrench. Heh heh heh...

Luciferin

Pulse of Light

Photo-detector

Luciferase

Gene Fusions

Reporter genes can be used to track the physical location of a segment of DNA or for more detailed genetic analysis. Suppose we have a gene which is complicated to assay, but we need to know how it is controlled. The target gene is cut between its regulatory region and coding region.

The same is done with a reporter gene. Then the regulatory region of the gene we are investigating is joined to the coding region of the reporter gene (Fig. 16.52). This is a gene fusion.

gene fusion hybrid in which the regulatory sequences from one gene are joined to the coding region of another gene

16.52 CONSTRUCTING A GENE FUSION

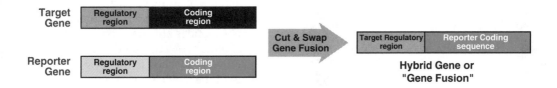

The gene fusion will be controlled the same way the target gene was controlled. But instead of making the original gene product that may be hard to assay, or even be unknown, it makes the enzyme belonging to the reporter gene. We can now grow our cell under a zillion different conditions and assay the reporter gene enzyme. This way we can monitor gene expression quickly and easily (Fig. 16.53).

16.53 MONITORING GENE EXPRESSION WITH FUSION

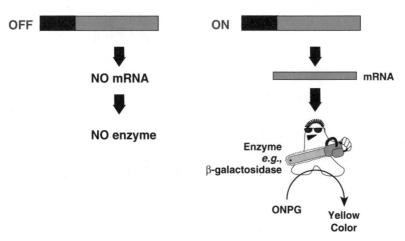

Gene fusions can be made between the coding region of, say, *lacZ* and the regulatory region of genes whose functions are completely unknown. They are then used to survey possible environmental conditions to see to what stimulus the target gene responds. This may give some clue as to the role of the unknown gene.

16.54 SOLID STATE DNA SYNTHESIS: SCHEME

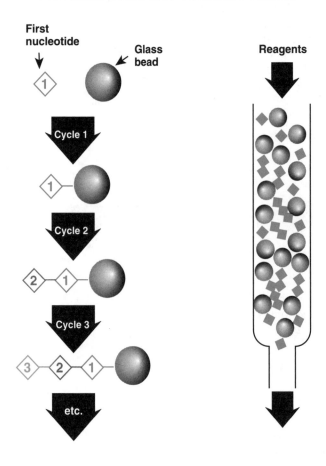

Chemical Synthesis of DNA

Artificially manufactured lengths of DNA are routinely used by molecular biologists for a variety of purposes. Short stretches of DNA for use as primers in PCR (see Ch. 17) and for DNA sequencing (see Ch. 23) are widely used today.

We begin by anchoring the first nucleotide to a solid support; often porous glass beads are used. The beads are packed into a column and the other reagents are poured down the column one after another. Nucleotides are added one by one and the growing strand of DNA remains attached to the glass beads until the synthesis is complete (Fig. 16.54).

The problem is that each deoxynucleotide has two hydroxyl groups, one for bonding to the next nucleotide and the other for bonding to the previous nucleotide. So each time a nucleotide is added, we must first block one of its hydroxyl groups and activate the one we want to react.

In practice, the first nucleotide has its 5'-hydroxyl blocked and is then bound to the glass bead via its 3'-hydroxyl group (Fig. 16.55).

16.55 SOLID STATE DNA SYNTHESIS – STEP I

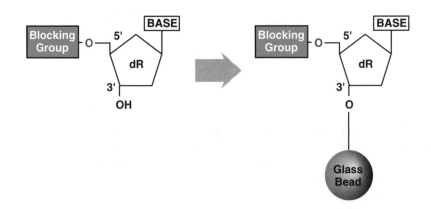

After the first nucleotide has been attached to the glass beads, acid is poured through the column to remove the blocking group and expose the 5'-hydroxyl group (Fig. 16.56).

16.56 **SOLID STATE DNA SYNTHESIS – STEP II**

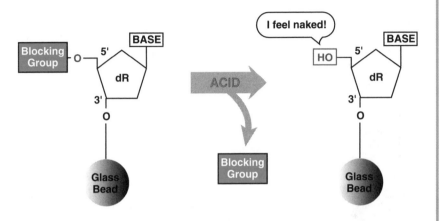

Next the acid is rinsed out. Then we add the next nucleotide which has had its 5'-hydroxyl group blocked but its 3'-phosphate group activated. The activated group of nucleotide No. 2 reacts with the exposed 5'-hydroxyl group of nucleotide No. 1 and a phosphate linkage is made (Fig. 16.57).

16.57 **SOLID STATE DNA SYNTHESIS – STEP III**

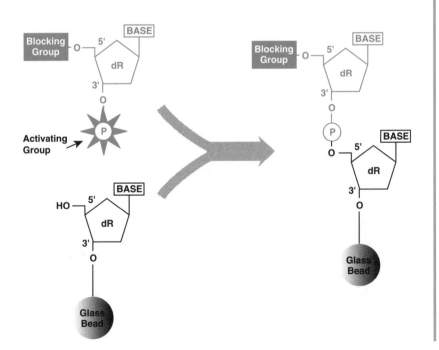

We now have a piece of DNA two units long with the 5'-hydroxyl of nucleotide No. 2 blocked. So we continue by exposing this with acid treatment and adding nucleotide No. 3. This cycle is repeated for each nucleotide to be added. Note that this artificial chemical synthesis proceeds in the 3' to 5' direction, whereas synthesis of DNA or RNA inside living cells always goes from 5' to 3'.

16.58 **DNA BY E - MAIL**

My lab tech is electronically enhanced. She can do chemistry via the internet!

Test tubes? We don't need test tubes!

Chemical synthesis of DNA is performed in practice by an automated machine. The molecular biologist just loads the machine with chemicals, types in the required sequence, and then waits. Gene machines take about five minutes per added nucleotide and can make chunks of DNA 100 nucleotides or more long.

Actually, most molecular biologists just write down the sequence they need and fax or e-mail it to their favorite biochemical supply company. It costs $2 to $3 per nucleotide for enough DNA to run a typical experiment a dozen or so times.

The Electron Microscope

With an ordinary light microscope, objects down to almost a micron (a millionth of a meter) in size can be seen. Typical bacteria are a micron or two long by about half a micron wide. Although they are visible under a light microscope, the internal details are too small to make out.

The resolving power of a microscope largely depends on the wavelength of the light (Fig. 16.59). If two dots are less than about half a wavelength apart they cannot be distinguished. Visible light has wavelengths in the range of 0.3 (blue) to 1.0 (red) micron, so bacteria are just at the limits of detection.

16.59 **RESOLVING POWER AND WAVELENGTH**

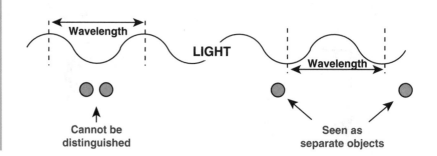

Wavelength

LIGHT

Wavelength

Cannot be distinguished

Seen as separate objects

A beam of electrons has a much smaller wavelength than does visible light and can distinguish detail far beyond the limits of resolution by light. Using an electron microscope allows us to visualize the layers of the bacterial cell wall and to see the folded-up bacterial chromosome as a light patch against a dark background. The electron beam is fired at the sample and materials that absorb electrons more efficiently appear darker. Because electrons are easily absorbed, even by air, the electron beam is used inside a vacuum chamber and the sample must be sliced extremely thin (Fig. 16.60).
To improve contrast, cell components are usually stained with compounds of heavy metals such as uranium, osmium or lead, all of which strongly absorb electrons.

Individual, uncoiled DNA molecules can be seen if they are shadowed with metal atoms to increase the absorption of electrons (Fig. 16.61). Shadowing is done by spreading the DNA out on a grid and then rotating it in front of a hot metal filament. Metal atoms evaporate and cover the DNA. Gold, platinum, or tungsten are typically used for shadowing.

16.60 ELECTRON MICROSCOPE

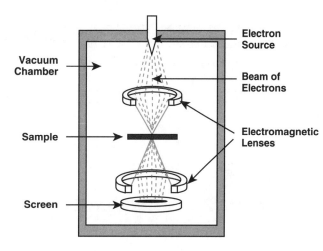

16.61 METAL SHADOWING

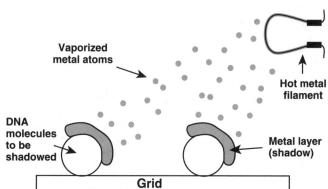

An example of the use of this approach was the direct visualization of the introns found in eukaryotic genes (see Ch. 11). The messenger RNA and the DNA both contain the exons that comprise the coding sequence, but the final mRNA lacks the introns (non-coding regions). If we hybridize mRNA to single stranded DNA from the corresponding gene, the results are regions of base pairing (the exons) interrupted by loops due to the extra intron sequences in the DNA. This is called R-loop analysis (Fig. 16.62).

intron segment of a gene that does not code for protein

exon segment of a gene that codes for protein

R-loop analysis when the DNA copy of a gene is base paired to the corresponding mRNA, the extra regions in the DNA, which have no partners in the mRNA, appear as loops

267

16.62 R - LOOP ANALYSIS

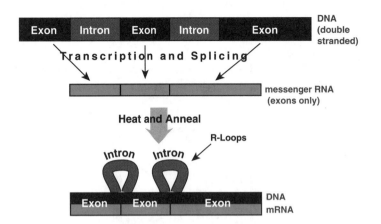

Exon | Intron | Exon | Intron | Exon

DNA (double stranded)

Transcription and Splicing

messenger RNA (exons only)

Heat and Anneal

R-Loops

Intron   Intron

Exon   Exon   Exon

DNA mRNA

X-ray crystallography (X-ray diffraction) determination of 3-D crystal structure by using X-rays

diffraction pattern array of spots formed by X-rays after traveling through a crystal

The resulting loops can be directly seen under the electron microscope.

X-ray Crystallography

X-ray crystallography, also known as X-ray diffraction, is used to work out the 3-D structure of molecules, in particular proteins and nucleic acids. It was X-ray diffraction that first revealed DNA was twisted into a double helix. Knowing their 3-D shapes allows us to understand how biological molecules fit together.

When a beam of X-rays is shone through a substance, the X-rays are scattered by the atoms they encounter. If the target substance is a crystal with a regular structure, the scattering of the X-rays will give rise to a regular, though complex, diffraction pattern (Fig. 16.63). In practice the crystal is rotated between a variety of positions on a computer controlled stage. The diffraction patterns are recorded and after computer analysis, are used to generate a 3-D atomic map of the protein molecule.

We need a large crystal of highly purified protein to put into the X-ray beam. Nowadays, molecular cloning and overexpression of the gene encoding it, allow us to obtain plenty of the protein under investigation. However, getting nice crystals is often tricky for such massive complex molecules as proteins, especially when their 3-D shapes are often irregular.

16.63 X - RAY DIFFRACTION OF PROTEIN CRYSTALS

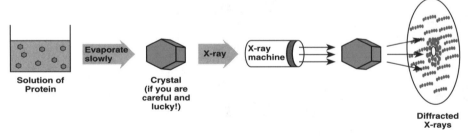

Solution of Protein

Evaporate slowly

Crystal (if you are careful and lucky!)

X-ray

X-ray machine

Diffracted X-rays

So dear reader, if your poor darling brain is overloaded from slogging through all these high-tech procedures, what do you think it was like writing them?! And we still haven't heard about the most famous technique, the one we actually hear about on TV - PCR. That's because PCR is so important it deserves its own chapter, so read on after you recover!

Additional Reading

Molecular Biology of the Gene by Watson JD, Hopkins NH, Roberts JW Steitz JA, & Weiner AM. 4th edition, 1987. Benjamin-Cummings, Menlo Park, California.

Molecular Biotechnology: Principles and Applications of Recombinant DNA by Glick BR & Pasternak JJ. 1994. American Society for Microbiology, Washington, DC.

PCR
The Polymerase Chain Reaction And Its Many Uses

17

Of all the recent technical advances in molecular biology, the polymerase chain reaction (PCR) has been by far the most useful. The PCR works by amplifying DNA sequences in a totally awesome way. Even if you have such a pathetically teeny weeny amount of a certain DNA molecule to begin with, so little it can scarcely be detected, you can generate literally zillions of copies by PCR. As a result there are almost as many uses for PCR as there are people in the PRC!

The PCR is used in clinical diagnosis, genetic analysis, genetic engineering and forensic analysis, and assorted practical examples are scattered around this book. In this chapter we will examine how PCR works.

Fundamentals of PCR

We begin with a segment of DNA we want to amplify in order to generate many copies. What do we need?

1) We need a few molecules of DNA which include the DNA segment we want to amplify (Fig. 17.2). Let's call this molecule the "template" and the segment of it we care about the "target sequence." We don't need much, just a trace amount will do.

17.1 LOTS OF USES FOR PCR

polymerase chain reaction (PCR) artificial amplification of a DNA sequence by repeated cycles of replication and strand separation

PRC Peoples Republic of China

17.2 DNA TEMPLATE FOR PCR

Double stranded DNA

Target sequence

2) We need two PCR primers (Fig. 17.3). These are short pieces of single stranded DNA that match the sequences at either end of our target DNA segment. They are needed to get DNA synthesis started.

primer short segment of DNA that binds to the longer template strand and allows DNA synthesis to get started

17.3 PRIMERS FOR PCR

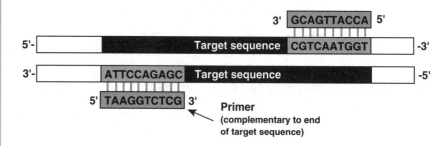

Taq polymerase heat resistant DNA polymerase from *Thermus aquaticus*

Thermus aquaticus a bacterium that only grows at very high temperatures and lives happily in hot springs

17.4 *THERMUS AQUATICUS* AT HOME

3) An enzyme to manufacture the DNA copies. The PCR procedure involves a couple of high temperature steps so we use a heat resistant DNA polymerase. This is extracted from heat resistant bacteria living in hot springs at temperatures up to 90°C. Most often we use Taq polymerase from *Thermus aquaticus*.

4) A supply of nucleotides is needed for the polymerase to use when making the new DNA.

5) Finally we need a PCR machine to keep changing the temperature (Fig. 17.5). The PCR process requires cycling around through several different temperatures. Because of this, PCR machines are sometimes called thermocyclers.

thermocycler machine used to rapidly shift samples between several temperatures in a pre-set order. Used for PCR

Before thermocyclers were invented, scientists had to move their tubes between three incubators at different temperatures every few minutes. At least they got some exercise back then!!

17.5 THERMOCYCLER FOR PCR

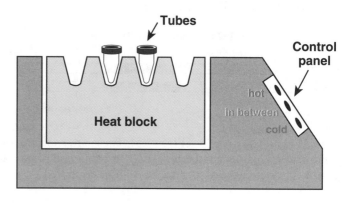

Cycling Through The PCR:

To separate the strands (Fig. 17.6), we start by heating our template DNA to 90°C or so for a minute or two.

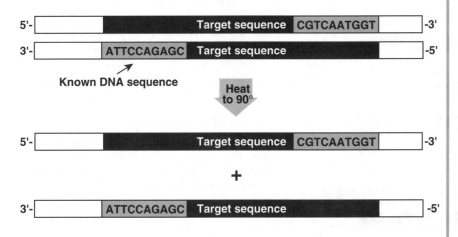

17.6 **PCR – STRAND SEPARATION 90°C**

Although the primers were present from the beginning, they cannot bind at 90°C. So, next we drop the temperature to around 50°C to 60°C, allowing the primers to anneal to the complementary sequences on the template strands (Fig. 17.7). We have shown 10 base primers for illustration purposes, though in real life they would be longer, say 15 to 20 bases. The longer the primer, the more specific will be its binding.

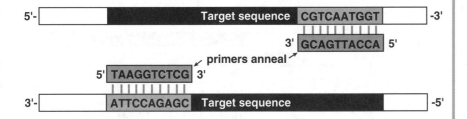

17.7 **PRIMERS ANNEALING 50°- 60°C**

Then we maintain the temperature at 70°C for a minute or two to allow the polymerase to elongate new DNA strands starting at the primers (Fig. 17.8). Note that DNA synthesis goes from 5' to 3' for both new strands.

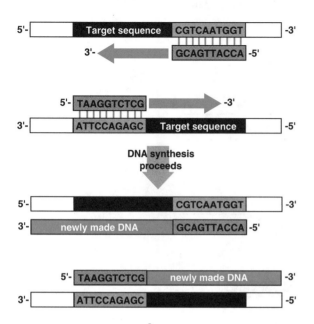

We now have two partly double stranded pieces of DNA. Note that the two new strands are not as long as the original templates, they are each missing a piece at the end where synthesis started. However they are double stranded over the region that matters, our target sequence.

We now repeat the cycle of events. The second cycle goes as follows (Fig. 17.9):

There are now four partly double stranded pieces of DNA. Note again that although they vary in length, they all have double stranded DNA throughout the target region.

As we continue to cycle through the PCR, the single strand overhangs are ignored and are rapidly outnumbered by segments of DNA containing only the target sequence. Lets do it one more time to show this (Fig. 17.10).

17.9 **PCR – SECOND CYCLE**

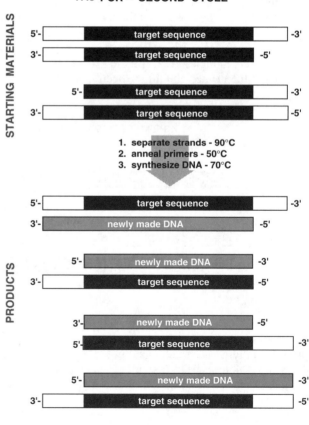

Notice that now we have our first two pieces of double stranded DNA which correspond exactly to the target sequence and do not have any dangling single stranded ends. The cycle is repeated over and over again. Once past the first two or three cycles, the vast majority of the product is double stranded target sequence with flush ends.

17.10 PCR – THE THIRD CYCLE

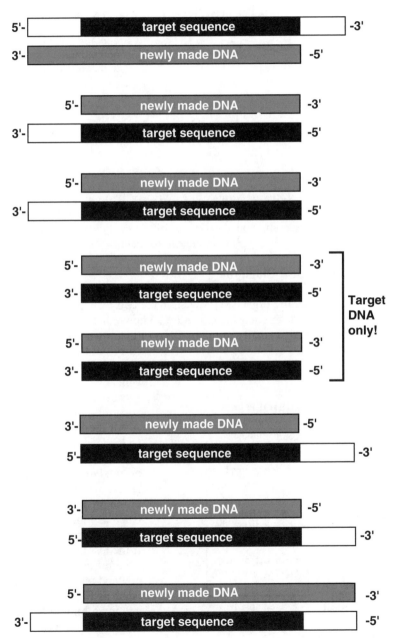

Each cycle doubles the amount of DNA. If we count up the numbers of double stranded DNA molecules (and ignore those molecules with dangling ends) we get:

Cycle Number	Target Molecules Made
1	0
2	0
3	2
4	8
5	22
6	52
7	114
8	240
9	494
10	1,004
15	32,738
20	1,048,536
25	33,554,382
30	1,073,741,764

... you get the idea.

When the PCR was invented by Kary Mullis in 1987, he used normal DNA polymerase. Since this is killed by the temperatures needed to separate DNA into single strands, he had to add a fresh dose of polymerase to each tube every cycle! What a right royal pain in the neck!

Luckily for the sanity of molecular biologists, or at any rate the sanity of their technicians, heat resistant DNA polymerase was purified from good old

J. Fonda PCaRobics: "...and separate
    and anneal
    and add polymerase
    and elongate
    and separate..."

Thermus aquaticus just a year or two later. This Taq polymerase can be thrown into the reaction mixture at the beginning and survives all of the heating steps. It actually requires a high temperature to manufacture new DNA.

Use of PCR in Medical Diagnosis

Suppose we already know the sequence of DNA characteristic for part of some gene in a specific organism. We can make two PCR primers corresponding to stretches of this sequence that are a known distance apart on the DNA. We can then take a small sample of DNA of unknown origin and carry out PCR using our primers. Next, the DNA generated is run on an agarose gel to separate it according to size.

If the DNA sample tested is indeed from another sample of the original organism, we will get a nice band of DNA of the predicted length (such as Fig. 17.12; unknown sample No. 2). If the test DNA is not from the same organism, no band will be generated (such as Fig. 17.12; unknown sample No. 1). If necessary, we can even

17.12 PCR DIAGNOSIS

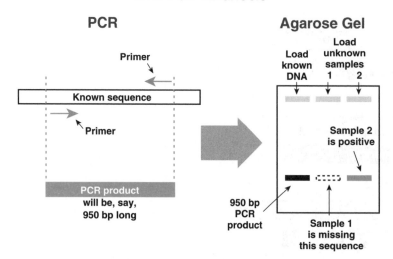

PCR

Primer

Known sequence

Primer

PCR product
will be, say,
950 bp long

Agarose Gel

Load known DNA

Load unknown samples 1 2

Sample 2 is positive

950 bp PCR product

Sample 1 is missing this sequence

sequence the DNA made by PCR to be absolutely sure. Clearly, PCR can be used in a variety of diagnostic tests.

For example, **AIDS** takes a long time after HIV retrovirus infection to produce visible symptoms, often several years. However, we can test DNA in blood samples by using PCR primers specific for sequences found only in the genes of this particular virus. Another example is tuberculosis. Unlike many bacteria, *Mycobacterium*, which causes this disease, grows very slowly. Culturing it to perform the classic diagnostic tests may take nearly a month, whereas PCR identification of mycobacterial DNA can be done in a day.

The DNA from 1/100th of a milliliter of human blood contains about 100,000 copies of each chromosome. Or, if you prefer, about one-tenth of a **picogram** by weight of a target sequence of, say, 500 base pairs. After a good PCR run we can get a whole microgram or more. A microgram is 10^{-6} gram or a millionth of a gram, which may not seem much but is plenty for complete sequencing or cloning. Obviously, it is possible to identify an organism from an extremely small trace of DNA-containing material. In fact, the DNA from a single cell is enough if you are careful (Fig. 17.13). The use of PCR in forensics is described in Chapter 18.

AIDS (acquired immunodeficiency syndrome) the disease caused by the HIV retrovirus

Mycobacterium the bacterial group whose members cause tuberculosis and leprosy

picogram 10^{-12} gram or a million-millionth of a gram

Human blood is not even a rich source of DNA for PCR since red blood cells lose their nuclei during development. Only the small percentage of white cells have nuclei with chromosomes

17.13 PCR IS VERY SENSITIVE

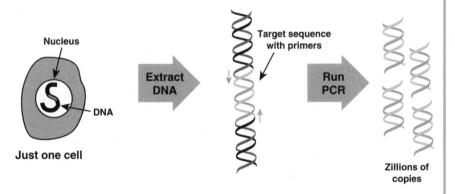

Nucleus

Just one cell

Extract DNA

Target sequence with primers

Run PCR

Zillions of copies

DNA

Degenerate Primers

The major snag with PCR is obvious. In order to make the PCR primers you need to know the sequence, at least at the ends, of your target sequence. There are several ways to deal with this, but in each case you need to know the sequence of something to get started!

Despite the name degenerate primers, these are not used for identifying sex offenders! They are used when there is some information but no complete sequence to go on. Suppose you have cloned and sequenced the gene for **insulin** from pigs and, for reasons best known to yourself, you want kangaroo insulin.

Thought 1: We presume that the sequence of the kangaroo insulin gene will be close to that of the pig gene.

Thought 2: We know that the genetic code is degenerate and several codons can encode the same amino acid (see Ch. 7 for details).

degenerate primer primer with several alternative bases at certain positions

insulin a hormone that controls sugar levels in the blood

degenerate when applied not to politicians but to genes, it refers to the use of more than one codon to encode the same amino acid

In particular, many codon families have the first two bases the same and vary in the third position.

Thought 3: The sequence of the protein is what is important for function, not the DNA sequence as such. So most of the variation between closely related genes is in this third codon position (Fig. 17.14).

17.14 DEGENERATE CODING OF PROTEIN

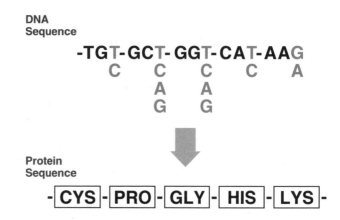

DNA Sequence

-TGT- GCT- GGT- CAT- AAG
   C   C   C   C   A
          A   A
          G   G

Protein Sequence

-| CYS |-| PRO |-| GLY |-| HIS |-| LYS |-

17.15 DEGENERATE PRIMER

T G Y G C N G G N C A Y A A R

N = any base
Y = either pyrimidine (T or C)
R = either purine (A or G)

Therefore we make degenerate DNA primers that have a mixture of all possible bases in every third position. A degenerate primer is actually a mixture of closely related primers (Fig. 17.15). We hope that one of the primers in our mixture will recognize the DNA of the gene we are looking for. We are also helped by the fact that a perfect match is not really necessary. If, say, 19 of 20 bases pair up, our primer will work quite well. Many segments of DNA have been PCR'ed successfully by using sequence data from close relatives.

Inverse PCR

Using degenerate primers is a statistical approach. Inverse PCR is a more sneaky way to do things. Here we start by knowing the sequence of part of a long DNA molecule, say a chromosome, and we want to continue exploring along this molecule into the unknown regions (Fig. 17.16).

17.16 EXPLORING THE UNKNOWN

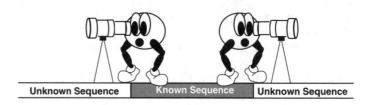

| Unknown Sequence | Known Sequence | Unknown Sequence |

To generate DNA by PCR, we need two regions of known sequence, for binding primers to on either side of the unknown target sequence. The present situation is exactly the opposite of that! So cleverly we convert our target molecule of DNA into a circle (Fig. 17.17). If we go round a circle we eventually get back to where we started. Pretty profound, huh?! In effect, the single known region is simultaneously on both sides of the target sequence!

To make our circle, let's choose a restriction enzyme that recognizes a six base sequence. If we travel far enough along the unknown DNA to either side of the known region, we will eventually come across sites for the chosen restriction enzyme. So we cut the DNA with this enzyme and ligate the ends together to make a circle of DNA, like so:

Any particular six base sequence will appear, on average, every 4 x 4 x 4 x 4 x 4 x 4 = 4,096 nucleotides.

17.17 INVERSE PCR – MAKING THE TEMPLATE

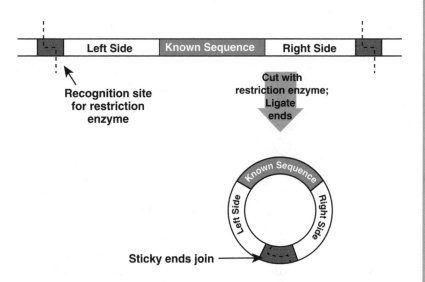

We use two primers corresponding to the known region and facing outwards round the circle (Fig. 17.18). Synthesis of new DNA will proceed around the circle clockwise from one primer and counter-clockwise from the other. Overall, the PCR reaction gives multiple copies of a chunk of DNA containing some DNA to the right and some DNA to the left of our original known region. These are the regions we want to explore.

17.18 INVERSE PCR – THE RESULT

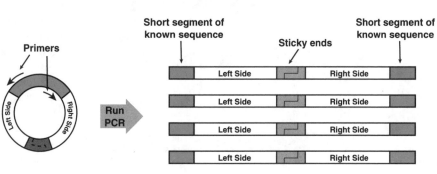

Randomly Amplified Polymorphic DNA (RAPDs)

randomly amplified polymorphic DNA (RAPD) method for testing genetic relatedness using PCR to amplify arbitrarily chosen sequences

Randomly Amplified Polymorphic DNA or RAPD is usually found in the plural as RAPDs and is pronounced "rapids," partly because it is a quick and dirty way to get a lot of information about the genes of the organism you are investigating.

The purpose of RAPDs is to test how closely related two organisms are. We may have a trace of blood from a crime scene and a suspect to compare it with, or we may be tracing an epidemic and need to sort out the disease-causing germs from innocent bystanders.

The principle is statistically based. Let's take any particular five-base sequence, such as ACCGA. How often will we find this exact sequence in any random length of DNA? The answer is that since we have four different bases to choose from, one of every 4^5 (or 4 x 4 x 4 x 4 x 4 = 1,024) stretches of five bases will be our chosen sequence. Any arbitrarily chosen 11 base sequence will be found once in approximately every 4 million bases. This is roughly the amount of DNA in a bacterial cell. In other words, any chosen 11 base sequence will happen by chance once only in the entire bacterial genome (Fig. 17.19). For higher organisms with lots more DNA per cell, we would need a longer sequence to be unique.

17.19 MOLECULAR IDENTITY CARDS

For RAPDs we do not want to be unique, just rare! We make PCR primers with our arbitrarily chosen sequence and run a PCR reaction using the total DNA of our organism as a template. Every now and then our primers will find a correct match, purely by chance, on the template DNA (Fig. 17.20). For PCR to happen, we need two such sites facing each other on opposite strands of the DNA. We also need the sites to be no more than a few thousand bases apart for the reaction to work well. The likelihood of two correct matches in this arrangement is quite low.

17.20 PRIMER SITES FOR RAPDS

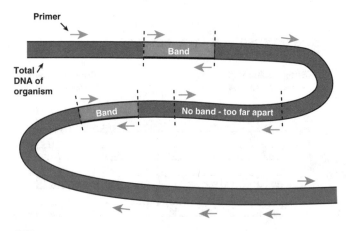

In practice, we make primers of such a length that we get five to 10 PCR bands. For higher organisms, primers of around 10 bases are typical. The bands from PCR are separated by gel electrophoresis (see Ch. 16) so we can measure their sizes.

We repeat this several times with primers of different sequence. The result is a diagnostic pattern of bands that will vary in different organisms, depending on how closely they are related (Fig. 17.21).

Although we do not know in which particular genes our PCR bands originate, this does not matter in measuring relatedness.

Adding Artificial Restriction Sites

Once we have made DNA by PCR what good is it? Among other choices we can sequence it (see Ch. 23) or we can clone it. However, to clone something we need convenient cut sites for restriction enzymes (see Ch. 9). Unless we are really lucky, we are unlikely to find such sites just at the ends of our PCR fragment.

So before we start our PCR reaction, we rig the outcome. When we design our primers, we add artificial restriction enzyme cut sites at the far ends of the primers (Fig. 17.22). As long as the primer has enough bases to match its target site, adding a few extra bases at the end will not affect the reaction. What does happen is that the bases making up the restriction site get copied and appear on the ends of all the newly manufactured DNA. This allows us to cut the PCR fragment with the restriction enzyme, and so generate sticky ends, and then clone it into a convenient plasmid.

PCR in Genetic Engineering

A whole plethora of modifications have been made to the basic PCR scheme. We can divide them into two broad categories:

    a) Deliberately changing one or two bases of a DNA sequence, and
    b) Deliberately rearranging large stretches of DNA.

Let's start with a gene we want to mutate by changing a single base. Suppose the sequence around this base is:

AAG CCG GTG GCG CCA

and we want to alter the T in the middle to an A. We simply make a PCR primer with the base alteration we want, that is,

AAG CCG GAG GCG CCA

We now use this mutant primer in PCR. As long as this primer is long enough to bind to the correct location on either side of the mutation, the DNA product will incorporate the change we made in the primer.

17.21 BAND PATTERNS

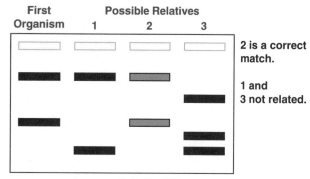

First Organism   Possible Relatives
1   2   3

2 is a correct match.

1 and 3 not related.

Agarose Gel of RAPDs

sequence   to find out the sequence of bases in a molecule of DNA

17.22 ADDING RESTRICTION SITES DURING PCR

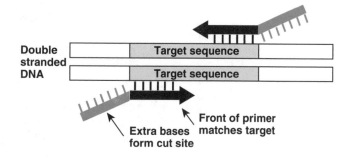

Double stranded DNA

Target sequence

Target sequence

Front of primer matches target

Extra bases form cut site

clone a gene   to obtain the DNA making up a particular gene and put it on a suitable vector such as a plasmid

As already illustrated, we can amplify any segment of DNA provided we have primers that match its ends. Suppose we want to make a hybrid gene. We may decide that we would like to join the front end of the growth hormone gene from a hippopotamus to the rear half of the equivalent gene from a gerbil. We'll skip the gruesome technical details. The crucial point is that we use an overlap primer that matches each part of both gene segments (Fig. 17.23).

17.23 OVERLAP PRIMER FOR FUSING GENE SEGMENTS

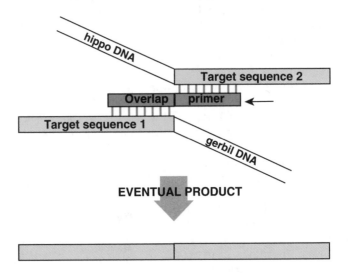

We run our reaction using a primer for the front end of the hippo gene, a primer for the rear end of the gerbil gene, and the overlap primer. And (hey presto!) PCR produces a hybrid gene. Some variants of this "molecular sewing" make the two halves separately and mix them later; other versions of this technique throw everything together in one big happy zoological mess.

This is not merely done to answer the profound question of whether we can make a miniature hopping hippo. Experiments like this may reveal which section of the growth hormone gene is most critical in determining the size of an animal. This sort of approach has been widely used to puzzle out in detail which regions of a gene are responsible for precisely which effects. It can also be used in biotechnology to construct artificial genes made up of chunks from different sources.

Tongue of frog,
tail of newt,
leg of dog,
let's hope its cute!

Reverse Transcriptase PCR

Suppose we want to clone genes from higher organisms. There is a big snag - most eukaryotic genes have intervening sequences in their DNA.

In other words, the coding sequence for the protein is interrupted by non-coding **introns** (see Ch. 11 for introns and RNA processing). If we clone the DNA and put it into a bacterial cell, the gene will not be expressed properly because the RNA will not be processed and the introns will not be cut out. As you remember, bacteria are not equipped for splicing mRNA.

It would be nice to get a copy of the gene without the introns. Eukaryotic cells remove introns during production of messenger RNA (Fig. 17.25). So if we obtain the mRNA, we have an intron-free sequence, except made of RNA not DNA.

intron segment of a gene that does not code for protein

17.25 INTRONS ARE REMOVED IN mRNA

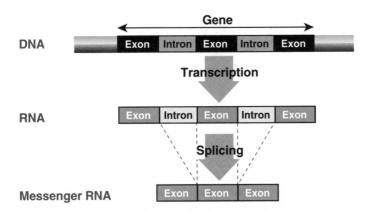

To convert RNA to DNA we use **reverse transcriptase**. (This enzyme originally comes from retroviruses; see Ch. 19.) Then we do PCR on the DNA to manufacture enough DNA copies for whatever experiment we have in mind (Fig. 17.26). This is referred to as **RT-PCR** and allows us to clone genes as intron-free DNA copies by starting with mRNA.

reverse transcriptase enzyme that starts with RNA and makes a DNA copy of the genetic information

RT-PCR combination of reverse transcriptase with PCR which allows DNA copies to be manufactured in bulk from messenger RNA

17.26 REVERSE TRANSCRIPTASE PCR

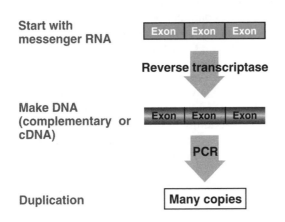

complementary DNA
(cDNA)   the DNA
sequence complementary to
an RNA sequence, usually
to messenger RNA

This DNA is known as complementary DNA (cDNA) because it is complementary to the message. Cloned eukaryotic genes are much easier to use for genetic engineering in their cDNA form. (Since bacterial genes do not have introns, this problem is irrelevant for bacteria).

RT-PCR has other uses. When a gene is expressed, the corresponding messenger RNA will be produced. In fact, there will be many more copies of the mRNA in the cell than of the original gene (Fig. 17.27). If we extract and purify the mRNA we will have several mRNA copies of every gene that was being expressed under the conditions in which we grew the cells.

cDNA is sometimes called "copy"-DNA by those who aren't sure whether its real name is spelt complementary or complimentary

oligo-dT   a short piece of DNA whose bases are all thymine, i.e.,
TTTTTTTTTTTTTT

the small "d" in front of the T indicates a deoxy nucleotide, that is, DNA, not RNA

17.27 mRNA CORRESPONDS TO EXPRESSED GENES

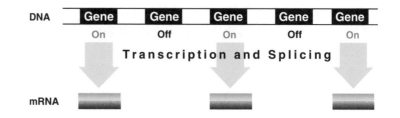

We now carry out RT-PCR on the mixture of mRNA using PCR primers that match some particular gene which we care about. If this gene was expressed we will get a PCR band, whereas if the gene was switched off, none of this particular mRNA will be present and no band will be generated (Fig. 17.28). So by carrying out RT-PCR on an organism under different growth conditions, we can see when the gene under scrutiny was switched on. So we can tell which environmental factors bring about expression of our favorite gene.

17.28 RT - PCR TO MONITOR GENE EXPRESSION

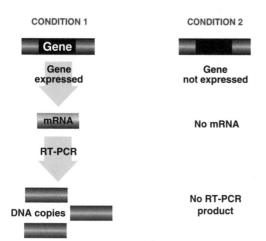

Differential Display PCR

One trendy variant of PCR is known as "differential display" PCR and is used to specifically amplify messenger RNA from eukaryotic cells. This technique is a combination of RAPDs (see above) with RT-PCR and has one clever little twist of its own, the use of oligo-dT primers. Since almost all eukaryotic mRNA molecules have a 3'-tail of poly-A, an artificial primer made only of T will base pair to this tail.

So we extract the RNA as before and use reverse transcriptase to make cDNA. Then we run a PCR reaction with two primers:

1) An oligo-dT primer that binds to the 3' end of all DNA copies of messenger RNA

2) As we do not know the sequences at the other end of the mRNAs, our second primer is actually a mixture of random primers similar to those used in RAPDs.

The result is that we end up with lots of DNA corresponding to each of the messenger RNA molecules in the original mixture. Finally, to separate the different components gel electrophoresis is used, as usual. This gives a series of DNA bands corresponding to each of the mRNAs being made in the cells we started with.

Jurassic Park PCR

Since any trace of DNA can be amplified by PCR and then cloned or sequenced, some scientists have looked for DNA in fossils. Stretches of DNA long enough to yield valuable information have been extracted from museum specimens such as Egyptian mummies, and fossils of various ages. This data has helped in studying molecular evolution (see Ch. 24).

Can we get enough DNA to resurrect *Tyrannosaurus rex*? Could a dinosaur amusement park really be put together? In the Sci-Fi best seller, Jurassic Park, the DNA was not obtained directly from dinosaur bones. Instead, it was extracted from prehistoric insects trapped in amber. The stomachs of bloodsucking insects would contain blood cells complete with DNA from their last victim, and if preserved in amber, this could be extracted and used for PCR (Fig. 17.29).

17.29 RESURRECTING OLD DNA

Amber

Ancient DNA

1. dissolve amber
2. extract DNA

This is fine in principle and DNA has indeed been extracted from insect fossils preserved in amber. The problem lies in the preservation of the fossil DNA. The older the fossil, the more decomposed the DNA will be. Normal rates of decay should break up DNA into fragments less than 1,000 bp long in 5,000 years or so. So, though we will no doubt obtain gene fragments from an increasing array of extinct creatures, you will have to be content with renting the video version of Jurassic Park for a while yet!

If you want to enjoy a present-day spectacle based on PCR, you can always watch a courtroom drama involving the use of DNA as forensic evidence. For this let's move on to the next chapter.

7.30 JURASSIC PORK PCR

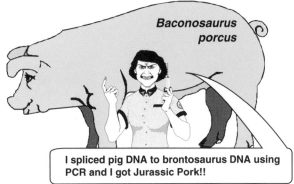

Baconosaurus porcus

I spliced pig DNA to brontosaurus DNA using PCR and I got Jurassic Pork!!

Additional Reading

Human Genetics: The Molecular Revolution by McConkey EH. 1993. Jones & Bartlett Publishers, Boston & London.

PCR: The Polymerase Chain Reaction by Mullis KB, Ferré F, and Gibbs RA. 1994. Birkhäuser, Boston, Basel & Berlin.

Whodunit?
Forensic Medicine and
Molecular Biology

One of the practical applications of DNA technology is its use to identify people from samples of blood or other tissues. This has been used by the legal profession to determine guilt or innocence in criminal trials or to determine a child's **paternity**. The history of DNA technology goes back to its first use in Britain in the mid 1980s; Americans caught on shortly afterwards. Lots of popular articles have appeared describing the use, methods and results of DNA testing in forensic medicine. While we don't like to admit it in public, we have all followed the O.J. Simpson "Trial of the Century" which publicized the use of DNA technology. Lets see what the lawyers must know about molecular genetics to try a case.

paternity fatherhood; that is, who is daddy?

Q. What do psychiatrists have in common with giant, 1,000-pound ducks??
A. Enormous bills!!

Who am I?

Psychologists all over the world claim they are helping people to "find themselves." More often than not the psychiatrists find their patient's wallets before the patients find themselves. We can do a much better job of helping people find themselves in the laboratory, with less expense and less grief. We can examine ourselves in various ways. For example, just a casual glance reveals major differences among people. Outward appearance is usually referred to as the **phenotype**. In the courtroom, the phenotype might be a verbal description or a drawing, or even pointing out the suspect in a "line-up."

Most physical differences between people are due to complex interactions of several genes during development. Some are obvious at a glance, such as webbed fingers or a cleft palate. Others are fine details like fingerprints, which are due to variations in the pattern of dermal ridges, the small skin elevations on our fingers (Fig. 18.2). The fingerprints of each individual are unique and were used for identification as early as the late 1800s.

phenotype characteristics due to the expression of our genes; usually refers to visible properties but may refer to characteristics revealed by laboratory tests

Geneticists who view DNA as all-supreme, hate to admit that fingerprints of identical twins are not quite identical, whereas their DNA *is* identical. There have been cases of parents of identical triplets losing track of which child is which. Only the FBI could tell them apart by comparing the fine details of their fingerprints with those taken at birth.

Fingerprint patterns depend on more than one gene (*i.e.*, they are multigenic). Although you might expect the fingerprints of identical twins to be the same, they are, in fact, not exactly identical. Minor variations in fingerprint patterns occur as a result of various factors influencing development.

Organizations obsessed with super secrecy take advantage of the unique pattern of blood vessels on the backs of our eyes. They make you look into a device that scans the retina of the eye for blood vessel patterns (retinal scans). A computer then compares your pattern with those on file to check your identity.

Another, more basic set of identifying features are the proteins made by all cells. Good examples of differences among proteins are the various blood types found in human populations (see below). But if we really, really, want to know ourselves, we must dig down to the level of the gene and determine our **genotype**. This is what is meant by DNA typing or DNA fingerprinting, techniques that are also described below.

18.2 FINGERPRINTS

Suspect 1
O. J.

Suspect 2
L.A.P.D. Police Officer

genotype the makeup of our DNA; the description of an organism at the genetic level

18.3 IDENTITY AT VARIOUS LEVELS

We can look at:	DNA	mRNA	proteins (antigens)	external characteristics
to obtain the:	genotype	genotype (message)	phenotype (lab tests)	phenotype (visible)
which is used legally in:	DNA typing/ DNA fingerprinting	not in practical use	blood typing	dermal fingerprint patterns

Blood, Sweat and Tears

All kinds of body tissues and fluids may be used to establish identity. While DNA technology is relatively new, it is a logical outgrowth of the work on blood typing that has been used in the courtroom for over 50 years. Although blood analysis is a relatively quick and easy method to identify individuals, sweat, tears, urine, saliva and semen also have cells with surface proteins that can be analyzed.

Terrorists mailing a bomb should never lick the stamps they stick on the envelope!

The proteins in blood are detected by using antibodies. So traditionally they are referred to as blood antigens (see Ch. 21 for antibodies and antigens). Binding of an antibody to its antigen is very specific. Consequently, two related proteins with only relatively small shape differences can be told apart because they will be bound by different antibodies. Let's begin by viewing the blood antigens as proteins expressed from our genes.

As humans, some of our blood antigens belong to the **ABO blood group system**. Although three letters, A, B and O, are used, there are only two antigens involved, the A antigen and the B antigen (both proteins).

ABO blood group system classic blood typing system for specific antigens found on red blood cells

If you have	you have blood type
A antigen alone	A
B antigen alone	B
A & B antigens	AB
neither A nor B	O

The A and B proteins are both coded for by different alleles of the same gene. Because we all have two copies of each gene, we all have two alleles for the ABO system protein. This pair of alleles may be the same or different in any given person. There are three alleles available, A, B, and O (no protein). The alleles for the A and B antigens are both dominant, so if you have at least one allele for either A or B, that antigen will be expressed. Thus, a mother with A and O alleles will make A antigen and a father with B and O alleles will make B antigen. They could still have a type O child because the child has a chance of inheriting the O gene on one chromosome from the heterozygous mother and another O gene from the heterozygous father (Fig. 18.4).

18.4 AO + BO PARENTS MAY HAVE AN OO CHILD

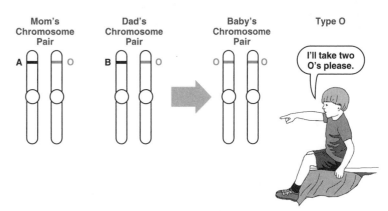

When two women both claimed the same baby, King Solomon, of Biblical times, suggested dividing the baby in half. When one woman withdrew her claim in order to save the child's life, Solomon knew who the real mother was.

Usually we know who the mother is. Mostly it is the father who may be difficult to identify. If you are accused of being a father in a paternity suit, the ABO typing will only exclude you in less than 15 to 20 percent of cases, even if you are innocent. There are several other blood and tissue antigen systems similar in principle to the ABO system that are also used in forensic medicine. Using the HLA system of white blood cells (see Ch. 21 for HLA system), the chance of exclusion is over 90 percent. When the HLA and ABO systems are combined the chances of exclusion are about 97 percent. Including the analysis of blood serum proteins with the others makes exclusion almost certain, if you really haven't been messing around!

Criminologists use the following formula for determining the combined probability (P) of exclusion based upon multiple tests with individual probabilities of P_1, P_2, P_3, P_4, etc.

$$P = 1 - (1\text{-}P_1)(1\text{-}P_2)(1\text{-}P_3)(1\text{-}P_4) \text{ etc.}$$

Juries have convicted suspects on ABO typing combined with other blood antigens giving overall probabilities as low as 25 to 50 percent that the suspect and the blood evidence matched. In many criminal cases, blood typing is the primary evidence.

DNA Testing

DNA tests alone, without supportive evidence, have sometimes been sufficient for conviction. DNA evidence is almost always sufficient for exoneration of mis-identified individuals who were wrongly convicted of committing a crime. DNA evidence can be obtained from any bodily tissue or secretion that has cell nuclei that contain DNA. There are two major types of testing used to determine if DNA found at the scene of a crime matches that of the suspect or the victim. One is popularly known as DNA fingerprinting and the other as PCR amplification.

DNA Fingerprinting

DNA fingerprinting relies on the unique pattern made by a series of DNA fragments after separating them according to their lengths by gel electrophoresis. Tissue samples taken from suspects may be compared with evidence obtained from a crime scene.

Like real fingerprints, DNA fragments show unique patterns from one person to the next. With the exception of identical twins, no two individuals have the same DNA, so they won't have similar DNA fingerprints either. Sets of individual chromosomes are distributed by parents to their offspring in so many possible combinations that it is incredibly unlikely that any two individuals will have the same DNA. Identical twins are the exception that proves the rule as they occur when the egg divides after fertilization has already happened. This makes it easier for twins to get away with murder!

Hang 'em both!

Just think if you had criminal tendencies and you were an identical twin. No one would know whodunit or would they take a less forgiving approach?

The fragments of different lengths are made by cutting the DNA from the sample with restriction enzymes (see Ch. 9). Consequently, the variation in the size of the fragments and therefore of their positions in the fingerprint pattern is due to differences in where cutting occurs. Such variations are known as <u>R</u>estriction <u>F</u>ragment <u>L</u>ength <u>P</u>olymorphisms or RFLPs (see Ch. 16).

DNA Fingerprinting - The Procedure

The steps involved in DNA fingerprinting are as follows:

1) Restriction enzymes are added to cut the DNA. Restriction enzymes (see Ch. 9) will cut DNA into fragments. These are very trustworthy enzymes and snip specific sequences of nucleotides every time they see them. There are many different restriction enzymes, most with unique cutting properties. The length of each fragment, and thus its molecular weight, depends on the location of the cut sites recognized by whichever restriction enzyme is being used. It is thought that there is one difference in every 1,000 nucleotides between non-related individuals.

For example, the restriction enzyme BamHI recognizes a six base pair sequence (at left) within any DNA double helix and cuts it as shown (right).

5'- GGATCC -3'   **5'- G   GATCC -3'**
3'- CCTAGG -5'   **3'- CCTAG   G -5'**

2) Separate fragments of DNA by gel electrophoresis (Fig. 18.7). This technique separates DNA fragments by their size or molecular weight (for details see Ch. 16). You may have many restriction fragments spread out on your gel, but they will be invisible.

18.7 GEL ELECTROPHORESIS OF RESTRICTION ENZYME CUT DNA

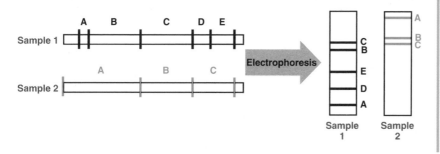

3) Visualize the fragments by Southern blotting. First transfer the separated fragments from the gel to nylon paper (Fig. 18.8). Next use a radioactively labeled DNA probe and see if it binds to any of the fragments (see Ch. 16 for labeling). The probe will bind with loving kindness to a DNA fragment if, but only if, the fragment has a DNA sequence complementary to the probe.

18.8 TRANSFER OF FRAGMENTS FROM GEL TO NYLON PAPER

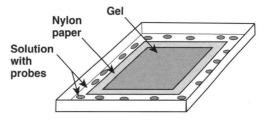

Even if mutations have changed a small percentage of the target sequence, there will usually still be enough similarity for binding to occur.

4) Make an autoradiograph by covering the blot with radiation sensitive film (Fig. 18.9). This will show the location of the DNA fragments that reacted with the radioactive probe.

18.9 AUTORADIOGRAPHY

Gel ⟶ Autoradiograph

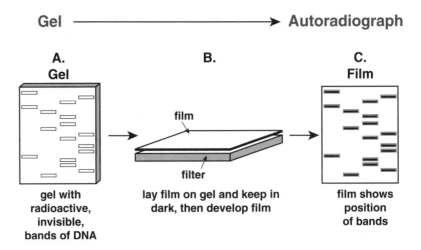

A.
Gel

B.

C.
Film

film

filter

gel with radioactive, invisible, bands of DNA

lay film on gel and keep in dark, then develop film

film shows position of bands

If a cutting site has been altered by changes in base sequence between two different people, a longer or shorter fragment will result and its location on the gel will change accordingly. This gives a Restriction Fragment Length Polymorphism (RFLP). The binding of the probe to a DNA fragment is expected, but where the fragment will appear on a gel after Southern blotting is variable. In practice, several different restriction enzymes are used and the positions of the fragments are compared for different people. RFLP patterns from different people can vary a lot. With luck it takes about six weeks to complete a DNA "fingerprinting" test.

18.10 FIVE ESSENTIAL LANES IN A DNA FINGERPRINTING AUTORADIOGRAPH

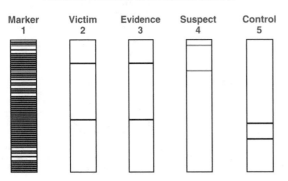

Marker 1   Victim 2   Evidence 3   Suspect 4   Control 5

How DNA Fingerprints Look

The final product of a DNA fingerprint is an autoradiograph that contains at least five essential lanes (Fig. 18.10).

The markers are standardized DNA fragments of known size which have been radiolabeled. The "control" is DNA from a source known to react positively and reliably to the DNA probes and shows if the test has worked as expected. We also need samples from the victim, the defendant and the crime scene. Since in this case the bands of the evidence and the suspect do not match on a horizontal row, the evidence DNA is not from the suspect. The evidence sample actually matches that of the victim.

An autoradiograph used in a criminal trial is shown below (Fig. 18.11). The blood-spattered clothing from a defendant charged with a murder was examined for RFLPs. The DNA from these blood stains did not match his own blood but that of the victim. In this instance, calculation of the frequency of a coincidental match gave one in 33 billion.

18.11 AUTORADIOGRAPH USED IN A CRIMINAL TRIAL

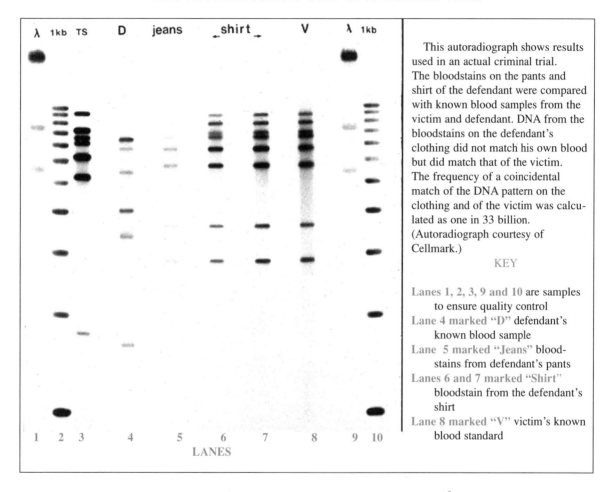

This autoradiograph shows results used in an actual criminal trial. The bloodstains on the pants and shirt of the defendant were compared with known blood samples from the victim and defendant. DNA from the bloodstains on the defendant's clothing did not match his own blood but did match that of the victim. The frequency of a coincidental match of the DNA pattern on the clothing and of the victim was calculated as one in 33 billion. (Autoradiograph courtesy of Cellmark.)

KEY

Lanes 1, 2, 3, 9 and 10 are samples to ensure quality control
Lane 4 marked "D" defendant's known blood sample
Lane 5 marked "Jeans" bloodstains from defendant's pants
Lanes 6 and 7 marked "Shirt" bloodstain from the defendant's shirt
Lane 8 marked "V" victim's known blood standard

Using Repeated Sequences

A variation of DNA fingerprinting is to look at regions of the DNA that contain variable number tandem repeats (VNTRs). This gobble-de-gook means that short sequences of DNA are repeated over and over and over and over and over and over, but that different people have different numbers of repeats. VNTRs usually occur in non-coding regions of DNA. They are visualized by using restriction enzymes to cut out the DNA segment containing the VNTR, followed by Southern blotting, as before.

variable number tandem repeats (VNTRs) sequences that are repeated multiple times at one location on the DNA and where the number of repeats varies from one individual to another

In Figure 18.12, a restriction enzyme is used to cut out a fragment of DNA containing a VNTR. Shown are DNA fragments from three individuals who differ in the number of repeats. Consequently the length of the fragment differs from person to person. Upon electrophoresis, these related fragments will move at different speeds through a gel and will end up in three different positions. The long pieces of DNA will move the least distance.

18.12 RESTRICTION ENZYME PRODUCES FRAGMENTS OF DIFFERENT LENGTHS IN DIFFERENT INDIVIDUALS

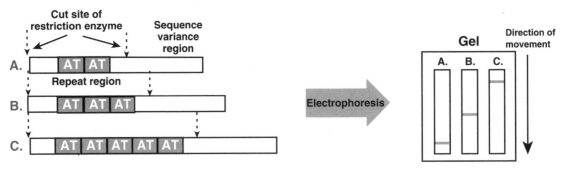

There is enormous variation between people in the number of repeats at any particular VNTR site in the DNA. So there is a very low frequency of two people matching exactly; or, if you prefer, a high probability they will differ. Some VNTRs have 100 to 200 different variants, making them very useful for forensic analysis. Although VNTRs are not genuine genes, their variants are often called alleles. Big words for this kind of complex system are "multi-allelic" or "hypervariable." The practical problem here is there may be so many closely packed bands that they cannot be easily told apart.

multi-allelic or **hypervariable** when a gene or site on the DNA has many variations or multiple alleles

a picogram (10^{-12}g) is one million millionth of a gram and a microgram (10^{-6}g) is one millionth of a gram; a gram is roughly one twenty-eighth of an ounce

Using the Polymerase Chain Reaction (PCR)

PCR is a procedure for amplifying tiny amounts of DNA and is used when there is too little DNA, or the DNA is too degraded for DNA fingerprinting. The details of PCR have already been discussed in the previous chapter (Ch. 17).

PCR machines can amplify a segment of DNA (100 to 3,000 bp long) in a few hours, starting from only a picogram (10^{-12}g), although microgram (10^{-6}g) quantities or larger are better. In fact, PCR can be used on DNA from a single cell such as an unfulfilled sperm cell. While DNA fingerprinting requires relatively long strands of DNA, PCR can be used on short segments of DNA. PCR is most useful for regions of the DNA with high individual variability. Small regions with high person to person variability are the best to amplify. If two samples match in several highly variable regions, they are probably from the same person.

Once the DNA from the forensic sample has been amplified, it is compared with DNA from the suspect, or suspects. Spots of both DNA samples are bound to a membrane and tested for binding to a DNA probe which is either radioactive or tagged with a fluorescent dye (see Ch. 16). The probe either binds or doesn't bind, so any spot is either positive or negative.

This kind of test is known as a dot blot (Fig. 18.13). Thus, the major difference is that DNA fingerprinting looks for differences in fragment sizes while PCR tests for the presence or absence of specific sequences.

Probability and DNA Testing

If two DNA samples are different, then they must have come from different people. Hence, DNA testing can readily exclude an individual from being suspected. But what if two DNA samples match for whatever tests we have run? To suggest that the individual was involved in a crime requires the use of probability. Exclusion means that the DNA pattern of the suspect and the evidence do not match. Inclusion depends on the probability of the DNA of the suspect and the DNA from the evidence matching being so great that it would not be expected to occur by chance in the general population, or more preferably, the ethnic subpopulation of the suspect, or even more preferably, the total population of the world.

The following general steps are important when determining the probability of a match:

1) From the same population of which the suspect is a member, select a random sample of individuals.
2) Determine the genotype of these randomly selected individuals and estimate the frequency of the alleles at the various loci.
3) Calculate the probability of the genotype of the suspect by assuming that this individual's alleles at each single locus represent a random selection from the population in general. (We also assume that alleles are not linked but independent of each other).
4) Multiply the frequencies that are determined from the various loci. The figure obtained represents the probability that the suspect's DNA would match the evidence.

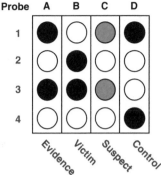

18.13 **THE DOT BLOT**

dot blot   a method used to recognize specific sequences of DNA with a tagged DNA probe

locus (plural, loci)   a site or location on a chromosome; it may be a genuine gene or just any site with variations in DNA sequence that can be measured, like RFLPs or VNTRs

Example: Suppose two loci, A and B, are examined. In one, the suspect was homozygous (a_1, a_1) and in the other he was heterozygous (b_1, b_2). The probability (p) for a homozygous match is given by $p(a_1)^2$ and for a heterozygous match by $2 p(b_1) \times p(b_2)$. Now assume the frequency of these alleles in the general population is 5 percent for a_1 and 10 percent each for b_1 and b_2. The overall frequency for this genetic combination in the population is calculated as follows:

$$p(a_1)^2 \times 2p(b_1)p(b_2)$$

$$[(0.05)^2 \times 2(0.10)(0.10)] = 0.000,05$$

$$\text{or 1 in 20,000}$$

So, using just two loci and alleles that are relatively rare, the probability that another person in the population has the same DNA characteristics is pretty low. Usually, three to five loci are examined to obtain odds that are much greater. In criminal proceedings frequencies of less than one in 100,000 are usually obtained and in many cases the frequencies were less than one in the total world population.

Some suggest that with the RFLP technique we have surpassed the level of mere probability and are now capable of absolute identity

Interestingly, convictions have been obtained using DNA evidence where the probability of a chance match was one in 100, but with the addition of supporting evidence. However, in cases where the evidence is primarily based on DNA testing, juries are more and more expecting astronomical odds such as one in a million or billion. Such was the case in the O. J. Simpson trial when a match between the evidence and O.J.'s blood was very probable, but the jury decided to ignore the blood evidence claiming instead that there was a police conspiracy. We should be cautious if close relatives are suspects in criminal proceedings since, for example, the probability that brothers may match each other is much greater than for the general population.

DNA Evidence and Convictions

The main impact of DNA technology has been the far greater certainty with which individuals can be associated with or excluded from a particular crime than was possible with traditional blood tests. Experience has shown that if DNA testing is given as evidence, there is a higher probability of conviction than if DNA testing is not used. DNA evidence is commonly used in cases of rape. However, in most cases of rape, the accused admits knowing the alleged victim and identity is not an issue. DNA testing can also be used by law enforcement to narrow the number of possible suspects, given that sex and racial characteristics can be determined by simple DNA testing protocols.

The most common criticism of DNA fingerprinting is the way in which statistics are used and the nature of the population from which the standard samples were taken. A panel from the U.S. National Academy of Sciences recommended that a more conservative statistical approach should be taken which places a ceiling on the odds that can be obtained by the match.

phosphoglucomutase (PGM) an enzyme present in blood whose different genetic forms can be easily analyzed

The British judicial system regards DNA testing highly. In one particular case, DNA testing was used to exclude the person first suspected of the sexual assault and murder of a young girl. But to find the real perpetrator, the police screened over 5,000 men in the village by blood testing (ABO and PGM or phosphoglucomutase), only to find no match with anyone. Ironically, the murderer was discovered because it was revealed that he had paid another man to give blood for testing. DNA testing subsequently confirmed that his DNA matched, to a high probability, that of the semen sample taken from the victim. A conviction was finally obtained.

In general, there is considerable public support in Britain to maintain DNA profiles on the entire population.

Admissibility of DNA Evidence

Criminals who did not attend law school may think that the Frye test means they will be examined for survival in the electric chair

The *Frye* test (Frye vs. United States, 1923) has nothing to do with the electric chair. The principle states that new scientific tests must be generally accepted in appropriate scientific circles before evidence from them is admissible in courts. In addition to the Frye rule, a "helpfulness" standard is applied in some states which involves the use of expert witnesses to assist the court in interpreting facts from scientific evidence. Recent court cases have almost all allowed DNA testing to be admitted into evidence, although there have been a few notable exceptions. By 1996, DNA evidence had been admitted in more than 2,500 criminal cases in the USA.

In the United States, there are presently few government labs doing DNA testing. Most forensic work is performed by accredited private labs and these services are available to both the prosecution and defense. All you have to have is $$$$.

Ethics and Forensic Genetics

Can DNA information be misused? The answer of course is yes, any information can be abused. People with different racial or genetic characteristics have been persecuted in the past. What reason is there to believe they won't be discriminated against in the future? Then again, RFLPs and VNTRs are hardly necessary for identifying people by race!

So, is DNA information being abused today? Most scientists believe the potential for abuse is low at present as we know so little about the human genome. Still, let's be brave and look into the future. If the genes being examined have medical, psychiatric, or social effects, this knowledge might be used to violate a person's privacy. If the sole purpose of genetic testing is for forensics, then there should be safeguards to limit the use of DNA testing to this and to keep the information confidential. We only need to listen to the current controversy over AIDS testing to see the potential for infringement of privacy as well as the fear of some people that their privacy will be violated. Imagine the hunger of health insurance companies to obtain genetic data that would let them predict the future health prospects of their applicants.

Since DNA is a blueprint of our identity, analysis of our genes could detect traits that affect our future. Privacy is an individual's own right to decide what should be known about oneself. Just as there has been a reluctance to issue a national identity card, there is a reluctance to use DNA data banks. It has been proposed that **non-coding regions** of the DNA should be used in forensics to avoid invading privacy. In fact, the very nature of repetitive sequences means that most VNTRs are in non-coding DNA.

non-coding regions
DNA sequences that do not code for proteins or functional RNA molecules

A national DNA data bank is presently being maintained by the FBI and is often screened by computer searches to find suspects. Although the paranoids among us shriek hysterically "Big Brother is coming!" those who feel that liberty includes the freedom to walk down the street without being assaulted may well consider this a positive innovation.

So much for justice. In the next chapter we will discuss creatures who practice mindless slaughter and care neither for truth nor justice, let alone the American way - the viruses!

Additional Reading

DNA Technology and Forensic Science by Ballantyne J, Sensabaugh G, and Witkowski J. 1989 Vol. 32 Banbury Report, Cold Springs Harbor Laboratory Press.

DNA Technology in Forensic Science. 1992 National Academy Press, Washington, D.C.

Human Genetics: The Molecular Revolution by McConkey EH. 1993. Jones & Bartlett Publishers, Boston & London.

Gene Creatures, Part 1: Viruses, Viroids and Plasmids

19

Dead or Alive?

Trying to define precisely what is living and what is non-living can be quite confusing. Take viruses for instance. Viruses can only multiply when they have entered a suitable host cell and taken over the cellular machinery. On the one hand, a virus cannot make its own proteins or generate its own energy. On the other hand, it is certainly not inert; it does replicate if it can subvert a host cell. So, is a virus truly alive?

Whatever answer you give to the question above, in practice viruses are discussed in biology books! In this chapter we will explore the twilight zone of biology and the **gene creatures** who live there. A whole range of such "subcellular life forms" or "genetic elements" exists. These gene creatures carry their own genetic information but they are not living cells. In order to replicate, gene creatures need to infiltrate the cell of another organism and use host cell facilities.

Gene creatures vary in their level of independence from viruses, which can survive outside the host cell as virus particles, down to mere stretches of DNA or RNA that replicate according to their own whims, yet are never found outside the cell where they live (Fig. 19.1). Many of these gene creatures carry out bizarre genetic maneuvers which would be regarded as "illegal" for a normal cell.

gene creature life-form consisting primarily of genetic information, sometimes with a protective covering, but without its own machinery to generate energy or replicate macromolecules

19.1 GENE CREATURES IN THE TWILIGHT ZONE

297

Viruses are Not Living Cells

Perhaps we should start by noting that being alive and being a living cell are not necessarily the same. We also need to define a living cell.

Multiple Choice: Which of the following apply to a living cell?
a) It contains genetic information as DNA
b) It uses RNA as a genetic messenger
c) It is capable of making its own proteins
d) It generates its own energy
e) It is surrounded by a cell membrane

Answer: ALL of the above. In other words, to qualify as a genuine cell you must send genetic messages (RNA) from your genes (DNA) to your own ribosomes to make your own proteins with energy you generate yourself (Fig. 19.2).

Virus particles do contain genes, either as DNA or RNA, although they do not contain both. However, many viruses do use both DNA and RNA during their replication cycle inside the host cell. Viruses flunk parts (c) and (d) of the above test as they are parasitic and rely on the host cell to provide both ribosomes and energy. As for (e), some virus particles are surrounded by genuine membranes, made mostly of material which they filched from the host cell, but many simple viruses have only a protein shell and no true membrane. Nonetheless, virus particles do have an outer covering and can survive on their own outside their host cells (admittedly without multiplying).

19.2 A MINIMAL LIVING CELL

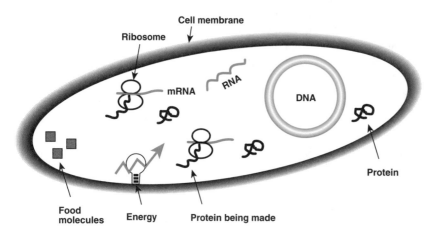

Cell membrane · Ribosome · mRNA · RNA · DNA · Protein · Food molecules · Energy · Protein being made

What Defines a Virus?

Viruses are all parasites that cannot multiply without a host cell. Furthermore, viruses are intracellular parasites; that is to say that they must actually enter the cells of the host organism to replicate. A virus alternates between two forms, an inert virus particle, the virion, which survives outside the host cell, and an active intracellular stage.

The virus particle consists of a protein shell, known as a capsid, surrounding a length of nucleic acid, either RNA or DNA, which carries the virus genes (Fig. 19.3). Many simple viruses have only these two components.

19.3 SIMPLE RNA AND DNA VIRUS PARTICLES

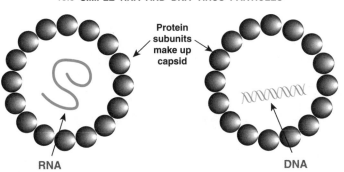

Protein subunits make up capsid

RNA   DNA

Note that although all viruses are parasites, not all parasites are viruses. A parasite is any life form that lives at the expense of another. Parasitic members are found in most groups of living things from disease causing bacteria to politicians

The nucleic acid with the genetic information is often referred to as the **viral genome**.

The life cycle of a typical virus has the following stages:
a) attachment to the correct host cell
b) entry of the virus genome
c) replication of the virus genome
d) manufacture of the virus proteins
e) assembly of new virus particles
f) release of new virus particles from the host cell.

Attachment of a virus means that a protein on the virus particle must recognize a molecule on the surface of the target cell. Sometimes this receptor is another protein, sometimes it is a carbohydrate. On some virus particles the recognition proteins form spikes or prongs sticking out from the surface (Fig. 19.4).

Many animal viruses have an extra envelope outside the protein shell (Fig. 19.5). This is made of membrane stolen from the previous host cell into which virus proteins have been stuck. These virus encoded proteins function to detect and bind to the next target cell.

viral genome the nucleic acid, either DNA or RNA, that carries the genetic information of a virus

19.4 RECOGNITION OF RECEPTOR BY VIRUS

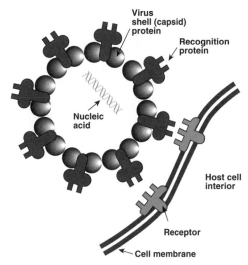

Virus shell (capsid) protein

Recognition protein

Nucleic acid

Host cell interior

Receptor

Cell membrane

19.5 ENVELOPED VIRUS PARTICLE

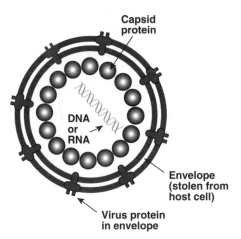

Capsid protein

DNA or RNA

Envelope (stolen from host cell)

Virus protein in envelope

When an enveloped virus enters a new animal cell, its envelope layer merges with the cell membrane and the inner protein shell containing the nucleic acid enters (Fig. 19.6). Once inside, the protein shell disassembles, exposing the genome.

19.6 ENVELOPED VIRUS PARTICLE ENTERS ANIMAL CELL

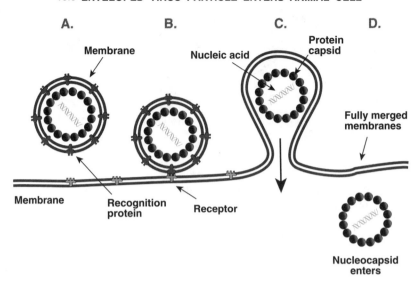

Bacteria have a cell wall protecting their cell membrane and so bacterial viruses cannot simply merge with the membrane as do animal viruses. Therefore, bacterial viruses do not bother with an outer envelope layer. They just have a protein shell surrounding the DNA or RNA. After binding to the cell surface they inject their nucleic acid into the bacterial cell and the outer protein coat of the virus particle is left behind (Fig. 19.7).

19.7 VIRUS PARTICLE ABANDONS COAT UPON ENTERING CELL

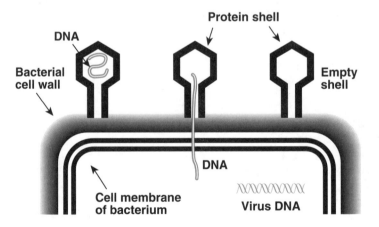

Once inside the host cell, the virus genome has two major functions. First, it must replicate to produce more virus genomes. Second, it must persuade the cell to manufacture lots of virus proteins so that new virus particles can be assembled. Note that viruses do not divide like cells. They are assembled from components manufactured by the host cell (Fig. 19.8).

19.8 VIRUS SUBVERTS CELLULAR MACHINERY

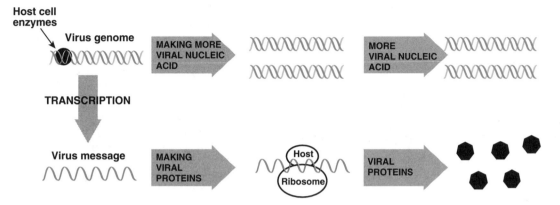

The Great Diversity of Viruses

In practice there is colossal variation in the structure of viruses and the detailed way in which they take over the cells they invade. As remarked above, some have DNA genomes while others have RNA. Furthermore, the nucleic acid may be either single or double stranded and either linear or circular. All of these possibilities exist (Fig. 19.9), though some are more common than others. Some viruses even have segmented genomes made up of several pieces of DNA or RNA.

The smallest viruses have only three genes, the largest have two or three hundred and can carry out some very slick maneuvers to outwit their host cells. Viruses have been found that attack animal cells, plant cells and bacterial cells. So no one is safe! Here we will discuss in detail a few of the most fashionable viruses. Note that bacterial viruses are often known as "bacterio-phages" from Greek meaning "bacteria eaters." For the sadists among you, some gruesome viral diseases afflicting humans are discussed in Chapter 21.

19.9 POSSIBILITIES FOR VIRUS GENOME

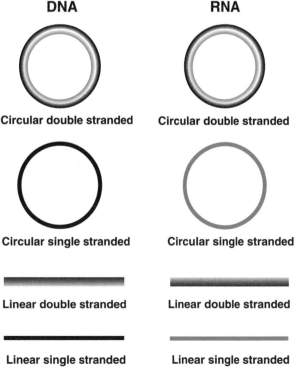

bacteriophage any
virus that infects bacteria

bacteriophage Qβ a
small spherical bacterial
virus containing single
stranded RNA and which
only infects male *E. coli*

Bacterial RNA Viruses Have Very Few Genes

A back-to-nature minimalist virus can get by with only three genes (Fig. 19.10): one to encode its protein coat, one to replicate its genome and one to burst the host cell so the newly made virus particles can get out. An example is the bacteriophage Qβ which has approximately 3,500 bases of single stranded RNA and infects the bacterium *Escherichia coli*.

Qβ and its relatives, such as MS2, are small spherical viruses with single stranded RNA and only three or four genes. They are "male-specific," which is to say they only infect bacteria carrying the F-plasmid (*i.e.*, male bacteria) because they attach to the sex pilus which is only found on the surface of F$^+$ cells (see Ch. 8). Very few bacterial viruses have double stranded RNA.

Bacterial Virus ΦX174 - A Small Single Stranded DNA Virus

Bacteriophage ΦX174 is a small simple virus which contains 5,386 bases of circular single stranded DNA. The virus is spherical in shape with protein spikes that recognize the receptors on the surface of its bacterial host cell. It looks rather like a World War II naval mine.

19.10 **VIRUS NEEDS ONLY THREE GENES**

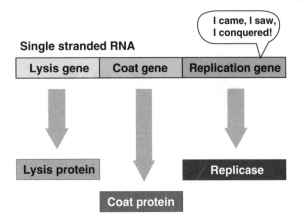

**male-specific bacterio-
phage** perhaps due to
their low information content, some bacterial viruses, such as Qβ or M13, are
feminists and only infect
male bacteria, *i.e.*, in the
particular case of *E. coli*
those bacteria containing
the F-plasmid which con-
fers the ability to conjugate

bacteriophage ΦX174
a small spherical bacterial
virus containing single
stranded DNA which
infects *E. coli*

eicosahedron solid
shape with 20 faces

19.11 **ΦX174 – A SMALL SPHERICAL VIRUS**

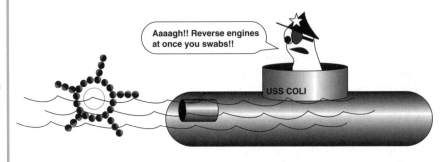

In fact, so-called spherical viruses are not truly spherical. They are actually geodesic dome structures. They have 20 triangular faces and are, strictly speaking, eicosahedrons. If you slice through an eicosahedron, the cross section is five- or six-sided depending on where you cut. At each of the 12 vertexes is a spike made of two different proteins involved in recognizing the host cell.

The most bizarre property of this teensy weensy parasite is that it has so little DNA that five of its 11 genes overlap others! For example, gene E is completely inside the DNA for gene D. Genes D and E are read in two different reading frames so they produce two totally different proteins (Fig. 19.12). D-protein helps in assembling the virus capsid, though it does not form part of the final structure, and E-protein destroys the cell wall of the host bacterium to allow the newly made virus to get out. A mutation in the DNA for gene E will also alter gene D. Although overlapping saves on DNA, the two genes are no longer free to evolve separately. Although overlapping genes are quite often found in small viruses, they are only found in real cells under exceptional circumstances.

19.12 ΦX174 HAS OVERLAPPING GENES

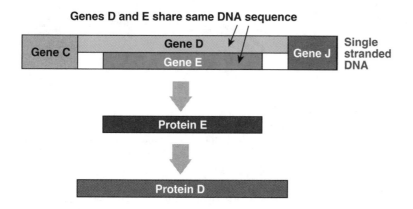

Some other bacterial viruses are filamentous rather than spherical. For example, bacteriophage M13 has single-stranded circular DNA like ΦX174 but the virus particle is a long thin filament. M13 is "male-specific" and only infects male bacteria carrying the F-plasmid (Ch. 8). M13 is most unusual in being released from the host cells without killing them and is used in molecular biology for sequencing DNA (see Ch. 23).

Complex Bacterial Viruses with Double Stranded DNA

These viruses all have a complex form made up of a head, tail, and tail fibers. The head of the virus particle contains a large molecule of linear dsDNA. They include bacteriophages T4, Lambda, P1 and Mu which are all used in bacterial genetics and molecular biology (see Ch. 8). The number of genes ranges from Mu with approximately 40 genes to T4 with nearly 200. T4 and its close relatives are some of the most complex types of virus known.

The head of the virus particle is more or less spherical in Mu but elongated to accommodate all the extra DNA in T4. Attached to the head is a tail with tail fibers that act as landing legs (Fig. 19.13). These viruses bind to bacterial cells by means of recognition proteins on the end of their tail fibers. After setting down like lunar landers, their tails contract and they inject their DNA like miniature hypodermic syringes.

bacteriophage M13 a small filamentous bacterial virus containing single stranded DNA that only infects male *E. coli* and is used to make single stranded DNA for sequencing

bacteriophage Lambda or λ a bacterial virus that infects *E. coli* and which is used both as a vector for cloning and in transferring bacterial genes by specialized transduction

bacteriophage P1 a bacterial virus that infects *E. coli* and that is used by geneticists to transfer bacterial genes by a mechanism known as transduction

bacteriophage Mu or μ a virus which infects *E. coli* and which causes mutations by inserting its DNA into the bacterial chromosome

Plant Viruses

Sad to say, even humble plants have their problems. Plant viruses containing DNA are relatively rare. One example is cauliflower mosaic virus which has circular DNA inside a small spherical shell and kills cauliflower and its relatives such as cabbages and Brussels sprouts - no great loss to the world! The promoters of genes from this virus have been used in plant genetic engineering to express insect killing toxins (see Ch. 15).

cauliflower mosaic virus   a double stranded DNA virus that infects cauliflower and related plants

tobacco mosaic virus a single stranded RNA virus that infects many plants, including tobacco; the most noticeable symptom is a mosaic of diseased blotches on the leaves

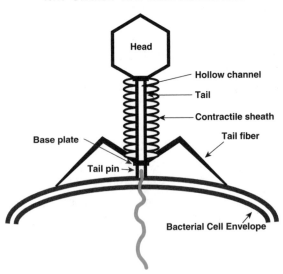

19.13 SYRINGE - LIKE VIRUS INJECTS DNA

- Head
- Hollow channel
- Tail
- Contractile sheath
- Tail fiber
- Base plate
- Tail pin
- Bacterial Cell Envelope

Most plant viruses are small with just a handful of genes and contain single stranded RNA. Some like cucumber mosaic virus are spherical. Others are rod-shaped, like tobacco mosaic virus, the most widespread plant virus, which attacks lots of plants including vegetables like the tomato, pepper, beet and turnip as well as tobacco. It makes yellowish blotches on the leaves (Fig. 19.15), hence, "mosaic."

The virus coat consists of 2,130 identical protein molecules arranged in a helix with the RNA in the center (Fig. 19.16). TMV and its relatives have around 10,000 bases of RNA which is enough for about 10 genes.

19.14 CAULIFLOWER MOSAIC VIRUS

Let's kick some veggie butt!

aaaaarrrrgh!! help! help!

19.15 MOSAIC VIRUS MAKES BLOTCHES ON LEAVES

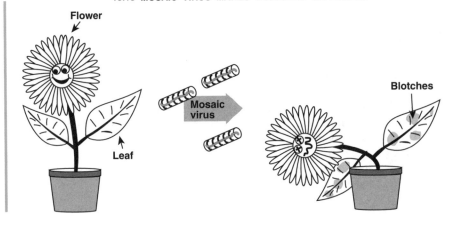

- Flower
- Leaf
- Mosaic virus
- Blotches

19.16 TOBACCO MOSAIC VIRUS STRUCTURE

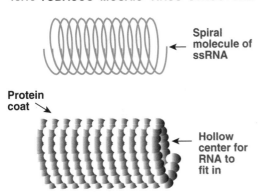

Spiral molecule of ssRNA

Protein coat

Hollow center for RNA to fit in

Viroids

Let's start with a virus whose genes are made of RNA. Now we take away its protein coat. We are left with a naked piece of RNA, shivering miserably in the cold, cheerless world - a viroid.

Viroids are small circular pieces of naked RNA which are only 250 to 400 bases long (Fig. 19.18). They are single stranded but base pairing occurs between bases on opposite halves of the circle to produce a rod-like structure.

19.17 VIROID SHIVERING IN THE COLD RAIN

At least I'm too small to catch cold!

19.18 VIROID RNA

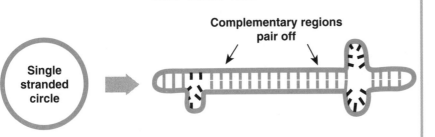

Complementary regions pair off

Single stranded circle

Rod shaped viroid

Viroids infect plant cells where they are replicated at the expense of the host cell. The viroid with the coolest name is coconut cadang-cadang viroid, which has only 246 bases worth of RNA (Fig. 19.19 next page). The viroid RNA does not even have any genuine genes that encode proteins, it merely carries signals for its own replication by the host machinery. Although the viroid codes for no protein enzymes, the viroid RNA is probably catalytically active itself, a ribozyme. Are viroids perhaps self-splicing introns that escaped (see Ch. 24)?

Because viroids have no protein coat, they cannot recognize and penetrate healthy cells as can a true virus. Viroids prey only on the weak and injured; they can only infiltrate a cell when its surrounding membrane is already damaged.

coconut cadang-cadang viroid   a particular viroid which infects coconut trees

ribozyme   RNA molecule which acts as an enzyme

19.19 COCONUT CADANG - CADANG VIROID

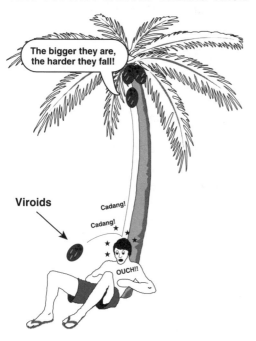

The bigger they are, the harder they fall!

Viroids

Cadang!
Cadang!

OUCH!!

this definition is no longer absolutely true. Recently some linear plasmids have been discovered, not in eukaryotic cells as you might have imagined, but in certain weird bacteria

Plasmids

Like viroids, **plasmids** are also naked molecules of nucleic acid without a protein coat. However, they normally stay inside cells and rarely cause disease. Plasmids are circular molecules of double stranded DNA that are separate from the chromosomes of a cell and are not needed for growth and division of the host cell, at least under normal conditions. Although they rely on the cell both for the machinery of DNA replication and the energy to make their own DNA, they are regarded as autonomous because they control their own replication (Fig. 19.20).

Plasmids were used first by researchers in bacterial genetics and have been promoted to become star players in molecular biology. They are now used to carry cloned genes and other engineered DNA (Ch. 9).

In Chapter 8 we talked about plasmids from the viewpoint of gene transfer in bacteria, as if plasmids were mere sex objects! However, a more sensitive approach is to consider plasmids as creatures in their own right. Just as fish live in the sea and monkeys hang out in the forest, so plasmids live inside their host cells. To a plasmid, the cell it inhabits is its warm, cozy home.

19.20 PLASMIDS ARE AUTONOMOUS CIRCLES OF DNA

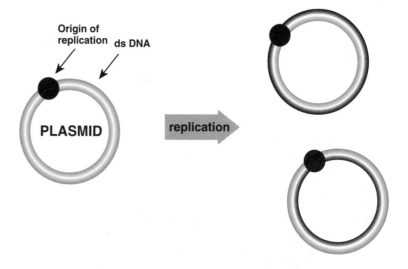

Origin of replication ds DNA

PLASMID

replication

Some cells may have plasmids living inside them, others may not. So, although the plasmid is not alive in the same sense as a cell, it is not just part of the cell. In some ways plasmids are like domesticated viruses that have lost the ability to move from cell to cell killing as they go. If you need to get sentimental and slushy, plasmids are like domestic dogs (Fig. 19.21), whereas viruses are more like wild jackals or wolves.

Plasmids resemble viruses in requiring a host cell. The plasmid genome is replicated by the host cell machinery and at the cell's expense in energy and raw materials. Unlike viruses, plasmids do not possess protein coats and since they cannot leave the cell they live in, they do not vandalize it. Viruses usually destroy the cell they are replicating in and are then released as virus particles to go in search of fresh victims. Plasmids replicate in step with their host cell (Fig. 19.22). When the cell divides, the plasmid divides and each daughter cell gets a copy of the plasmid.

Plasmids Help Their Host Cells

If plasmids are not an essential part of the cell, why do cells put up with them? Then again, why do people keep dogs? Many dogs are useful to their human partners. Guard dogs keep watch, sheep dogs round up sheep, and guide dogs lead the blind. Other dogs are useless ornaments,

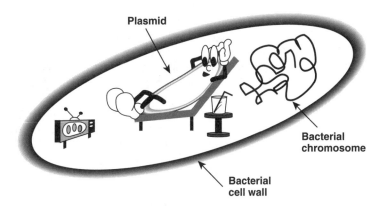

19.21 **PLASMID AT HOME**

Plasmid

Bacterial chromosome

Bacterial cell wall

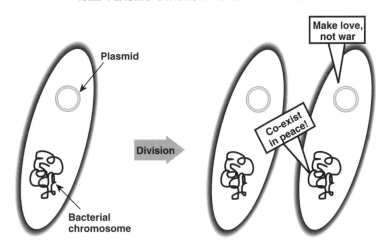

19.22 **PLASMID DIVIDES IN STEP WITH HOST CELL**

Plasmid

Bacterial chromosome

Division

Make love, not war

Co-exist in peace!

19.23 **PEKING DOG ONCE USED AS FOOD**

MMMM Peking Dog!

like poodles, or merely irritating like Pekinese. Although its podgy, yipping, western descendants have evolved to resemble their owners, the Pekinese was originally bred by the Chinese to cook and eat. Plasmids are much the same. Most provide useful properties to their host cells while a few are just useless, welfare molecules.

The first plasmids to be discovered were found living in

bacteria in Japan. The bacterium, *Shigella*, causes dysentery (see Ch. 8) and was being treated with the antibiotic sulfonamide. Then, suddenly, strains of *Shigella* appeared that were resistant to sulfonamide treatment. The genes providing resistance to sulfonamide were carried on a plasmid. What's more, this particular plasmid was able to transfer copies of itself from one bacterial cell to another (see below) so that the sulfonamide resistance spread rapidly from *Shigella* to *Shigella*. Although this is no doubt a good thing from the viewpoint of the *Shigella*, it is extremely dangerous from a human medical viewpoint.

Plasmids often protect bacteria from human medicine by carrying genes for resistance to antibiotics (Fig. 19.24). They also protect bacteria from industrial pollution by carrying genes for resistance to toxic heavy metals such as mercury, lead or cadmium. Some plasmids provide genes allowing bacteria to grow by breaking down herbicides, certain industrial chemicals or the components of petroleum. Such bacteria are sometimes a nuisance but may be useful in cleaning up oil spills or other chemical pollution.

19.24 PLASMID PROTECTS BACTERIA AGAINST ANTIBIOTICS

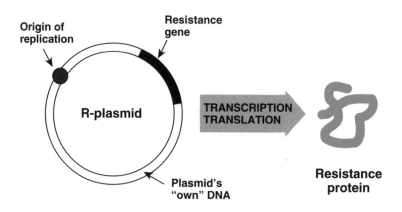

Some plasmids provide their bacterial owners with aggressive weaponry. Bacteriocin plasmids allow bacteria to kill other bacteria (see Ch. 20). Virulence plasmids help bacteria infect humans, animals or even plants, by a variety of mechanisms. Some virulence factors are toxins that kill animal cells, others help bacteria to invade animal cells, whereas others protect bacteria against retaliation by the immune system. The Ti-plasmid carried by soil bacteria of the *Agrobacterium* group confers the ability to infect plants and produce tumors, inside which the bacteria grow and divide happily. The Ti-plasmid and its use in plant genetic engineering are discussed in Chapter 15.

Plasmid DNA Replicates by Two Methods

Plasmids use two mechanisms for replicating their DNA, though not both at once!! Most plasmids replicate like miniature bacterial chromosomes.

They have an origin of replication where the DNA opens and replication begins. Then two replication forks move around the circular plasmid DNA in opposite directions until they meet (Fig. 19.25; see Ch. 5 for details of replication forks). A variation on this theme is that some very tiny plasmids have only one replication fork which moves around the circle till it gets back to the origin.

rolling circle replication mechanism of replicating double stranded circular DNA that starts by nicking and unrolling one strand and using the other, still circular strand, as a template for DNA synthesis

19.25 BIDIRECTIONAL REPLICATION OF PLASMID

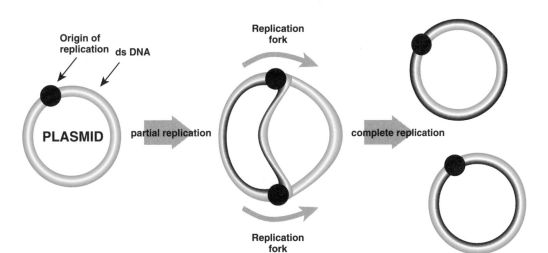

The other replication mechanism is known as "rolling circle" replication. The rolling circle mechanism is shared by some plasmids and quite a few viruses. At the origin of replication, one strand of the double-stranded DNA molecule is nicked (Fig. 19.26). The other, still circular strand, starts to roll away from the broken strand. This results in two single stranded regions of DNA, one belonging to the broken strand and one that is part of the circular strand.

DNA is now synthesized starting at the end of the broken strand, which is therefore elongated (Fig. 19.27). The circular strand is used as a template and the gap left where the two original strands

19.26 ROLLING CIRCLE STARTS TO ROLL

19.27 ROLLING CIRCLE CONTINUES TO ROLL

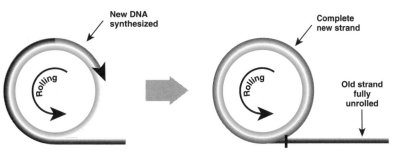

rolled apart is filled in. This process of rolling and filling in continues. Eventually the original broken strand is completely unrolled and the circular strand is all paired with a newly made strand of DNA.

We now have a single strand of DNA, equal in length to the original DNA circle, hanging loose. What happens next varies, depending on the circumstances.

Some plasmids, such as the famous F-plasmid, can transfer themselves from one bacterium to another. Such plasmids have two separate origins of replication. They divide by bidirectional replication when their host cell divides but use the rolling circle mechanism if they move from one cell to another. During plasmid transfer, the broken strand moves from the original host cell into another bacterial cell. Only after the broken strand has entered the new host cell is its opposite strand made. (The details of plasmid transfer are described in Ch. 8.)

In some ways, this transfer process is similar to virus infection, except that the two bacteria must form a bridge as the plasmid cannot survive outside its host cell. Perhaps when plasmids move from cell to cell like this, they are dreaming of the good old days when they used to be viruses and could travel freely around killing their host cells!

Rolling circle replication is also used by many viruses (Fig. 19.28). Some want to manufacture lots of double stranded molecules of virus DNA.

19.28 ROLLING CIRCLE VIRUS FACTORY

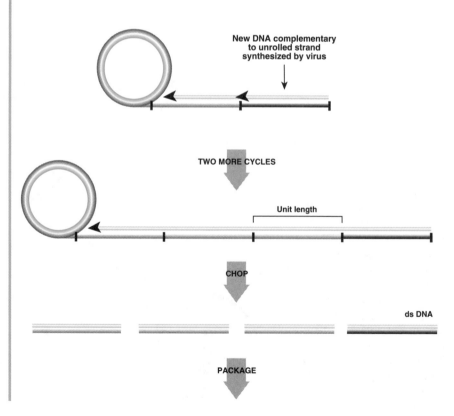

These viruses use the dangling strand as a template to synthesize a new strand of DNA. They just keep rolling and synthesizing and end up with a long linear double stranded DNA many times the length of the original DNA circle. This is chopped into unit lengths and packaged into virus particles. (Some of these viruses convert the DNA into circles before packaging, whereas others package linear DNA and only circularize their DNA after infecting a new cell when it is time to replicate again.)

Other viruses contain single stranded molecules of virus DNA. These viruses leave the unpaired, dangling strand alone. They just keep rolling and end up with a long linear single stranded DNA (Fig. 19.29). This is chopped into unit lengths and packaged as above. When these viruses infect a new cell, the first thing they must do is synthesize the opposite strand, so converting their single strand to a double stranded DNA molecule.

19.29 ROLLING CIRCLE MAKES SINGLE STRANDS

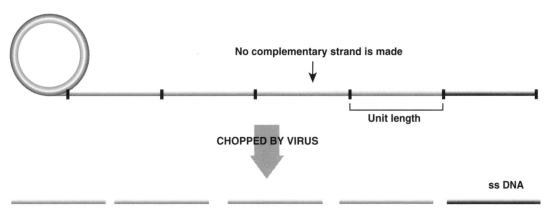

No complementary strand is made

Unit length

CHOPPED BY VIRUS

ss DNA

PACKAGED BY VIRUS

Lysogeny and Latency

There are other resemblances between the behavior of plasmids and viruses. In fact some schizoid circles of DNA can choose to live as either a plasmid or a virus. The bacterial virus P1 is a good example. It can indeed behave as a virus, in which case it destroys the bacterial cell, replicates by rolling circle mode and manufactures large numbers of virus particles to infect more bacterial cells (Fig. 19.30, next page). This is known as lytic growth since the host cells are "lysed," Greek for broken.

On the other hand P1 can choose to live as a plasmid. In this case the P1 DNA divides using bidirectional replication, and this only happens

lytic growth   mode of virus growth that results in destruction of the infected host cell

19.30 LYTIC GROWTH OF BACTERIOPHAGE P1

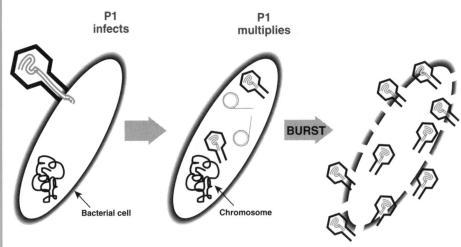

lysogeny mode of
virus growth in which
duplication of the virus
genome keeps in step with
division of the host cell
which is therefore not
destroyed

lysogen cell containing
a virus in harmless, lysogenic mode

when the host cell divides (Fig. 19.31). So, each descendant of the infected
bacterial cell gets a single copy of P1 DNA. The cell is unharmed and no
virus particles are made. This state is known as lysogeny and a host cell
containing such a virus in its plasmid mode is called a lysogen.

19.31 LYSOGENIC GROWTH OF BACTERIOPHAGE P1

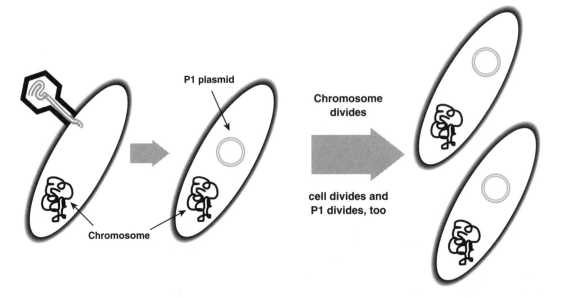

Changing conditions may stimulate a lysogenic virus to stop being
Dr. Jekyll, the plasmid, and go back to being Mr. Hyde, the virus. In particular, this tends to happen if the host cell is injured. The virus decides
that it may as well abandon ship and makes as many virus particles as

19.32 **P1 LEAVES THE SINKING SHIP**

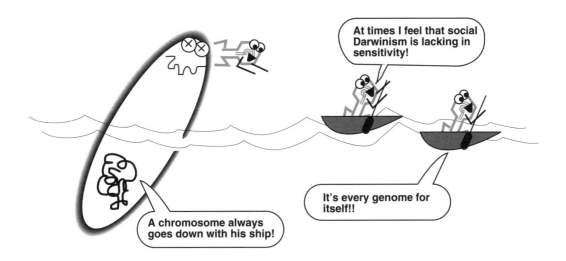

possible before the cell dies. If, on the other hand, the host cell is doing well, growing and dividing, the virus will decide its best bet is to go along for the ride and it stays in plasmid mode, a passenger on the train!

When a virus infecting humans or other higher organisms replicates its DNA in step with its host cell, it is called latency, rather than lysogeny. However, the principle is the same.

Herpesvirus (famous for cold sores and genital warts, etc.) is notorious for latent infections. The virus stays "hidden" in cells of the nervous system. It is difficult to detect because only one or two copies of virus DNA are present in each infected cell. No virus particles are made and for a while the patient shows no symptoms. Then, one day, the herpesvirus may decide to change into viral mode and symptoms of the disease emerge again. In the case of herpes, such outbreaks are often caused by "stress." In other words, cells of the nervous system have a rough time and the virus decides to quit while it's ahead, just as in the case of a lysogenic bacterial virus.

latency another name for lysogeny, normally applied to animal viruses

herpesvirus a family of DNA-containing animal viruses often causing latent infections

Lysogeny or Latency by Integration

Lysogeny, or latency, means that the virus has decided to divide in step with the host cell instead of killing it. It does not necessarily mean the virus has decided to live as a plasmid. Some cases of lysogeny or latency are caused by integration of the virus DNA into a host cell chromosome (Fig. 19.33, next page). Such an integrated virus is known as a provirus. The virus DNA becomes a physical part of the chromosome and is replicated when the chromosome divides.

The bacterial virus lambda (λ) operates like this. It recognizes a special sequence of DNA on the chromosome of its host cell, the bacterium *E. coli*, and integrates at this site, known as *att*λ (attachment for λ).

provirus form of a virus in which the viral DNA is integrated into the host chromosome

19.33 LYSOGENY / LATENCY BY INTEGRATION

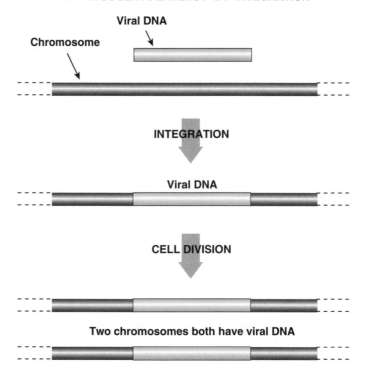

Some animal viruses insert themselves into the chromosomes of their host cells. Some have special sites, while others insert at random. Retroviruses which use RNA to carry their genes in the virus particle, must first make a DNA copy of themselves which they can then insert into the host chromosome (see Ch. 20).

Replicons

Replicons are molecules of nucleic acid which are in charge of their own destiny, at least in the sense of controlling their own replication. Chromosomes, plasmids, virus genomes (both DNA and RNA) and viroids are all replicons. Not all gene creatures are replicons; some, in particular the various kinds of **transposable elements**, cannot replicate themselves. By definition, replicons all contain an origin of replication where synthesis is initiated. Transposable elements do not have their own replication origin and are only duplicated when replication starting somewhere else on the same DNA molecule proceeds through them, as we shall see in the next chapter.

Additional Reading

Biology of Microorganisms by Brock TD, Madigan MT, Martinko JM, & Parker J. 8th edition, 1997. Prentice Hall, Englewood Cliffs, New Jersey.

Microbial Genetics by Maloy SR, Cronan JE, & Freifelder D. 2nd edition, 1994. Jones & Bartlett Publishers, Boston & London.

---

replicon molecule of DNA or RNA which contains an origin of replication

transposable element segment of DNA that can move from one location to another, but which always remains part of another DNA molecule

Gene Creatures, Part II: Jumping Genes and Junk DNA

20

As mentioned in the previous chapter, there is a whole slew of genetic elements lacking a replication origin of their own and which therefore do not qualify as replicons. Consequently they can only get replicated by integrating themselves into other molecules of DNA, such as chromosomes or plasmids. These genetic elements are not merely dependent on a host cell, they are dependent on a host DNA molecule! Some of them can move around from one host DNA molecule to another. These are the transposable elements or transposons (Fig. 20.1). Some of these have alter egos as viruses, whereas others only exist as chunks of integrated DNA.

20.1 REPLICONS VERSUS TRANSPOSONS

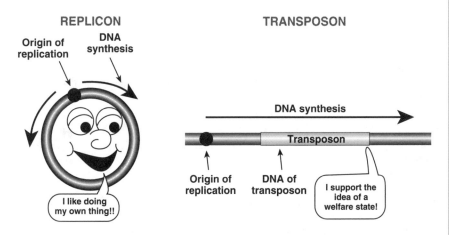

Other stretches of parasitic DNA are stuck permanently where they are and are probably the remains of once mobile gene creatures. They have degenerated into junk DNA.

Transposons

Transposable elements, or transposons, are segments of DNA that can move as a unit from one location to another. They are always inserted

gene creature life form consisting primarily of genetic information, sometimes with a protective covering, but without its own machinery to generate energy or replicate macromolecules

replicon molecule of DNA or RNA which contains an origin of replication

transposable element or transposon segment of DNA that can move as a unit from one location to another, but which always remains part of another DNA molecule

20.2 JUMPING GENES

into other DNA so they are never free as separate molecules. They are sometimes called "jumping genes" because they hop around from place to place on the chromosome. The process of jumping from one DNA molecule to another is called transposition.

Simple transposons cannot replicate themselves. So a transposon is even less in control of its own destiny than is a plasmid (Fig. 20.3). As long as the DNA molecule of which the transposon is part gets replicated, the transposon will also be replicated. If the transposon guesses wrong and inserts itself into a DNA molecule with no future, the transposon dies with it.

transposition process by which a transposon moves itself from one host molecule of DNA to another

20.3 TRANSPOSONS ARE NEVER FREE

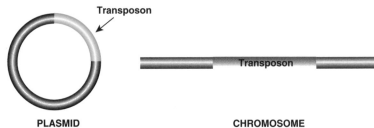

PLASMID   CHROMOSOME

The Essential Parts of a Transposon

insertion sequence the simplest type of transposable element, consisting only of two terminal inverted repeats and a gene for the transposase enzyme

inverted repeats two DNA sequences that are the same except that one is inverted relative to the other

The simplest transposons, known as insertion sequences (IS), were first found in bacteria. They have two vitally important characteristics. First they have inverted repeats at either end. This means that the sequence of the DNA at one end is the same as that at the other end as long as you read it backwards and on the other strand, as indicated by the arrows in Figure 20.4. Second, insertion sequences have just one gene that encodes the transposase, the enzyme needed for movement.

20.4 PARTS OF AN INSERTION SEQUENCE

transposase the enzyme that carries out the transposition process

Typical insertion sequences are 750 to 1,500 base pairs (bp) long with terminal inverted repeats of 20 to 40 bp. Insertion sequences are found in the chromosomes of bacteria and also in the DNA of their plasmids and viruses. For example, several copies each of the insertion sequences IS1, IS2, and IS3 are found in the chromosome of *E. coli*. The F-plasmid has zero copies of IS1, one copy of IS2, and two copies of IS3. When plasmid and chromosome possess identical IS sequences this allows integration of the plasmid into the host chromosome. This, in turn, allows transfer of chromosomal genes by the F-plasmid as explained in Chapter 8.

Just as human faces are somewhat lopsided rather than being truly symmetrical, so the inverted repeats at the ends of insertion sequences are not quite exact repeats. For example, the inverted repeats of IS1 match in 20 out of 23 positions.

To qualify as a genuine transposon, as opposed to a mere insertion sequence one needs "character." In other words, you must have some extra genes which encode useful characteristics inside the inverted repeats

20.5 PARTS OF A TRANSPOSON

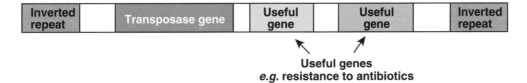

Useful genes
e.g. resistance to antibiotics

20.6 OVERALL RESULT OF TRANSPOSITION

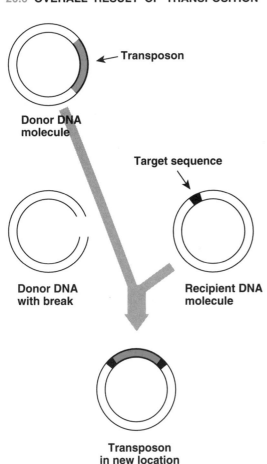

(Fig. 20.5). When the transposon moves, these internal genes will of course move as part of it. The first true transposons to be analyzed carried genes for antibiotic resistance and so protected the bacteria that hosted them from attack by human medicine.

Movement of Transposons

The transposase is responsible for moving the transposon around (Fig. 20.6). The transposase has two things to do. First, it recognizes the inverted repeats at the transposon ends and this tells it which piece of DNA must be moved. But to where will the transposon be moved? The transposase must also recognize a specific sequence on the DNA molecule it has chosen

317

20.7 MECHANISM OF TRANSPOSITION I:
TRANSPOSON LEAVES ITS OLD HOME

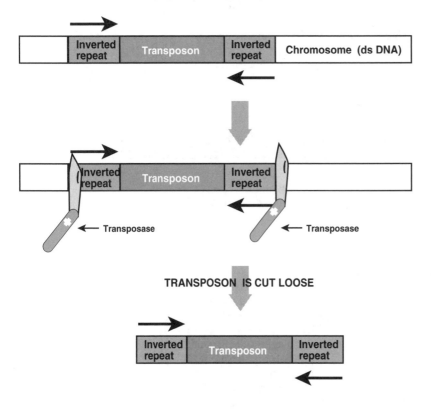

as its future home. This is known as the **target sequence** and is quite short, often only five to nine base pairs long. In fact, transposases will often accept a target site with a sequence that is a near match to the pre-ferred target sequence. So any DNA molecule of reasonable length will have quite a few possible target sites for all but the fussiest transposons.

The transposase starts the movement process by cutting the transposon loose from its original site (Fig. 20.7). Next, the transposase grabs the molecule of DNA that will be the transposon's new home. It makes a staggered cut that opens the target sequence to give over-hanging ends (Fig. 20.8). Then it sticks the trans-poson into the gap. Finally, the single stranded stretches of tar-get sequence are filled in to make double stranded DNA.

target sequence short sequence of DNA which is recognized by the trans-posase and into which it inserts the transposon when it moves

The net result is that the transposon has moved, and the target sequence has been duplicated in the process. Transposons can move from one place to another on the same DNA molecule or that can move between two separate DNA molecules. The DNA molecules into which a transposon jumps can be a plasmid, a virus, or a chromosome, any DNA molecule will do.

conservative transposi-tion the version of trans-position in which the trans-poson is removed from its original location, so leaving a gap in the DNA, and is inserted unaltered into a new site

The transposition process described above is known as **conservative transposition** because the transposon DNA is not altered during the move. When the transposon has cut itself out of its original home, it leaves a dou-ble stranded break in the DNA. There is a high likelihood that this dam-aged DNA molecule will not get itself back together again and is doomed. Life sucks!

Replicative Transposition

Conservative transposition leaves behind damaged DNA when the transposon moves. Is there not, you ask, with tears of sympathy misting your eyes, a kinder and gentler form of transposition? Yes, dear reader, there is. Some transposons are capable of replicative transposition during which the transposon creates a copy of itself. Consequently, both the

20.8 MECHANISM OF TRANSPOSITION II: TRANSPOSON FINDS A NEW HOME

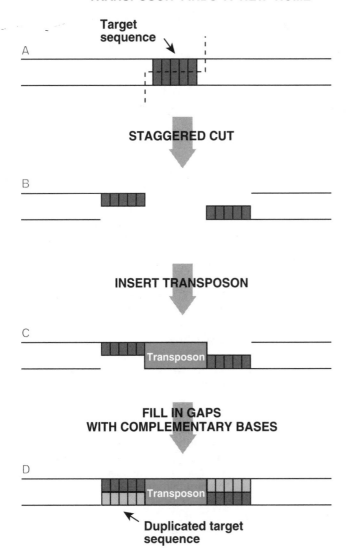

original home site and the newly selected location end up with a copy of the transposon. The original home DNA molecule is not destroyed.

These sweethearts are known as complex transposons because, unfortunately, the way they move is horribly complex. For this they need an extra enzyme, resolvase, and an extra DNA sequence, the internal resolution site (IRS) which is recognized by the resolvase (Fig. 20.9).

Complex transposons have a transposase that recognizes their inverted repeats and the target sequence just as for the other kinds of transposon (Fig. 20.10).

complex transposon type of transposon that leaves behind one copy in its original location when it moves, and inserts a second copy into the new location

resolvase enzyme needed by complex transposons to cut apart the fused intermediate formed during the transposition process into two separate DNA molecules

internal resolution site (IRS) DNA sequence which is recognized by the resolvase

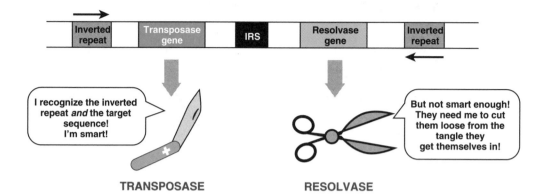

TRANSPOSASE

RESOLVASE

20.10 REPLICATIVE TRANSPOSITION – PRINCIPLE

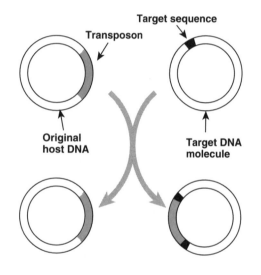

Although complex transposons manage to get replicated while moving, they are not replicons as they have no origin of replication. Instead they trick the host cell into duplicating them. First the transposase makes single stranded nicks at the ends of both the transposon and the target sequence. Next it joins the loose ends to create a tangled mess in which both DNA molecules are linked together via single strands of transposon DNA. This alarms the host cell so that it mends the single stranded regions, thus duplicating the transposon.

The function of resolvase is to resolve the mess and get the two DNA molecules separated again. It does this by recognizing the two IRS sequences and carrying out recombination between them (Fig. 20.11).

20.11 RESOLVASE SEPARATES THE FUSED DNA

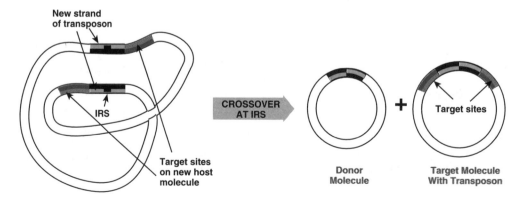

Although the net result is that the transposon has been duplicated, please note that each "copy" has half of the original transposon. The transposon did NOT make a new copy of itself which was liberated to wander round the cell and find a new home.

Tn1 and Tn3 are complex transposons carrying resistance to antibiotics of the penicillin family and are found in both the plasmids and chromosomes of many bacteria.

Composite Transposons

Lets surround a segment of DNA at both ends by two identical insertion sequences. When transposition occurs we have two possibilities (Fig. 20.12). First, the insertion sequences may move one at a time on their own. Second, and much more interesting, is that they may cooperate. In this case the whole structure moves as a unit and is referred to as a composite transposon.

composite transposon segment of DNA flanked by two insertion sequences that can jump as a complete unit

20.12 PRINCIPLE OF THE COMPOSITE TRANSPOSON

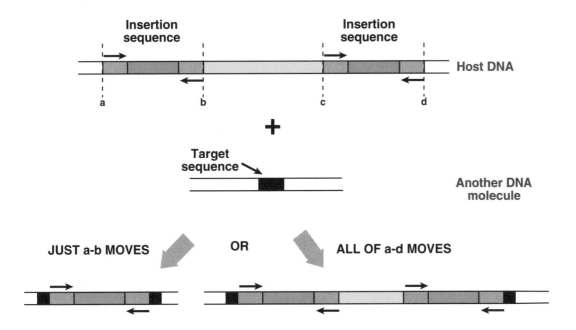

Many of the best known bacterial transposons which carry genes for antibiotic resistance or other useful properties are actually composite transposons. Perhaps the best known is Tn10 which is 9,300 base pairs long and confers resistance to tetracycline.

Once a useful composite transposon has evolved, it is silly for its parts to wander off by themselves. What usually happens is that mutations inactivate the innermost pair of inverted repeats which prevents the insertion sequences from jumping independently. Often, one of the two transposase genes is also lost. The result is that the two ends and the middle are now stuck with each other permanently and always move as a unit (Fig. 20.13). In practice, all stages from newly formed to fully united composite transposons are found in bacteria. For that matter, novel composite transposons can be assembled in the laboratory by genetic manipulation!

20.13 COMPOSITE TRANSPOSITION – EVOLVED VERSION

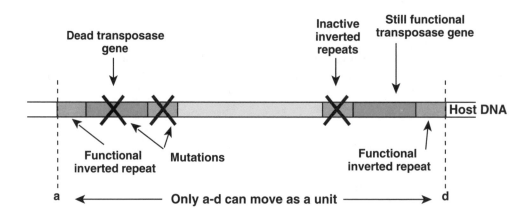

Transposons Come in 57 Different Flavors

There is a whole slew of transposable elements that vary in their mechanisms of movement, and whether or not they have an RNA phase. They tend to sound like subatomic particles, for example, transposons, retroposons and retrons (Table 20.1).

Transposons in Higher Life Forms

Although we have so far used examples from bacteria to illustrate how they work, there are transposons scattered through the DNA of all forms of life. In fact, the first jumping genes were observed by Barbara McClintock during genetic crosses in maize (corn) plants. She worked before the DNA double helix was even discovered but nonetheless realized that segments of the plant chromosomes must be moving around. When technology caught up, Barbara McClintock was proven right and she got her Nobel prize in 1983.

The Ac/Ds family of transposons in corn are

Ac/Ds family a family of transposons found in multiple copies, some of which are damaged, in cells of the maize plant

Table 20.1 **The Variety of Transposable Elements**	
Insertion Sequence	Simplest kind of transposable element, with two terminal inverted repeats and a gene for transposase
Composite Transposon	Segment of DNA flanked by two insertion sequences which jumps as a complete unit
Complex Transposon	Type of transposon which gets duplicated during transposition
Retrotransposon or retroposon	Transposon resembling a retrovirus in having reverse transcriptase and having an RNA phase
Retron	Element found in bacteria which has reverse transcriptase and makes an RNA/DNA hybrid molecule

simple and conservative. They leave behind double stranded gaps in the DNA when they move. They have inverted terminal repeats of 11 base pairs and insert at an 8 bp target sequence. The Ac element is 4,500 bp long and is a fully functional transposon with the ability to move itself. The Ds elements vary in size and are defective. They are derived from Ac by deletion of part or all of the transposase gene and so they cannot move by themselves. To remain mobile, the Ds elements must keep the inverted repeats, otherwise the Ac transposase will not recognize them (Fig. 20.14).

If a cell contains an Ac element anywhere in its DNA, then the transposase enzyme made by Ac can also move the Ds elements around (Fig. 20.15). The Ac and Ds elements do not need to be on the same chromosome for transposition to occur.

If a Ds element has been inserted into the gene for purple kernels of corn, the gene is disrupted and the kernels are white. If there is no Ac element in the cells the white color is stably inherited. If an Ac element is also present in such a

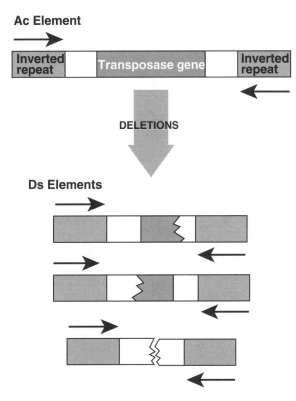

20.14 Ac / Ds TRANSPOSONS OF CORN – STRUCTURE

Ac Element

Ds Elements

20.15 Ac / Ds TRANSPOSONS OF CORN – MOVEMENT

A) Ac and Ds on same chromosome

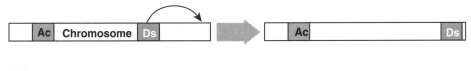

OR

B) Ac and Ds on different chromosomes

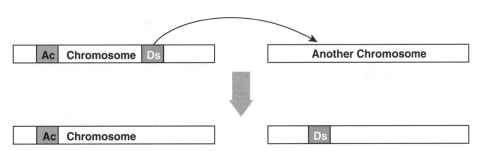

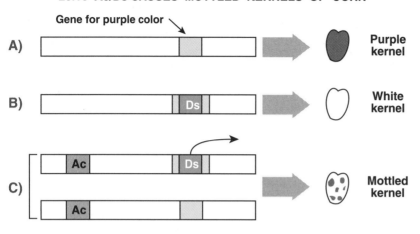

20.16 Ac/Ds CAUSES MOTTLED KERNELS OF CORN

A) Gene for purple color → **Purple kernel**

B) Ds — **White kernel**

C) Ac / Ds, Ac — **Mottled kernel**

cell, it allows the Ds element to jump out again and the result is that the original cell divides to form a mixture of purple and white cells. This produces a mottled kernel of corn (Fig. 20.16).

Animals and plants frequently contain transposon families with both active and defective members. Not only are there defective members that need help to move, but we also find dead transposons. These have suffered mutations in their terminal repeats and can no longer even be recognized by the transposase, so they cannot move.

Retro(trans)posons

Retrotransposons or retroposons, are transposable elements that replicate via an RNA intermediate. They are found in eukaryotes and resemble retroviruses, except that they can't get out of the cell because they don't have the ability to make virus particles. (The relationship to retroviruses is even more marked in some retroposons which pack their RNA in defective virus-like particles, but which still cannot get out of the cell.)

The TY-1 (Transposon Yeast No. 1) retrotransposon of yeast is around 6,000 bp long, contains a gene encoding reverse transcriptase and has long terminal repeats like those of a retrovirus. When moving, its first step is to make a single stranded RNA copy of itself by transcription (Fig. 20.17). Next, the reverse transcriptase makes a double stranded DNA copy, going via an RNA/DNA hybrid just as in the case of a retrovirus. Finally, the DNA is inserted into a new site within the host cell DNA. There are 30 to 40 copies of TY-1 per yeast cell.

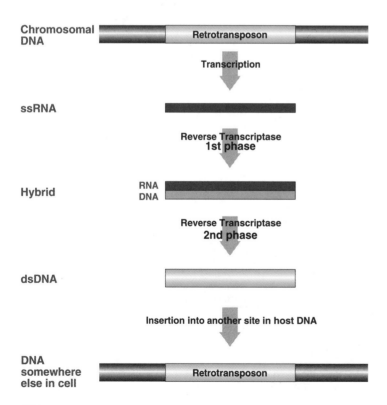

20.17 MOVEMENT OF A RETROPOSON

Chromosomal DNA — Retrotransposon

Transcription

ssRNA

Reverse Transcriptase 1st phase

Hybrid — RNA / DNA

Reverse Transcriptase 2nd phase

dsDNA

Insertion into another site in host DNA

DNA somewhere else in cell — Retrotransposon

Moderately Repetitive DNA of Mammals

A substantial portion of the DNA of both animals and plants consists of repeated sequences that may be derived from retrotransposons. In mammals there are two classes, short interspersed sequences or SINES, and long interspersed sequences or LINES (Fig. 20.18).

retrotransposon transposon that resembles a retrovirus in having reverse transcriptase and which moves by making an intermediate RNA copy

TY-1 or **Transposon Yeast No. 1** a retrotransposon found in multiple copies in yeast cells

SINES the short interspersed sequences making up much of the highly repetitive DNA of mammals

LINES the long interspersed sequences making up much of the highly repetitive DNA of mammals

20.18 FAMILY OF LINES

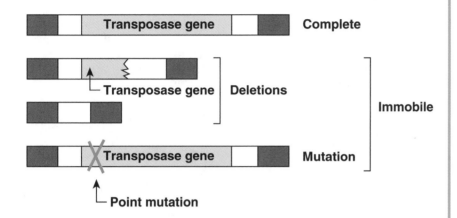

Many of the repeats are defective and finding the complete version of a repeated sequence with a fully active reverse transcriptase is not always easy. For example, in humans the LINES-1 element is present in 50,000 to 100,000 copies and makes up about 5 percent of the total DNA! The complete LINES-1 sequence is 6,500 bp long and contains a gene for reverse transcriptase. However, only about 3,000 of the LINES-1 sequences are full length and most of these are crippled by point mutations.

Very rarely LINES-1 makes a new copy of itself and inserts it somewhere else in the DNA. As shown in Figure 20.19, this may disrupt one of the genes of the host chromosome.

Hemophilia is an inherited condition caused by a defect in blood clotting factors. A few very rare cases of hemophilia are due to the

20.19 NOW AND THEN LINES-1 MOVES

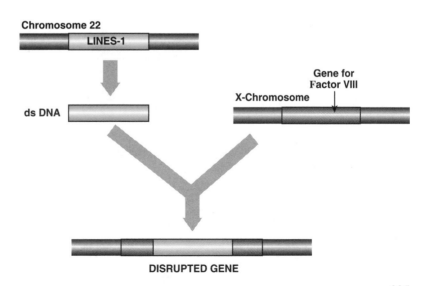

20.20 GORILLAS HAVE LINES-1 TOO

I can remember my LINES too!!

insertion of a LINES-1 sequence into the gene for blood clotting factor VIII on the X-chromosome. The culprit, the intact LINES-1 sequence that jumped, came from chromosome 22.

This still active copy of the LINES-1 sequence has been found in the same location in the DNA of our cousin the gorilla, implying that it has been lurking in the same place for millions of years of primate evolution.

Retrons

Retrons are found in bacteria. They are shorter, stranger, relatives of the retroposons of higher cells. They have just one gene which, when transcribed, makes a stretch of untranslated RNA followed by the coding region for reverse transcriptase (Fig. 20.21).

retron genetic element found in bacteria whose mechanism of movement is still unknown but which has reverse transcriptase and uses it to make a bizarre RNA/DNA hybrid molecule

selfish DNA any region of DNA that manages to replicate, but which is of no use to the host cell it inhabits

20.21 TRANSCRIPTION OF A RETRON

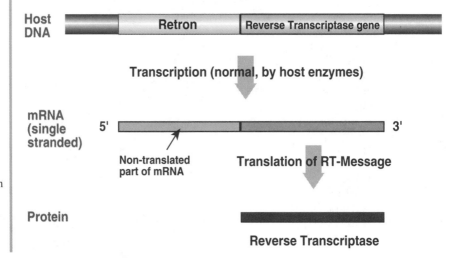

Retrons use their reverse transcriptase to manufacture lots of copies of a bizarre molecule that is part RNA and part DNA (Fig. 20.22). As far as is known, this does not reintegrate into the chromosome, and there is only a single copy of each retron in the bacterial chromosome.

No one knows yet how retrons move around or even if they still do. They are found in relatively few bacteria and are often found inserted into the DNA of bacterial viruses which have then inserted into the bacterial chromosome. Where did the virus pick them up? Who knows? Maybe the copies observed so far are defective and the master copy has not yet been found.

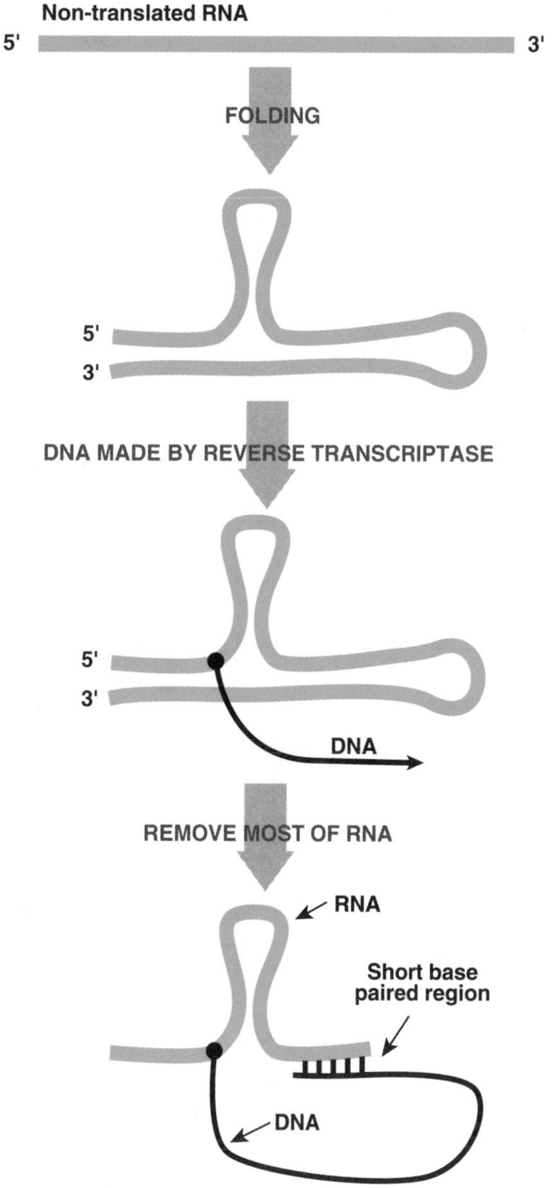

Junk DNA and Selfish DNA

DNA sequences that perform no useful function but merely "inhabit" the chromosomes of other organisms, can be regarded as genetic parasites of a very degenerate kind. These include the LINES, SINES and assorted retroposons discussed above. This general type of DNA has been named "selfish DNA" since it looks out for its own best interests, not for that of the host DNA. The selfish DNA multiplies inside its host DNA molecule just like a virus replicating inside a cell, or an infectious bacterium multiplying inside a patient (Fig. 20.23).

In most higher organisms, a substantial proportion of the DNA consists of multiple copies of such "selfish DNA." Why doesn't the host cell purge its chromosome of these parasites? Easier said than done. Remember that more rats than people live in New York City, and we aren't just referring to lawyers! In small, efficient, fast-growing cells, like bacteria or even yeasts, there is much less "selfish DNA." In large, slow-growing cells the parasites replicate faster relative to the host DNA and gradually increase.

Probably most "selfish DNA" is the remains of viruses or transposons that inserted into the chromosome long ago. Over long periods of time the copies diverge due both to single base mutations and deletions. Eventually most of the copies become defective and lose the ability to form virus particles or to move around; they degenerate into mere "junk DNA" (Fig. 20.24).

20.22 RETRONS MAKE A WEIRD RNA / DNA MOLECULE

Non-translated RNA

FOLDING

DNA MADE BY REVERSE TRANSCRIPTASE

REMOVE MOST OF RNA

RNA

Short base paired region

DNA

junk DNA defective selfish DNA of no use to the host cell it inhabits and which is no longer capable of either moving or expressing its genes

327

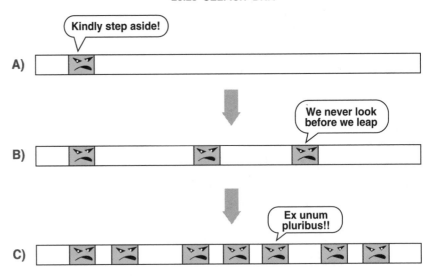

A smaller proportion of the junk DNA is the cell's own genetic trash. Pseudogenes are defective copies of cellular genes that sometimes arise by gene duplication followed by mutation (see Ch. 11). Although present in relatively few copies, they qualify as junk DNA.

pseudogene defective copy of a genuine gene

20.24 **MUTATIONS PRODUCE JUNK DNA**

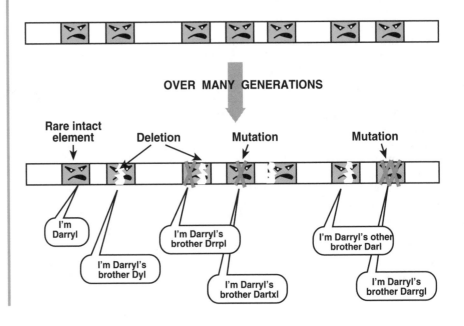

Gene Creatures and the Evolution of Higher Organisms

Though tiny, the various gene creatures have had a major influence on the evolution of higher organisms. We may consider three main effects.

1) Diseases caused by viruses and viroids, as well as those due to bacteria, have killed large numbers of higher organisms. Individuals resistant to these diseases have been selected and left descendants. In animals this has lead to the development of a complex immune system (see Ch. 22), and a whole range of modifications to avoid infection.

2) In a somewhat more positive vein, useful genes may be provided to their host cells by gene creatures. The best known examples are actually the R-plasmids that make bacteria resistant to antibiotics and so frustrate the control of human disease!

3) As discussed above, the DNA of higher organisms is mostly non-coding, in mammals, anywhere from 95 to 99 percent! Could things be any worse? Yes they could! Some newts have 20 times as much junk DNA as you do, some ferns have 50 times more and a few pathetic amoebas have almost 200 times as much. Politicians have yet to be tested. Most of this junk DNA consists of repetitive sequences, probably derived from transposons and integrated virus genomes, in particular from retroviruses and retrotransposons.

Sad to say, most of the DNA on your supposedly human chromosomes is merely the crippled remains of defunct gene creatures. You are merely a mobile mausoleum of yesterday's retroviruses! If you are starting to feel sorry for yourself, remember this, most species which have evolved on earth have become extinct. Whether you deserve it or not, you belong to one of the few species that has muddled through, so far. Let's move on to the next chapter where we can brood over possible future means of death, disease and decay.

Additional Reading

Biology of Microorganisms by Brock TD, Madigan MT, Martinko JM, & Parker J. 7th edition, 1994. Prentice Hall, Englewood Cliffs, New Jersey.

Molecular Biology of the Gene by Watson JD, Hopkins NH, Roberts JW Steitz JA, & Weiner AM. 4th edition, 1987. Benjamin-Cummings, Menlo Park, California.

Biological Warfare

21

Biological warfare has been practiced throughout history by organisms at all levels on the evolutionary scale. Before getting down to the nitty-gritty of how to get rid of your mother-in-law, let's look at some other life forms and their conflicts with each other and with us.

Lethal Proteins Made by Bacteria

When closely related bacteria are interested in occupying the same habitat or consuming the same resources, what do they do to settle the matter? Peaceful coexistence? No way! They kill each other, just like cowboys and Indians.

Bacteria make a variety of toxic proteins. Some are targeted at humans, some at other animals or plants, and some at other bacteria. Generally speaking, bacteria are most enthusiastic about killing their close relatives. The reason for this is that the more closely related they are, the more likely other bacteria will compete for the same resources.

Proteins made by bacteria to kill their relatives are known as **bacteriocins** (not bacterial <u>sins</u>). For example, many strains of *Escherichia coli*, the favorite bacterium of molecular biologists, deploy a wide variety of bacteriocins, referred to as **colicins**, intended to kill other strains of the same species.

21.1 BACTERIA COMPETE FOR RESOURCES

Hey! That's my burger!

No way! I saw it first!

Bacterial cell

bacteriocin a toxic protein made by bacteria to kill other, closely related, bacteria

colicin a bacteriocin made by some strains of *Escherichia coli* to kill other strains of *E. coli*

Yersinia pestis, the bacterium that causes Black Death (bubonic plague), is credited with wiping out nearly a third of the human population of Europe, and probably most of Africa and Asia too, on several occasions. You would think that this was enough to make anyone proud of their cultural heritage. But no, *Yersinia pestis,* a true professional, also makes bacteriocins, called <u>pesti</u>cins in this case, designed to kill competing strains of its own species.

21.2 COLICIN PLASMID

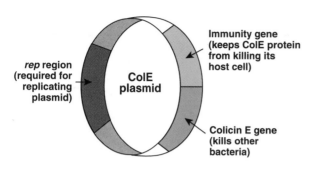

rep region (required for replicating plasmid)

ColE plasmid

Immunity gene (keeps ColE protein from killing its host cell)

Colicin E gene (kills other bacteria)

21.3 COLICIN E1 MAKES AN ION CHANNEL

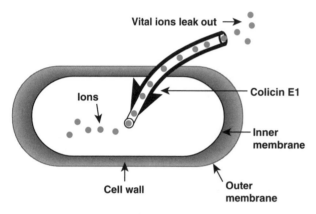

Vital ions leak out →

Ions

Colicin E1

Inner membrane

Cell wall

Outer membrane

Bacteriocins are Usually Coded for by Plasmids

The ability to make bacteriocins is usually due to the presence of a plasmid in the producer cell. Most famous by far are the three related ColE plasmids of *Escherichia coli*, ColE1, ColE2 and ColE3 (Fig. 21.2). These ColE plasmids exist in 50 or more copies per cell and have been used to derive many of the major genetic engineering plasmids, all of which have the actual colicin genes removed (see Ch. 9).

The original ColE plasmids allow the strains of *Escherichia coli* that possess them to kill other strains. There are two basic approaches to this. The first is to drill holes in a victim's cell membrane. A gene on the ColE1 plasmid encodes the colicin E1 protein that inserts itself through the membrane of the target cell and creates a channel allowing vital cell contents to leak out (Fig. 21.3). A single molecule of colicin E1 is enough to kill the target cell.

The second approach is to chop up the nucleic acids of the victim. The ColE2 and ColE3 plasmids both encode nucleases, enzymes that degrade nucleic acids. Colicin E2 is a deoxyribonuclease which hacks the chromosome of the target cell into pieces (Fig. 21.4). Colicin E3 is a ribonuclease which delicately snips the 16s rRNA of the small ribosomal subunit, releasing a fragment of 49 nucleotides from the 3' end. This abolishes protein synthesis and is just as lethal.

plasmid circular molecule of double helical DNA that replicates independently of the bacterial cell's chromosome

nuclease an enzyme that cuts nucleic acids into shorter pieces

deoxyribonuclease an enzyme that cuts DNA into shorter pieces

ribonuclease an enzyme that cuts RNA into shorter pieces

How Do Bacteria That Make Colicins Avoid Killing Themselves?

Bacterial cells which produce a particular colicin are immune to their own brand, but not to other brands. Immunity is due to specific immunity proteins that bind to the corresponding colicin proteins and cover their active sites (Fig. 21.5). If you possess the ColE1 plasmid, you have genes for both colicin E1 and the immunity protein that binds colicin E1.

21.4 COLICIN E2 DEGRADES DNA

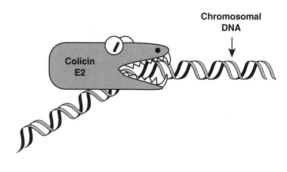

Chromosomal DNA

Colicin E2

This immunity protein will not protect you from colicin E2 or E3. All forms of immunity are based on the ability of immune system proteins to recognize and neutralize the bad guys. The immune systems of animals are very complex (see Ch. 21); the colicin immune system of bacteria is about as simple as you can get.

Actually, in a population of Col plasmid-carrying bacteria most cells do not produce colicin. Every now and then one of them gets the urge to be a hero. It goes berserk and manufactures large amounts of colicin, then it bursts and releases the colicin into the medium. Note that it is the burst and release mechanism that kills the berserker cell, not the actual colicin. All sensitive bacteria in the area are wiped out, but those with the Col plasmid have immunity protein and survive.

21.5 ACTION OF IMMUNITY PROTEINS

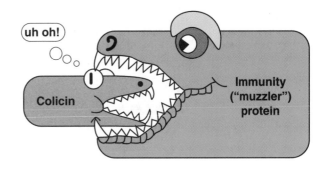

Bacteria Make Proteins to Kill People

When the poisonous proteins produced by bacteria are used to attack higher organisms rather than fellow bacteria, we get uptight and refer to them as **toxins**. But let's look at this from a broader, more multicultural, viewpoint. Bacteria that deploy bacteriocins against their fellow bacteria do it with the deliberate intention of killing them. In contrast, disease-causing bacteria do not really mean to kill the people they infect. The longer you stay alive, the longer you provide a home for them. If death occurs, it's merely environmental mismanagement (Fig. 21.6). Don't take it personally!!

Do we see here the beginnings of altruism at the single cell level? One cell sacrifices itself so that its relatives carrying the same Col plasmid can take over the neighborhood.

toxin a toxic protein that acts against higher organisms

Who is morally responsible when bacteria carrying a colicin plasmid kill their plasmid-less relatives? Since plasmids may be regarded as gene creatures inhabiting cells, perhaps we should think more in terms of a takeover attempt by the plasmid. Either bacteria provide a home for the plasmid or they are killed by the colicin.

21.6 DISEASE AS ENVIRONMENTAL MISMANAGEMENT

There are many different bacterial toxins, just as there is a wide variety of bacterial diseases. Let's take **choleratoxin** as an example. *Vibrio cholerae* (the cholera bacterium), *Shigella* which causes dysentery, and those *E. coli* strains causing food poisoning and diarrhea, all produce variants of the same toxin. This family of toxins causes loss of water and ions from intestinal cells due to massive overproduction of the control molecule, cyclic AMP.

These toxins each comprise three proteins: A1, A2 and B. The A1 and A2 proteins are the two halves of an original A protein that was cut in two. A1 and A2 remain linked by a disulfide bond. Five B proteins form a donut-like structure into which one A1-A2 unit is inserted (Fig. 21.7).

21.7 STRUCTURE OF CHOLERATOXIN

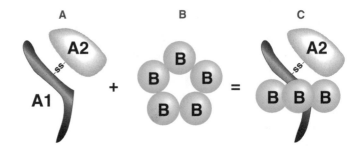

The B protein of choleratoxin recognizes and binds to the sugars attached to a grease molecule (ganglioside GM1) in the animal cell membrane (Fig. 21.8). The A1 protein is the lethal component. It is released from A2 by breakage of the disulfide bond and enters the cell. The ring of five B proteins and A2 are abandoned outside.

Choleratoxin A1 protein blocks the correct regulation of an enzyme found in the membrane of animal cells. This enzyme is known as **adenylate cyclase** because it makes **cyclic AMP**. The cyclic AMP in turn controls the cell's balance of water and essential mineral ions. Adenylate cyclase consists of two parts, the enzyme itself and a regulatory subunit, the G-protein, which acts as a switch.

21.8 ENTRY OF CHOLERATOXIN

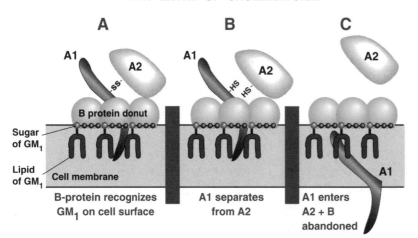

Choleratoxin A1 protein sticks a chemical group onto the G-protein and this jams the switch permanently in the "ON" position (Fig. 21.9). The adenylate cyclase then goes wild and makes several hundred times too much cyclic AMP. Affected cells leak their ions and then lose water. Victims of cholera lose their bodily fluids in a deluge of dilute diarrhea and may die of dehydration.

21.9   ACTION OF CHOLERATOXIN ON G - PROTEIN

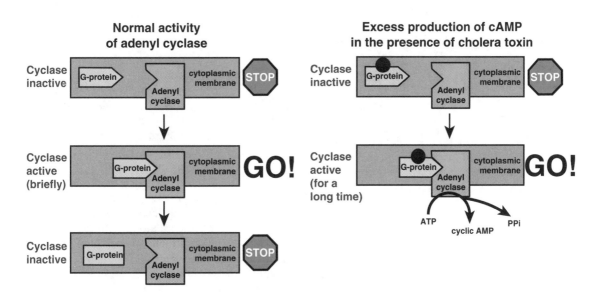

Choleratoxin A1 protein actually gets the chemical group, ADP-ribose, that it uses to modify G-protein by ripping it off from NAD (nicotinamide adenine dinucleotide), a component used by all cells to carry hydrogen atoms during energy generation. Several bacterial viruses inactivate the bacteria they infect in a similar way. They possess toxins that split NAD and then attach the ADP-ribose fragment to bacterial proteins. It seems likely that the bacteria borrowed this idea from their viruses, but modified the toxin to damage proteins in animal cells. When you have a kindness shown, pass it on!

Killer *Paramecium*

Paramecium is a protozoan that is a microscopic single celled animal. It cruises around, feeds by swallowing bacteria whole, and is generally viewed as being fairly laid back (Fig. 21.10).

21.10 RELAXED LIFESTYLE OF A *PARAMECIUM*

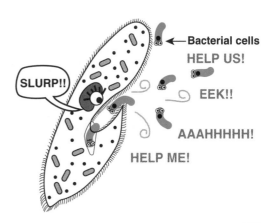

335

Despite this, some strains of *Paramecium* practice germ warfare against others. The bad guys are known as killers. Those who are killers will kill any non-killers who get too close. Killers are immune to their own poison. However, different brands of killer *Paramecium* exist that kill each other.

21.11 KAPPA PARTICLES IN *PARAMECIUM*

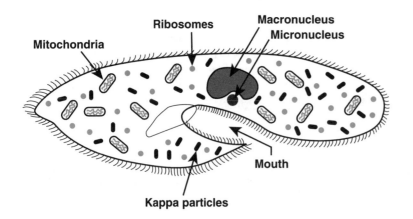

In killer *Paramecium*, the biological weapon consists of **kappa particles** found floating inside the cell (Fig. 21.11). Kappa particles are actually symbiotic bacteria (*Caedibacter*) that grow and divide inside the larger, eukaryotic, *Paramecium* cell.

From the viewpoint of the killer *Paramecium*, bacterial kappa particles are inherited like mitochondria (Fig. 21.12). This is known as cytoplasmic or maternal inheritance since you get your mitochondria or kappa particles from your mother cell, along with the chunk of cytoplasm in which they are floating. Unlike mitochondria, which have lost their identity and become mere organelles (see Ch. 24), kappa particles are still obviously bacteria. Some of these symbiotic bacteria can still live outside on their own, if necessary, although most types cannot.

21.12 INHERITANCE OF KAPPA PARTICLES IN *PARAMECIUM*

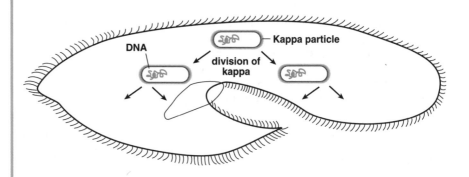

The ability to kill also depends on a gene found in the nucleus of the *Paramecium*, which has two alternate forms: K and k. *Paramecium* is diploid so any individual has two copies of each gene and may therefore be KK, Kk or kk. Possession of one or two K genes (*i.e.*, KK or Kk) allows the kappa particles to grow and divide inside the *Paramecium*. However, kk individuals lose the kappa particles. Thus to be a killer, you need both the cytoplasmic kappa particle and at least one copy of the nuclear K gene.

All *Paramecium* cells without kappa particles are sensitive, whether they are KK, Kk or kk. Killing occurs when a few kappa particles are liberated from killer strains into the culture medium. The sensitive *Paramecium* strains swallow a kappa particle, thinking they are going to get a tasty snack. Inside the kappa particles are toxic coils of protein called R-bodies. When the kappa particle is digested by a sensitive *Paramecium* the R-body is liberated, uncoils and kills the *Paramecium* (Fig. 21.13). "One man's meat is another man's poison!"

R-body, refractile body a toxic protein found as a coiled body inside the kappa particles of killer *Paramecium*

21.13 KAPPA KILLS

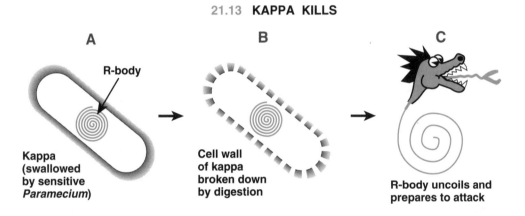

A

R-body

Kappa (swallowed by sensitive *Paramecium*)

B

Cell wall of kappa broken down by digestion

C

R-body uncoils and prepares to attack

The R-body protein is in, fact, encoded by a gene belonging to a bacterial plasmid or a defective bacterial virus, not by the kappa bacterial chromosome. So a toxin made by a virus infecting the kappa particles has been domesticated and diverted to the higher purpose of killing *Paramecium*.

The War on Bugs!!! Eradication of Insect Pests

Insects eat a substantial portion of the food humans grow for themselves. Some insects also spread disease, such as mosquitoes which carry malaria and yellow fever. Chemical insecticides have been used in vast amounts since World War II. The two main classes are the chlorinated hydrocarbons (DDT, dieldrin, lindane) and the organophosphates (malathion, parathion). Appropriately enough, the first generation organophosphates were initially developed as chemical warfare agents.

Apart from losing the occasional farmer to organophosphate poisoning, there are two general problems. First, many insect pests have now developed resistance to chemical insecticides. Second, chemical insecticides do not discriminate; they practice chemical affirmative action and poison

chlorinated hydrocarbons class of organic chemical compounds containing chlorine which are often toxic and difficult to break down

organophosphates class of organic chemicals with phosphate groups and which are often neurotoxic

beneficial insects and pests equally. In some cases, chemical treatment has even increased the number of pests by killing their natural predators!

In addition, the chlorinated hydrocarbons, especially DDT, have caused a major pollution problem. These compounds decay very slowly and accumulate in the fatty tissues of animals. The higher up the food chain, the higher the concentration of DDT. In fact, present-day Americans contain DDT levels above the U.S. government's legal limit for food products. No wonder cannibalism is so rare nowadays!

Bacillus Toxins and Insects

The recent growth of biotechnology has provided some alternative approaches to insect control. The bacterium *Bacillus thuringiensis* has become famous due to its production of a toxin that kills insects. Different strains of *Bacillus thuringiensis* kill different types of insects.

When *Bacillus thuringiensis* forms spores, it also makes protein crystals. After being eaten by an insect, the crystals dissolve into pairs of proteins linked together by disulfide bonds. When the pairs are split apart they give single protein molecules. These are still not the actual toxin but are "pro-toxin" molecules, *i.e.,* they are precursors to the toxin (Fig. 21.15). For reasons of safety, toxins are often made as inactive precursors and are only activated later after they have left the cell that made them! In this case digestive enzymes in the insect gut convert the protoxin to the actual toxin.

The toxin then inserts itself into the membranes of cells making up the insect's gut wall where it forms a channel (Fig. 21.16). Vital ions and ATP, the cells' main energy supply, escape through the channel. About 15 minutes later, the insect loses its appetite, becomes dehydrated and eventually dies.

To work, the toxin must be eaten by the insect larva. So, cultures of *Bacillus thuringiensis* that have just formed spores are sprayed onto insects in the larval stage (such as caterpillars). However, although cute, this toxin has some practical drawbacks. These have mostly been surmounted by genetic engineering as described below:

21.15 **CRYSTALS CONVERTED TO TOXIN**

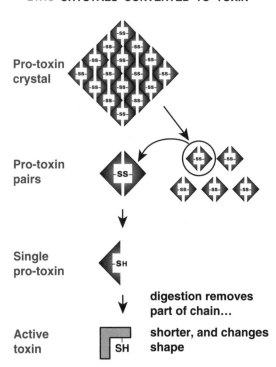

Pro-toxin crystal

Pro-toxin pairs

Single pro-toxin

digestion removes part of chain...

Active toxin

shorter, and changes shape

21.16 TOXIN KILLS INSECTS

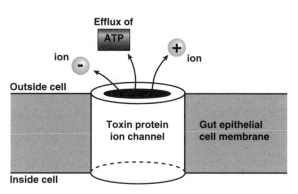

Problem: The toxin is normally formed only when the bacteria are in the spore-forming stage. *Solution:* Cloned toxin genes have been placed under the control of promoters giving constant gene expression. So toxin is made all the time.

Problem: The toxin from any particular strain of *Bacillus thuringiensis* kills only a restricted range of insects.

Solution: Several cloned toxin genes from different strains of bacteria can all be inserted into the same strain of *Bacillus thuringiensis*. These engineered strains kill a much wider range of insects.

Problem: Bacteria sprayed over a field do not protect the roots of plants against root-boring insects.

Solution: Toxin genes from *Bacillus thuringiensis* have been inserted into other kinds of bacteria. For example, a toxin gene was inserted into the chromosome of *Pseudomonas fluorescens* which often lives on the surface of corn roots. Such engineered strains killed root-eating insect larvae under laboratory conditions but still need to be tested in the field.

21.17 NEW IMPROVED BACTERIA KILL CATERPILLARS

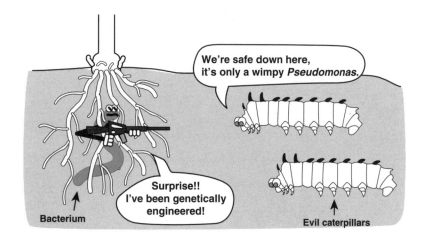

Warfare Against the Insects Using Viruses

The evil tobacco budworm and its comrade the cotton bollworm (alias tomato fruitworm or corn earworm) are not really worms, they are insects. They have worm-like larvae (mere "maggots" to the uneducated) that munch on their plant victims.

It has been known for quite a while that certain parasitic wasps lay their eggs by

21.18 TOBACCO BUDWORM EATING TOBACCO

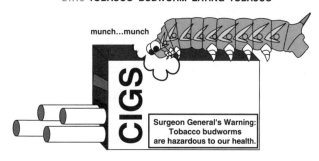

339

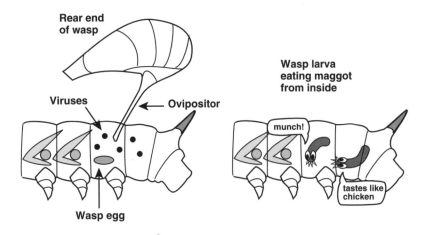

Rear end of wasp

Viruses

Ovipositor

Wasp egg

Wasp larva eating maggot from inside

munch!

tastes like chicken

injecting them into these plant eating maggots (Fig. 21.19). The eggs hatch and the newborn wasps eat the living maggots from inside.

The maggots are eventually killed and a new generation of wasps is released to continue the crusade against these pests.

The secret to the wasp's success is germ warfare. The wasps inject a virus along with their eggs. The virus, a member of the adenovirus family, heads for the maggot's "fat body" (equivalent to the liver of higher animals). The virus also trashes the maggot's primitive immune system and messes up its hormone control system. The result is that the maggot loses its appetite for plants and is prevented from moulting and turning into a pupa, the next stage on the way to an adult insect capable of laying more eggs. There are many types of plant-eating maggots and many types of wasps that attack them, and therefore many types of virus, each designed to soften up a particular plant pest.

anthrax a disease of cattle caused by the spore-forming bacterium *Bacillus anthracis*

Humans and Germ Warfare

O.K., so after all the chintzy diversions you really want to know about human biological warfare. Germ warfare has been frequently used throughout human history. During the Middle Ages, cattle with anthrax were sometimes catapulted over the walls into castles under siege. Anthrax bacteria usually cause disease in cattle, but they are highly infectious and give a high death rate among humans. To be fair, if you were stuck in a medieval castle under siege, you would have been at least as likely to die from the plague, typhoid, or smallpox already inside as from anthrax thrown over the wall!

Those of you who think that democracies would never practice biological warfare should be aware that they already have. During the eighteenth and nineteenth centuries, measles virus was deliberately spread among the American Indians by settlers in the USA. Measles is a mild disease among Europeans, but is often fatal to American Indians. Blankets used by white children who had just had measles were given as gifts to the Indians. This proved a cost effective way to clear the environmental niche of North America for newcomers of the same species.

So then, the idea is we pick a human disease and use it to eliminate our human competitors. Perhaps we could even improve it by genetic engineering. Before we rush to get started, there are several factors to consider:

Time: One major problem with biological warfare is that even if you go to all the trouble of creating a designer germ and spraying it all over the bad guys, they will take a while to die. Even the best diseases, such as Lassa fever or the pneumonic form of Black Death, take at least 24 hours to kill. And 24 hours is plenty of time for a couple of retaliatory nuclear exchanges.

Distribution: Another drawback to germ warfare is the problem of delivery. Suppose that your submarine pops up out of the sea just off the enemy coast and sprays the germs. First, you need a breeze and second, the wind needs to blow in the *right* direction!

According to the U.S. Army, a biological warfare agent should fulfill the following requirements:

-It should consistently produce death, disability or damage.

-It should be capable of being produced economically and in militarily adequate quantities from available materials.

-It should be stable under production and storage conditions, in munitions and in transportation.

-It should be capable of being disseminated efficiently by existing techniques, equipment or munitions.

-It should be stable after dissemination from a military munition.

-All munitions containing lethal biological warfare agents must carry a warning notice from the Surgeon General. (Just kidding on this last point!)

21.21 SUBMARINE SPRAYING GERMS

During the 1950s the British government conducted field tests with harmless bacteria. When the wind blew the germs over "healthy" farmland, many of the airborne bacteria survived the trip and landed alive and well. In contrast, when the wind blew the bacteria over industrial areas, especially oil refineries or similar installations, the airborne bacteria were wiped out. In the event of aerial delivery, even if the wind is in the right direction, most of the population of an industrial nation will be found in cities where they will be protected from airborne germs by air pollution!

Persistence of the Agent: Once you have successfully sanctioned the hostile units, you presumably want to move in and take over the territory yourself. Therefore, you do not want the germs you used to hang around forever as they will then infect your own men.

Anthrax was the favorite choice in the Dark Ages because it acts rapidly and is highly lethal. Unfortunately, *Bacillus anthracis*, which causes this disease, spreads itself around by spores which are tough, difficult to destroy, and last for a very long time. When suitable conditions return, they hatch out and resume growth as normal bacterial cells.

21.22 AIRBORNE BACTERIA

Oil refinery

Off Scotland, there is a tiny island which was sown with anthrax around the time of World War II. Although this island has been fire-bombed and disinfected, it is still uninhabitable, even today, because of the survival of anthrax spores in the soil.

biosafety levels a series of levels of increasingly strict laboratory containment intended for dealing with dangerous agents; from 1 (mild) to 4 (bad!!)

Which Disease? Bacteria or Virus? Viral diseases seem more scary in some ways. In particular, they cannot be cured by antibiotics. One practical drawback is that viruses can grow only inside host cells. Culturing animal cells is far more difficult than growing bacteria and the yields are much lower. Large scale manufacture of viruses is a real pain.

Bacteria are much easier to culture. Black Death is not a bad choice. It has a snappy name, it is highly infectious, the death rate is high, and it works fast. In the years just after World War II, the British kept large scale cultures of Black Death constantly on the go, rather like an automatic coffee machine always keeps a nice hot drink ready in case you need one.

High Containment Laboratories: Before you start monkeying around, it's not a bad idea to blow a few million on a high containment laboratory. If you can afford it, go for Biosafety Level Four. All your operations will be done inside safety cabinets with glove ports. Your whole lab is sealed off and kept at a little less than normal atmospheric pressure. That way, if there is a leak, air will leak *in*! (Smart, huh!) If you are the romantic sort, you can get dressed to kill by wearing one of those cool spacesuit-type rigouts you see in sci-fi movies!

When you enter Biosafety Level Four, you take off your outside clothes and put on a separate set of lab clothes, not just a lab coat but lab underwear too! When you exit, you take a disinfectant shower and leave all the lab clothes behind. Some high containment labs are designed so that you can only get out by totally submerging in a pool of disinfectant (Fig. 21.23), remember to shut your eyes first! If your budget runs to it, an airlock at the door with an ultraviolet light is a good idea for working with the worst viruses.

21.23 DISINFECTANT DIP FOR BIOSAFETY LEVEL 4

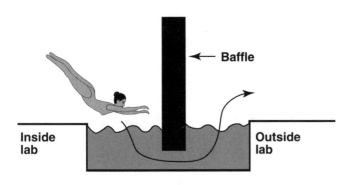

Baffle

Inside lab

Outside lab

If such a fuss each time you come out seems burdensome, remember this - if you stick yourself with a contaminated needle while you are inside, they won't *let* you come out until a quarantine period has passed. This is for the greater good of mankind, in particular of the administrators whose offices you would pass before moving on to infect the world at large.

Genetic Engineering of Diseases:

So let's take a harmless laboratory bacterium, such as *Escherichia coli*, and turn it into something nasty. We insert genes for binding the bacteria to the gut wall after being swallowed and for invading human cells.
We provide genes for tearing vital supplies of iron away from blood cells. And finally, we add genes for a toxin so potent that even a very tiny dose is fatal.

What have we got? An unstoppable disease which will wipe the human race from the face of the Earth? Actually, all we have done is to convert *Escherichia coli* into its near relative, *Yersinia pestis*, the Black Death organism. The truth is that Mother Nature has long ago beaten us to the punch. The reason we are not all dying of Black Death today is not due to dereliction of duty on the part of *Yersinia pestis,* but to the existence of modern hygiene. Given the right conditions, diseases that already exist are quite capable of decimating human populations. The "improvement" of diseases by genetic engineering is really a minor threat.

So What is the Real Threat Today?

Most bacterial diseases can be taken care of by using antibiotics. Although human supremacy is threatened by **R-plasmids** which confer antibiotic resistance on bacteria (Chs. 8 and 19) biotechnology can be used to make improved antibiotics (Ch. 10). Some tropical diseases caused by eukaryotic parasites can also be treated with chemical agents such as quinine for malaria. However many parasites are problematic because they confuse the immune system (see Ch. 22).

Nonetheless, at least in temperate climates, most incurable diseases are due to viruses. Vaccination (see Ch. 22) may protect you from catching many viral diseases such as mumps, measles, smallpox, etc., however, no actual cure exists for these viral diseases.

The fundamental problem is that viruses are parasites at the genetic level (see Ch. 19). Viruses are not themselves living cells but rely on the host cells they infect to assemble new virus particles. Consequently, chemical agents that prevent virus replication will usually kill the host cells too! So let's indulge the pessimists amongst us by brooding over the misery and death caused by viral diseases.

As discussed in Chapter 19, viruses may contain DNA or RNA in a variety of conformations, single or double stranded and linear or circular. The RNA viruses constitute a bigger threat because genes made of RNA mutate much faster than DNA and so RNA viruses are constantly changing. Let's start with the less alarming DNA viruses.

DNA Viruses of Animals

Most DNA viruses of animals contain double stranded DNA. **Simian Virus 40 (SV40)** is a smallish, spherical virus that causes cancer in monkeys by inserting its DNA into the host chromosome (see Ch. 14).

The **Herpesviruses** are spherical with an extra outer envelope of material stolen from the nuclear membrane of the host cell. The internal nucleic acid with its protein shell is referred to as the **nucleocapsid**. This family includes cold sores and genital herpes as well as chickenpox and infectious mononucleosis.

Poxviruses are the most complex animal viruses and are so large they may be seen with a light microscope (Fig. 21.24). **Vaccinia** is a poxvirus that causes cowpox. It measures 0.4 by 0.2 microns, compared to 1.0 by 0.5 microns for bacteria like *E. coli*. Unlike other animal DNA viruses which all replicate inside the cell nucleus, poxviruses replicate their dsDNA in the

R-plasmids (or **R-factors**) plasmids that confer antibiotic resistance on their bacterial hosts

Simian virus 40 (SV40) a cancer-causing virus of monkeys which contains dsDNA

herpesviruses family of viruses with dsDNA and an outer envelope surrounding the nucleocapsid

nucleocapsid innermost protein shell of a virus plus the DNA or RNA inside it

poxviruses family of viruses with dsDNA carrying up to 200 genes and an outer protein layer surrounding the nucleocapsid

vaccinia member of poxvirus family that causes cowpox

21.24 IDEALIZED POX VIRUS

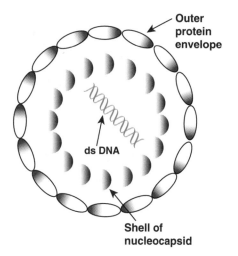

Outer protein envelope

ds DNA

Shell of nucleocapsid

cytoplasm of the host cell. They build little subcellular factories known as inclusion bodies, inside which virus particles are manufactured. They have 150 to 200 genes, about the same number as the T4 family of complex bacterial viruses.

Double Stranded RNA Viruses of Animals

Double stranded RNA is relatively uncommon among animal viruses. Reoviruses are spherical, with two concentric protein shells but no envelope. They contain a dozen or so separate dsRNA molecules, each coding for a single virus protein. They are tasteless and messy rather than deadly; their most famous member is the rotavirus (Fig. 21.25) which causes infant diarrhea, the longest running disease in human history!

Blue tongue disease of sheep is due to another reovirus with a coarse sense of humor.

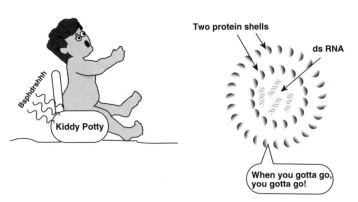

21.25 **ROTAVIRUS**

Two protein shells

ds RNA

Bsphdrshhh

Kiddy Potty

When you gotta go, you gotta go!

Plus and Minus Strands of RNA

Let's consider a gene made of double stranded DNA. When we make messenger RNA we use one of the DNA strands as the template strand. The mRNA will be complementary in sequence to the template strand and identical in sequence to the other strand of DNA, except that we use U instead of T in RNA. The non-transcribed strand of DNA is thus the coding strand (see Fig. 21.26). Let us designate coding strands as "plus" and the other strand as "minus." The non-transcribed strand of DNA and the mRNA, which encodes the protein, are both "plus" strands and the template strand is a "minus" strand.

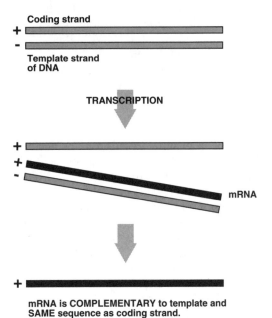

21.26 **POSITIVE AND NEGATIVE STRANDS OF DNA AND RNA**

Coding strand

+

−

Template strand of DNA

TRANSCRIPTION

+

+

−

mRNA

+

mRNA is COMPLEMENTARY to template and SAME sequence as coding strand.

344

Suppose that we have a virus particle containing single stranded RNA (ssRNA). We have two alternatives, the virus RNA could be either the plus strand or the minus strand. If the virus contains the plus strand it can use this RNA directly as a messenger RNA to make proteins immediately upon entering the host cell. Such viruses are called positive-strand RNA viruses. Conversely, if the virus particle contains the minus strand, it must first make the complementary, plus strand before moving on to manufacture proteins. These are called negative-strand RNA viruses.

Positive-Stranded RNA Viruses of Animals

Picornaviruses are small, spherical ssRNA viruses. They include polio, common cold, hepatitis A and foot and mouth disease. Their genome is long enough for about a dozen genes. Since they are positive-strand RNA viruses, their RNA can be directly used as mRNA. But, there is a technical problem. Unlike bacteria where a single mRNA molecule may code for several proteins (an operon; see Ch. 6) in higher organisms each molecule of mRNA only encodes a single protein (see Ch. 11). Eukaryotic ribosomes will only translate the first reading frame on an RNA message, even if it carries several.

Picornaviruses do indeed use their single stranded virus RNA molecule directly as a messenger RNA. They weasel around the problem by using the RNA to code for a single giant protein that uses all of their genetic information (Fig. 21.27). This "polyprotein" is then chopped up into 10 to 20 smaller proteins.

ssRNA single stranded RNA

positive-strand RNA virus RNA virus that contains a single plus, or coding, strand of RNA

negative-strand RNA virus RNA virus that contains a single minus, or non-coding, strand of RNA

picornaviruses family of positive-stranded RNA viruses with a single protein shell surrounding the ssRNA

polyprotein large protein made by stringing several genes together and which is cut up into individual proteins after synthesis

21.27 ONE GENE – LOTS OF PROTEINS

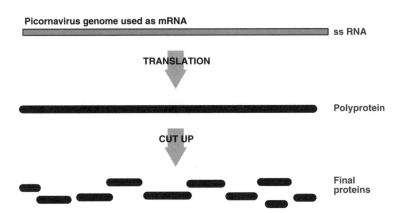

Picornavirus genome used as mRNA — ss RNA

TRANSLATION

Polyprotein

CUT UP

Final proteins

Influenza and Other Negative-Strand RNA Viruses

The negative-strand RNA viruses are divided into several families. These families include such wonderful contributions as rabies, mumps, measles and influenza. In these, the ssRNA in the virus particle is complementary to the messenger RNA and is therefore the minus strand. These viruses are similar, having an outer envelope actually derived from the membrane of the host cell where they were assembled.

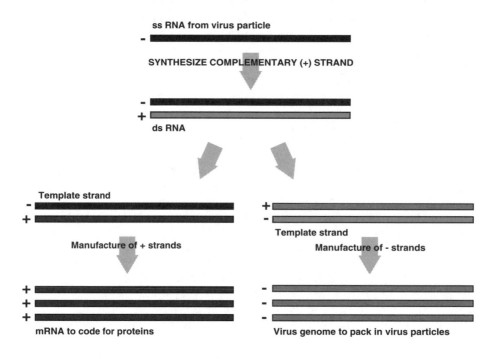

ss RNA from virus particle

SYNTHESIZE COMPLEMENTARY (+) STRAND

ds RNA

Template strand

Manufacture of + strands

Template strand

Manufacture of - strands

mRNA to code for proteins

Virus genome to pack in virus particles

After infiltrating the cell, the first mission of a negative-strand RNA virus is to make its RNA double stranded by synthesizing the corresponding positive RNA strand. Once it has the two strands it uses them both as templates. The plus strand is used as a template to manufacture lots of negative strands for the next generation of virus particles. The minus strand is used as a template to manufacture lots of plus strands which act as mRNA molecules (Fig. 21.28).

Let's take as our example the **influenza virus** which is an **orthomyxovirus**. The flu virus particle contains eight separate pieces of single stranded RNA ranging from 890 to 2,341 nucleotides long. These are packed into an inner nucleocapsid surrounded by an outer envelope (Fig. 21.29).

influenza virus member of the orthomyxovirus family with eight separate ssRNA molecules

orthomyxoviruses family of negative-stranded RNA viruses with an outer envelope surrounding the nucleocapsid which contains several pieces of ssRNA

21.29 **STRUCTURE OF INFLUENZA VIRUS**

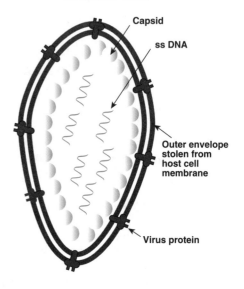

Capsid

ss DNA

Outer envelope stolen from host cell membrane

Virus protein

21.30 ENTRY OF INFLUENZA VIRUS

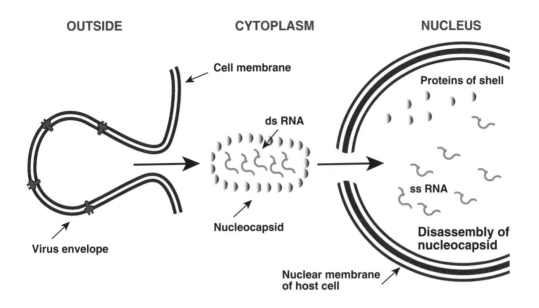

OUTSIDE CYTOPLASM NUCLEUS

Cell membrane

ds RNA

Proteins of shell

Nucleocapsid

ss RNA

Disassembly of nucleocapsid

Virus envelope

Nuclear membrane of host cell

The outer envelope is abandoned as the virus particle enters the host cell. The capsid with the RNA invades the cell nucleus where it disassembles releasing the RNA molecules (Fig. 21.30).

Replication of the influenza RNA occurs in the nucleus. The viral mRNA exits the nucleus just like normal cellular mRNA and travels to the ribosomes in the cytoplasm. Here the proteins for the new virus particles are made (Fig. 21.31).

21.31 MANUFACTURE OF NEW FLU VIRUS

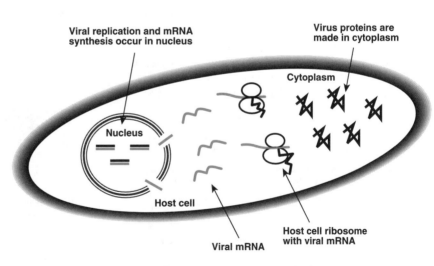

Viral replication and mRNA synthesis occur in nucleus

Virus proteins are made in cytoplasm

Cytoplasm

Nucleus

Host cell

Viral mRNA

Host cell ribosome with viral mRNA

Because influenza virus has its genes scattered over eight separate molecules of RNA, different strains of flu can trade pieces of RNA and form new combinations (Fig. 21.32). In addition, mutations occur at a higher rate in RNA than in DNA. These two factors result in a lot of genetic variation.

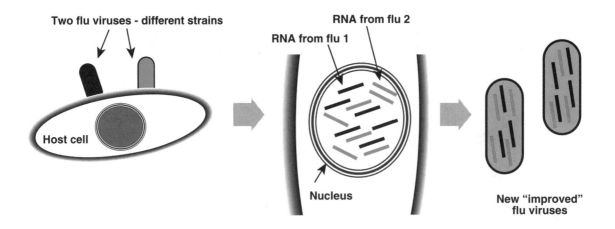

Two flu viruses - different strains

RNA from flu 2

RNA from flu 1

Host cell

Nucleus

New "improved" flu viruses

Still worse, flu is shared by people, pigs and poultry. When two strains of flu from humans are both caught by the same pig, they swap RNA segments and mutate. A variant strain of flu may then be selected which survives better in the pig. Later, when this strain returns to a human host, it may be sufficiently altered so it is no longer recognized by the immune system. The result is a major epidemic on a yearly basis.

21.33 **FLU VIRUS TRAVEL TO OTHER ANIMALS**

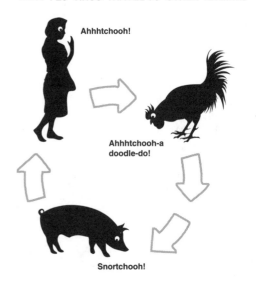

Ahhhtchooh!

Ahhhtchooh-a doodle-do!

Snortchooh!

Flu strains vary greatly in their virulence. For example, influenza virus achieved the honor of being public enemy No. 1 in 1918-1919 when it caused an epidemic that killed more humans than World War I. Mankind felt so embarrassed about being outperformed by a mere virus that it felt obliged to hold World War II not long after, just to show who was smarter.

1918 WORLD SERIES

	Jan	Feb	Mar	Apr	May	Jun	Jul	Aug	Sep	Oct	Nov	Dec
Humans	10	12	6	15	20	12	23	10	4	19	21	25
Flu	100	221	104	212	400	666	365	358	104	269	340	180

Filoviruses - Ebola and Marburg

Every now and then an unknown virus hits the headlines by killing a few people in a gruesome way in some obscure part of the world. Instead of "dog bites mailman," what happens is "monkey bites missionary." Lassa fever (No, not Lassie fever!) appeared this way in the 1960s.

Ebola virus has caused outbreaks in Sudan and Zaire with 80 to 90 percent fatality. In 1989, an Ebola outbreak occurred among monkeys in a research facility in Reston, Virginia. The monkeys, long-tailed macaques, had been freshly imported from the Philippines. Although the incident gave rise to the exaggerated hype of the book, *"The Hot Zone,"* which was followed by a whole slew of B grade plague movies, the Reston Ebola virus was not even lethal to humans.

Ebola, and its buddy Marburg, are negative-strand RNA viruses like flu. They were named filoviruses because they form long, thin filaments. The virus spreads through the blood system and causes vomiting of blood in doomed patients. However, it is relatively difficult to catch by casual exposure and transmission requires substantial exposure to infected body fluids.

If you feel obliged to worry yourself sick over the possibility of a new plague devastating the human race, remember this. Diseases that kill their victims fast do not have much time to find new hosts and so the outbreaks usually burn out quickly. The reason why the AIDS epidemic is so much more dangerous than Ebola virus or Lassa fever is precisely because the disease develops slowly and may therefore be passed on before the first victim is even aware of having it.

Retroviruses and AIDS

Retroviruses are another group of single stranded RNA viruses taking yet a third approach (Fig. 21.35). They convert their ssRNA back into a double stranded DNA copy (Fig. 21.36). Then they insert this DNA into the host cell DNA which makes them impossible to get rid of completely.

Retrovirus particles carry ssRNA of the plus conformation. Although this RNA has a sequence identical to a messenger RNA, it is not used as an mRNA.

What makes a retrovirus "retro" is that upon entering a host cell it reverses the normal flow of genetic information by making a DNA copy of its genes! The process of making a DNA copy from an RNA sequence is known as reverse transcription (Fig. 21.36) and the enzyme which carries out the reaction is reverse transcriptase.

Lassa fever new and very virulent virus disease which appeared in Africa in the 1960s

Ebola virus new and very virulent member of filovirus family which appeared in Africa in the 1980s

filoviruses family of negative-stranded RNA viruses with a filamentous structure

AIDS (acquired immuno-deficiency syndrome) a disease caused by the HIV retrovirus which slowly undermines the immune system by destroying helper T-cells

retroviruses family of RNA viruses with an outer envelope surrounding the nucleocapsid and which convert their ssRNA backwards into DNA by reverse transcriptase

reverse transcription process in which a single-stranded RNA molecule is converted into a double-stranded DNA version

reverse transcriptase enzyme that starts with RNA and makes a DNA copy of the genetic information

21.35 **STRUCTURE OF A RETROVIRUS**

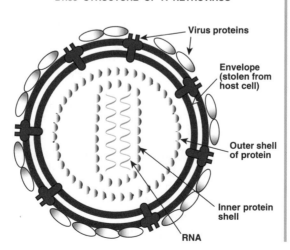

Virus proteins

Envelope (stolen from host cell)

Outer shell of protein

Inner protein shell

RNA

21.36 TRANSCRIPTION AND REVERSE TRANSCRIPTION

NORMAL TRANSCRIPTION

DNA

Template
strand

RNA

mRNA

REVERSE TRANSCRIPTION

RNA

MAKE COMPLEMENTARY
DNA STRAND

DNA

DEGRADE ORIGINAL
RNA

MAKE ds DNA

The retrovirus particle is surrounded by an envelope made from the cell membrane of its previous victim. This membrane layer has retrovirus proteins both inserted through it and covering its surface. Inside is a core particle made of protein and RNA.

Although a retrovirus genome is made of single stranded RNA, the virus particle actually contains two identical ssRNA molecules (Fig. 21.37). These are bound together by base pairing with two molecules of transfer RNA stolen from the previous host cell.

When a retrovirus enters a new cell, the core particle is released into the cytoplasm and disassembles, liberating the ssRNA. One of the two ssRNA molecules is then used by the reverse transcriptase to make a complementary DNA strand.

The first DNA strand is then used as a template to make a second strand. We now have the double stranded DNA form of the retrovirus.

21.37 RETROVIRUS CONTAINS TWO ssRNA MOLECULES

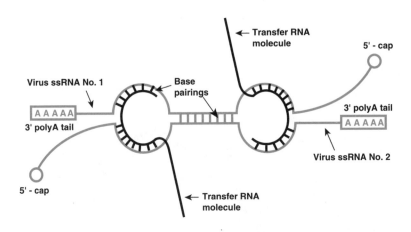

Transfer RNA molecule

5' - cap

Virus ssRNA No. 1

Base pairings

A A A A A
3' polyA tail

3' polyA tail
A A A A A

Virus ssRNA No. 2

5' - cap

Transfer RNA molecule

long terminal repeat (LTR) repeats at end of retrovirus DNA necessary for integrating it into the host cell chromosome

provirus the form in wich a retrovirus exists when its DNA is integrated into the host cell chromosome

This dsDNA now enters the nucleus of the host cell. The dsDNA has a repeated sequence at each end, the long terminal repeats or LTRs. These are required for the integration of the retrovirus DNA into the host cell DNA (Fig. 21.38). The site of integration is more or less random and once integrated, the retrovirus DNA is there to stay. It is called a "provirus" and has become a stable part of the host cell chromosome.

21.38 RETROVIRUS INTEGRATION

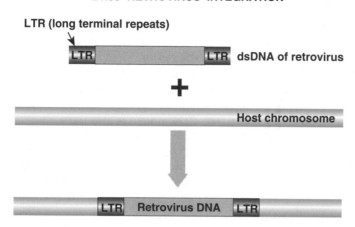

LTR (long terminal repeats)

For a sexually transmitted disease like AIDS, this sadly gives fresh meaning to the phrase "Till death us do part."

The integrated retrovirus DNA is transcribed to give messenger RNA molecules which are capped and tailed just like the mRNA of a typical eukaryotic cell (Fig. 21.39; see Ch. 11 for eukaryotic mRNA). The retrovirus RNA molecules exit the nucleus to the cytoplasm. Some are translated to produce viral proteins and others are packaged into virus particles. Thus the virus particle actually contains mRNA molecules. However, when infecting a new cell, this mRNA is used as a template to make DNA instead of being used as a message. A couple of molecules of reverse transcriptase are packaged along with the RNA.

The first retrovirus to be studied, **Rous Sarcoma Virus**, or RSV, causes tumors in chickens. (See Ch. 14 for cancer-causing viruses.) But not all retroviruses cause cancer. AIDS is caused by **Human Immunodeficiency Virus**, (HIV), which does not cause cancer, but damages the immune system. As a consequence, cancers triggered by other causes often grow out of control. In practice, most AIDS patients die of **"opportunistic" infections**. These are infections only seen in patients with defective immune systems and are caused by assorted viruses, bacteria, protozoans and fungi which are normally harmless but take the opportunity to attack when host defenses are down.

HIV infects circulating blood cells belonging to the immune system, the **T-cells**. Although these cells are not directly killed they are prevented from dividing, which seriously weakens the immune system over a period

A provirus which has inserted into the DNA of a germ line cell will be stably inherited over many generations, just like any other gene. This assumes that the virus is only mildly harmful or has been inactivated by mutation, and does not kill the cell it inhabits. Much of the junk DNA in our human genome may consist of defunct retrovirus DNA (see Ch. 19)

Rous sarcoma virus (RSV) a retrovirus that causes cancer in chickens

human immunodeficiency virus (HIV) the retrovirus that causes AIDS

21.39 RNA SYNTHESIS BY RETROVIRUS

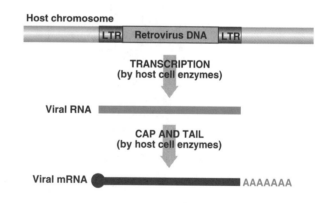

opportunistic infection a disease that does not infect healthy people but which attacks patients with immune system defects

T-cell type of immune system cell responsible for cell-mediated immunity and which makes T-cell receptors instead of antibodies

21.40 CD4 IS RECOGNIZED BY gp120

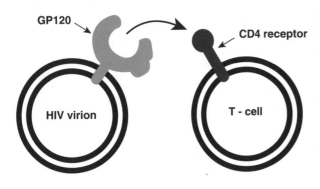

of time. The CD4 protein is found on the surface of many T-cells, where it acts as an important receptor during the immune response (see Ch. 22). Unfortunately, the HIV particle also uses the CD4 protein as a receptor (Fig. 21.40). The gp120 protein in the outer envelope of HIV binds to CD4. This results in entry of the virus.

The CD4 protein is also found on the surface of some other immune system cells, the monocytes and macrophages. These two cell types are not seriously harmed by HIV but act as reservoirs to spread the virus to more T-cells. It is the damage to the T-cells that matters.

CD4 protein a protein found on the surface of many cells of the body's immune system

gp120 or gene product 120 a protein on the surface of HIV which binds to the CD4 protein on a susceptible host cell

Infected T-cells carry the gp120 protein from the HIV particle in their surface membrane. This will then bind to the CD4 protein on other T-cells. The result is that many T-cells clump together and fuse (Fig. 21.41). The giant, multiple cell soon dies. About 70 percent of the body's T-cells carry the CD4 receptor. As they gradually die off, the immune response fades away over a 5- to 10-year period.

No cure or vaccine yet exists for AIDS. The base analog azidothymidine (AZT or Zidovudine), appears to increase the survival time of AIDS patients in some cases. AZT resembles thymidine closely enough to be incorporated in the growing DNA chain during reverse transcription (Fig. 21.42). Because AZT lacks a 3' hydroxyl group, the DNA chain cannot be extended. AZT is thus a DNA chain terminator.

One problem with AZT is that it also inhibits host DNA synthesis in

azidothymidine (AZT) base analog that terminates a growing DNA chain and is used to treat AIDS

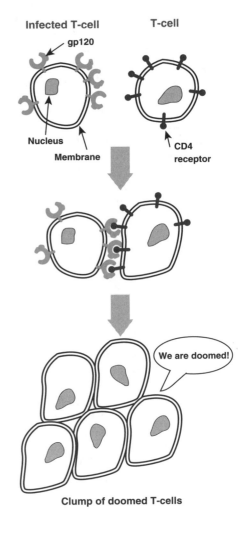

21.41 **CLUMPING OF T-CELLS**

352

uninfected cells of the body. In particular it is toxic to bone marrow cells (B-cells) which are another part of the immune system.

However, the fundamental problem is that HIV is an RNA virus and so has a relatively high mutation rate. HIV mutates at a rate of approximately one base per genome per cycle of replication. Even within a single patient, HIV exists as a swarm of closely related variants known as a "quasi-species." Consequently, strains of HIV resistant to AZT and other similar drugs appear at a relatively high frequency. Other attempts to control AIDS, whether by using vaccines, protein processing inhibitors or antisense RNA all face the same problem, the HIV will mutate to produce resistant variants.

21.42 AZIDOTHYMIDINE VS. THYMIDINE

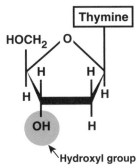

AZT

Thymine

$HOCH_2$

H   H

H   H

N_3 H

↖ Azido group

Thymidine

Thymine

$HOCH_2$

H   H

H   H

OH   H

↖ Hydroxyl group

quasi-species group of related RNA-based genomes that differ slightly in sequence, but arise from the same parental RNA molecule

Puzzle of the Prion

Until recently all infectious diseases were thought to be caused by germs that carry at least some of their own genes. Some diseases are due to living cells like bacteria, while others are due to viruses or viroids (see Ch. 19), but all contain their own genetic information in the form of DNA or RNA. However, several weird diseases of the nervous system are caused by infectious agents containing no nucleic acid at all. These bizarre diseases are due to rogue proteins known as "prions" and include both inherited diseases and infectious diseases.

The prion protein is actually coded for by a gene belonging to the victim. This gene is transcribed and translated normally and produces a protein found attached to the outside of nerve cells, especially in the brain. Its proper function in the brain is still unknown.

The critical property of the prion protein is that it has two alternative structures, good and bad (Fig. 21.43). Occasionally the normal, properly folded form rearranges to produce the rogue form of the protein which somehow messes up nerve cell operations.

In the inherited form of prion disease, for example, Creutzfeldt-Jacob disease, a mutation in the gene results in a mutant prion protein that changes more often into the bad form. The overall result is little different than a typical inherited disease.

prion distorted disease-causing form of a normal brain protein that can transmit an infection

Creutzfeldt-Jacob disease inherited brain degeneration disease of humans caused by prion

21.43 TWO FORMS OF THE PRION PROTEIN

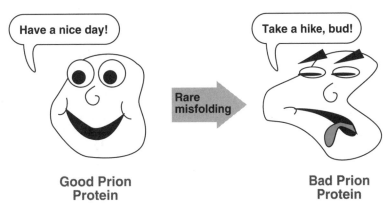

Have a nice day!

Take a hike, bud!

Rare misfolding

Good Prion Protein

Bad Prion Protein

Good **Bad**

The interesting prion diseases are the infectious ones. When a bad prion protein infects a healthy nerve cell, it finds its good relatives and convinces them to go bad, too. Presumably the bad prion protein binds to the good version and somehow twists it into the bad shape (Fig. 21.44). The protein is not chemically altered, it simply changes shape. Good prions consist mostly of α-helical chunks, whereas bad prions have less helix and lots of β-sheet structures instead (for protein structures see Ch. 7).

Infection of a new victim by prions is quite difficult, and requires transfer of bad prion proteins from infected brain tissue. The best known of these diseases are:

1) **Scrapie**, a disease of sheep and goats

2) **Mad Cow Disease** - the official name, Bovine Spongiform Encephalopathy, sounds even worse!

3) **Kuru**, a disease of cannibals

Mad cow disease is not really a "natural" disease since cows do not usually munch on each other's brains. What happens is that animal remains, including the brains, are ground up and incorporated into animal feed. In fact, the present epidemic of mad cow disease which started in England in 1986, is thought to have originated from sheep suffering from scrapie, a disease identified in 1738. Sheep remains were processed into feed for cows and hence, the disease was transmitted. The prion proteins of sheep and cows differ by only half a dozen amino acids and are close enough in shape for bad sheep prions to turn good cow prion proteins bad. Whether mad cow prions can cause disease in people is still unknown. As the name suggests, the symptoms are that the infected animals go wacko.

Kuru was transmitted by ritual cannibalism before Eurocentric intolerance led to a decline in such cultural practices. Kuru used to be endemic among the Fore tribe of New Guinea. The women had the honor of preparing the brains of dead relatives and participating in their ritual consumption. As a result, 90 percent of the victims were women and younger children who accompanied them. It may take 10 to 20 years for symptoms to develop, but once they do, the progression from headaches to difficulty walking, to death from neural degeneration, takes from one to two years. No one born to the Fore since 1959, when cannibalism stopped, has developed kuru.

scrapie brain degeneration disease of sheep caused by prion

mad cow disease brain degeneration disease of cattle caused by prion which originally came from sheep with scrapie

kuru brain degeneration disease of cannibals caused by prion

Now that we're feeling happy about the demise of kuru, let's move on and see how our own genetically encoded strategic defense initiative, the immune system, stomps on wicked invading germs.

Additional Reading

Bacterial Pathogenesis: A Molecular Approach by Salyers AA & Whitt DD. 1994. American Society for Microbiology, Washington, DC.

The Hot Zone by Richard Preston 1994. Random House, New York.

The Molecular Defense Initiative:
Your Immune System At Work

As you probably realize, the world is full of nasty germs, all conspiring to infect you. Every second of every day, bacteria, viruses and other microorganisms with evil in their tiny hearts, are attempting to gain entry into your tissues. Some land on your skin, others float in the air entering your lungs, and yet others sleaze by hidden in your food or water. Others, even less polite, may sneak in when you are happily cooperating with another person to extend the human species!

If nothing was done about these constant attempts at invasion, you would be long gone. In fact, your body surfaces, both outside and inside, as well as your internal tissues and bloodstream, are patrolled by a counter-insurgency system that makes the KGB look like amateurs.

The cells of the immune system constantly patrol the body looking for outsiders. Invading microorganisms will have their own distinctive proteins which are different in sequence and therefore in 3-D structure from those of the host animal. Any foreign macromolecules that are not recognized as being "self," trigger an immune response.

22.2 **LOOKING FOR STRANGERS**

357

22.3 OVERVIEW OF THE IMMUNE SYSTEM

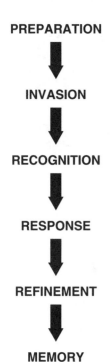

PREPARATION

> Immune system generates zillions of antibodies by gene shuffling. Each B-cell knows how to make antibody of just one shape.

INVASION

> A disease causing agent invades the body.

RECOGNITION

> A few among the zillions of waiting antibodies fit the surface molecules (antigens) of the invader well enough to bind them.

RESPONSE

> Those B-cells which make the chosen antibodies divide and produce mass quantities of antibody.

REFINEMENT

> Mutations occur which improve those antibodies which bind the antigen.

MEMORY

> Special memory cells remember antigens which have invaded the body in the past.

What happens when we are invaded by bacteria or viruses that cause disease? The exposed molecules on the surfaces of these enemy germs will be what the immune system makes contact with. These molecules will be different in their detailed shapes from the molecules that are part of the body itself, so they will be regarded as signs of an intrusion. These foreign molecules are the antigens and the immune system molecules that recognize and bind to them are the antibodies (see Fig. 22.3).

You may think that our story begins when an evil germ invades the body and the immune defense forces rush to stomp its microscopic butt. Not so. In fact, the clever stuff has already been done before our invader shows up. The body is lying in wait. What has it been doing? How do you recognize something you have never met before?

Although we often speak sloppily of how the body makes **antibodies** in response to invasion by a foreign **antigen**, this is a bit misleading. In fact, long before the infection the immune system generates zillions of different antibodies and keeps a few cells on standby that know how to make each one. This happens before encountering the antigens and without knowing which antibodies will actually be needed later. If enough different antibodies are available, at least one or two should match an antigen of any conceivable shape, even if the body has never come into contact with it before.

Eventually, the evil foreign antigen appears (Fig. 22.4). Among the zillions of waiting antibodies, at least one or two will fit the antigen reasonably well. Each immune B cell knows how to make just one type of antibody.

22.4 ARE YOU FEELING LUCKY?

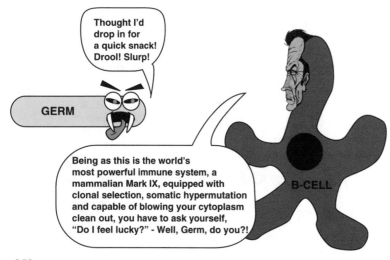

Those B-cells that make antibodies that recognize the antigen now divide rapidly and go into mass production. Once invading antigens have been bound by the corresponding antibody, the body brings other mechanisms into play to destroy the invaders.

A little later there is a stage of refinement during which those antibodies that bound to the invader are modified by mutation to fit the antigen better. In addition, the immune system keeps a special record of antibodies that are actually used. If the same invader ever returns, the corresponding antibodies can be rushed into action, faster and in greater numbers than before. And now for the nitty gritty…

Antibodies, Antigens and Epitopes

Foreign molecules detected by the immune system are called antigens. In practice, most antigens are proteins made by invading bacteria or viruses. However other macromolecules such as **polysaccharides**, often found as surface components of infiltrating germs, can also work as antigens. The antigens exposed on the surface of an alien microorganism will usually be detected first by the immune system (Fig. 22.5).

Several types of immune system proteins recognize and bind to antigens; the antibodies and the **T-cell receptors** are the most famous. Antibodies bind to whole proteins, whereas T-cell receptors bind to chewed up fragments of protein.

When an antibody binds to a protein it recognizes a relatively small area on the surface of the protein. Such recognition sites are known as **epitopes** (Fig. 22.6). Just as we recognize other people by their ugly noses or projecting ears, so antibodies bind best to lumps and projections sticking out from the protein surface.

antigen a molecule that causes an immune response and which is recognized and bound by an antibody

antibody protein of the immune system that recognizes and binds to foreign molecules, *i.e.*, antigens

polysaccharide polymeric molecule made of sugar subunits

22.5 FOREIGN MOLECULES ON SURFACE OF CELL

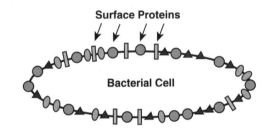

Surface Proteins

Bacterial Cell

T-cell receptor protein found on the surface of T-cells that recognizes and binds to foreign protein fragments when they are displayed by the MHC proteins

epitope localized region of an antigen to which the antibody binds

22.6 ANTIBODY BINDS TO EPITOPE ON ANTIGEN

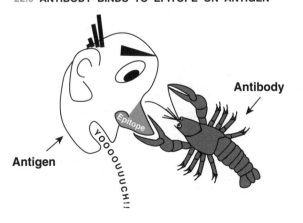

Antibody

Epitope

Antigen

YOOOOUUUCH!!

Since intact proteins are large molecules, they may have suitable recognition sites for antibodies at several places on their surfaces (Fig. 22.7). Consequently, several different antibodies may be able to bind the same protein, although not usually all at once as they get in each other's way!

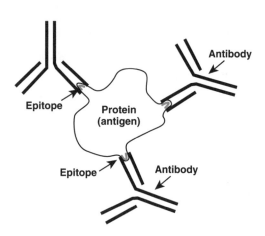

Epitope

Protein
(antigen)

Antibody

Epitope

Antibody

Diversity of Antibodies

Antibodies are protein molecules that recognize and bind to alien molecules. Since there is an almost infinite variety of possible alien molecules, we need a correspondingly colossal number of antibody molecules. Fortunately, the amino acids making up protein molecules can be arranged to give an almost infinite number of different sequences and therefore, of different shapes.

This leads to a major problem. If each antibody protein were encoded by a separate gene, we would need a gigantic number of genes and a correspondingly vast amount of DNA. Unfortunately, calculations based on the number of different antibodies produced by humans or other animals, show that the amount of DNA available is nowhere near sufficient to code for all of them.

The answer? Simple, once you have thought of it and won your Nobel prize. Instead of complete genes for each antibody we have a collection of partial genes. By shuffling our set of partial genes and joining the lengths of DNA together to make complete antibody genes, we can generate an immense variety of different antibodies.

In Figure 22.8 this idea is illustrated using three alternative front ends and three rear ends. Combining them all possible ways allows us to make nine different genes.

Suppose in our collection of gene parts we have a set of 10 front ends and a set of 10 back ends, all different. We can combine any front end with any rear end. Therefore we can make 10 x 10 = 100 combinations. Suppose we have front, middle and end pieces we could make 10 x 10 x 10 = 1,000 combinations. This is getting closer to what really happens with antibody genes. Note that to generate our 1,000 possible proteins we need only 30 lengths of genetic information to store on our chromosome.

light chain   the shorter of the two pairs of chains comprising an antibody molecule

heavy chain   the longer of the two pairs of chains comprising an antibody molecule

disulfide bond   chemical linkage between two cysteine residues that binds together two protein chains

constant region   region of an antibody protein chain remaining the same for all chains

variable region   region of an antibody varied by gene shuffling in order to provide many alternative antigen binding sites

Antibody Structure

So now that we have the general idea, let's take a look at real antibodies. Each antibody consists of four protein subunits, two light chains and two heavy chains, arranged in a Y-shape (Fig. 22.9). The chains are held together by disulfide bonds between cysteine amino acid residues.

22.8 MODULAR GENE SHUFFLING

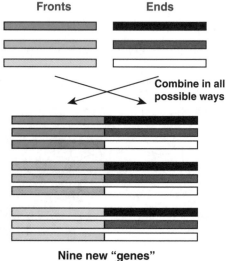

Fronts   Ends

Combine in all possible ways

Nine new "genes"

22.9 ANTIBODY STRUCTURE

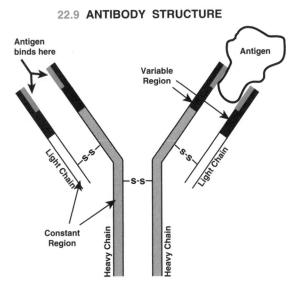

Each of the light and heavy chains consists of a **constant region** and a **variable region**. The constant region is the same for all chains of the same class. The variable region binds to the target molecule, the antigen. There are millions of different variable regions which, as we shall see, are generated by genetic shuffling.

If an antibody is broken at the "hinge"

Fab fragment antigen binding fragment of an antibody

Fc fragment the stem region of an antibody, *i.e.*, the fragment not binding the antigen

λ (Lambda) chain one of the two alternative types of antibody light chain

κ (Kappa) chain the other of the two alternative types of antibody light chain

where the heavy chains bend, we get three chunks, two identical **Fab fragments** and one **Fc fragment** (see Fig. 22.10). Fab, meaning "fragment, antigen binding," consists of one light chain plus half of a heavy chain. Fc, meaning "fragment, crystallizable" contains the lower halves of both heavy chains. Other components of the immune system often recognize and bind to the Fc region of an antibody (see below).

Two Types of Light Chain

There are actually two types of light chains. The human genes for making up the λ (lambda) chain are on chromosome 2 and the genes for the κ (kappa) chain on chromosome 22. Any particular antibody will have two identical λ chains or two identical κ chains, never a mixture.

The genetic information for a κ chain consists of three types of modular chunks, designated V-, J- and C-regions, used to make the final product. The variable region of the κ light chain is made by shuffling the V- and J- regions, while the constant region is encoded by the single C-region. These gene segments are originally

22.10 Fc AND Fab FRAGMENTS OF AN ANTIBODY

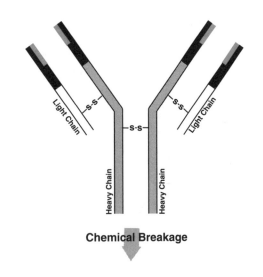

Chemical Breakage

Two Fab Fragments   **One Fc Fragment**

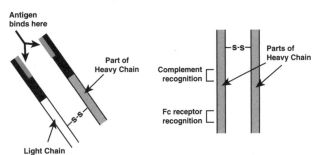

22.11 GENE RE-ARRANGEMENTS FOR KAPPA CHAIN

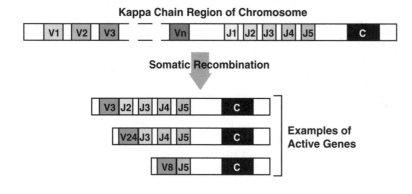

Kappa Chain Region of Chromosome

Somatic Recombination

Examples of Active Genes

arranged as follows: first, approximately 300 alternative V-regions, then five alternative J-regions and finally the single C-region (Fig. 22.11).

During development of an immune cell, DNA is deleted by a process known as somatic recombination. The first step is to choose one each of the V- and J-regions. The DNA between the chosen V-region and the chosen J-region is then deleted. This results in an active gene consisting of the chosen V-region fused to the chosen J-region, together with the single constant region (Fig. 22.11). Note that the V- regions upstream of the chosen V- region are not deleted, they are merely ignored when making the light chain, and are not even transcribed. Some unused J-segments between the chosen J-segment and C-region remain at this stage and are transcribed when messenger RNA is made (Fig. 22.12).

Only the J-region fused to the chosen V-region will appear in the final κ chain. The extra J-regions will be removed after transcription and during RNA processing (Fig. 22.12). We've illustrated this for the first active gene from Figure 22.11 (above) which has the DNA sequence V3-J2-J3-J4-J5-C and the mRNA structure V3-J2-C (plus poly-A tail).

somatic recombination
gene rearrangement by recombination which occurs in somatic (non-reproductive) cells, in this case B-cells or T-cells of the immune system

22.12 RNA PROCESSING FOR KAPPA CHAIN

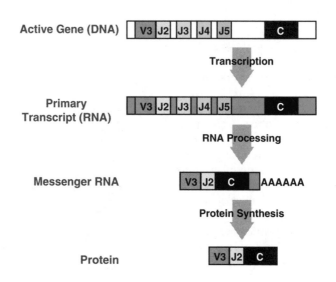

Active Gene (DNA)

Transcription

Primary Transcript (RNA)

RNA Processing

Messenger RNA

Protein Synthesis

Protein

This leaves a messenger RNA that is translated into a κ chain containing a single V-region linked to a single J-region and the constant region. There are approximately 300 x 5 = 1,500 different possible κ chains. Wow! There are about the same number of possible λ chains which are generated in much the same way. Double wow!!

Heavy Chains and Antibody Classes

Now let's deal with the heavy chains. The principle is the same but the details are more complicated and we need to polish our Greek. First, there are five heavy chains: μ (mu), δ (delta), γ (gamma), ε (epsilon), and α (alpha). These give rise to five corresponding antibody classes (M, D, G, E and A) each playing different roles in immunity. Antibodies are often referred to as immunoglobulins (Ig), so these antibody classes are called IgM, IgD, IgG, etc. (A globulin is simply a soluble globular protein found sloshing around in blood or other body fluids.)

The heavy chain genes are all on human chromosome 14 and consist of about 100 V-segments, 20 D-segments, and four J-segments and then the constant chains in the order μ – δ – γ – ε – α. The variable region of the heavy chain results from choosing one each of the V-, D-, and J-segments at random and fusing them together. This gives 100 x 20 x 4 = 8,000 possible combinations. Wow cubed!!! During immune cell development, a single active gene is produced containing the fused V/D/J-segment attached to all five possible constant regions (Fig. 22.13). Antibody-producing cells are known as B-cells and each individual, original B-cell contains a different V/D/J combination.

antibody classes there are five classes of antibody, each depending on which of the corresponding five types of heavy chain is present

immunoglobulin another name for an antibody protein

B-cell the type of immune system cell that produces antibody

22.13 **HEAVY CHAIN GENE SHUFFLING**

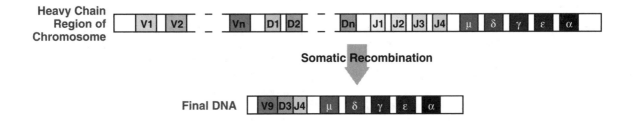

This leaves the question of which heavy chain constant region to use. In fact the different C-regions are all used, one after the other. Early in the life of an antibody-secreting B-cell, the μ (mu) chain is used. Transcription at this stage produces a primary RNA transcript containing the randomly chosen variable region, consisting of the fused V/D/J-segments, separated from the μ segment by intervening DNA (Fig. 22.14). RNA processing splices the μ segment to the variable region (the fused V/D/J segments) and allows production of the final IgM heavy chain.

junctional flexibility increasing antibody diversity by alterations introduced when joining together the separate segments making up the variable region

terminal deoxynucleotide transferase an enzyme that adds a few extra bases when junctions are made between the gene segments for the V-, D-, and J-regions of antibody heavy chains

The M-class antibodies (immunoglobulin M or IgM), are found in the membrane of the B-cell that makes them. Each B-cell produces only one type of antibody. However, there are billions of B-cells, each carrying on its surface an IgM class antibody with a different variable region.

22.14 SYNTHESIS OF IMMUNOGLOBULIN

DNA | V9 | D3 | J4 | μ | δ | γ | ε | α |

Transcription

Primary RNA Transcript | V9 | D3 | J4 | μ |

RNA Processing

Messenger RNA | V9 | D3 | J4 | μ | AAAAAAAA

RNA Processing

μ Heavy Chain Protein | V9 | D3 | J4 | μ |

22.15 JUNCTIONAL FLEXIBILITY: EXAMPLES

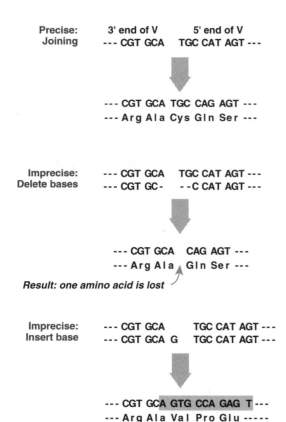

Precise:
Joining

3' end of V   5' end of V
--- CGT GCA   TGC CAT AGT ---

--- CGT GCA TGC CAG AGT ---
--- Arg Ala Cys Gln Ser ---

Imprecise:
Delete bases

--- CGT GCA   TGC CAT AGT ---
--- CGT GC-   - -C CAT AGT ---

--- CGT GCA   CAG AGT ---
--- Arg Ala Gln Ser ---

Result: one amino acid is lost

Imprecise:
Insert base

--- CGT GCA   TGC CAT AGT ---
--- CGT GCA G   TGC CAT AGT ---

--- CGT GCA GTG CCA GAG T ---
--- Arg Ala Val Pro Glu -----

Result: change of reading frame beyond insertion

Junctional Flexibility

So far we have approximately 8,000 possible heavy chains and 3,000 possible light chains (1,500 each κ or λ) giving 24,000,000 total possible combinations. Wow to the fourth power!!!! This might seem like quite a few, but more variation is introduced by sloppily joining the V-, D-, and J-segments.

The mechanism for joining the variable region DNA segments is unusual in being imprecise. There may be up to four bases of "slop" at the V/D and D/J junctions (Fig. 22.15). This process is known as junctional flexibility and enormously increases the number of possible combinations. Wow wo woww wow woow!!!!

Normal genetic recombination is very accurate and joining occurs precisely at equivalent bases in both strands of DNA. However, in forming the variable regions of both the antibody light and heavy chains, one or more nucleotides are often lost from the ends of the segments when they are joined. Moreover, in the case of the heavy chains, up to half a dozen nucleotides are sometimes inserted. The enzyme terminal deoxynucleotide transferase adds these random nucleotides when V/D and D/J junctions are being made.

Clonal Selection

After reaching the stage of making surface-bound IgM, the antibody producing B-cells hang around in the spleen and lymph nodes, waiting. B-cells with IgM on their surface are referred to as **virgin B-cells**. Further development only occurs if the IgM on the cell surface recognizes and binds to an alien molecule, the antigen. (In fact, stimulation of a B-cell to divide is a bit more complicated. It also requires permission from a T-helper cell. We'll worry about that below.)

Since there are so many different antibodies, whenever an alien molecule infiltrates the body, somewhere there will be a B-cell with an antibody to match. This happy event will, of course, only happen to a tiny minority of the zillions of different IgM antibodies. Most never achieve fulfillment.

Those B-cells whose IgM finds a matching antigen are stimulated to divide (Fig. 22.16). This is known as "clonal selection," as only the B-cell matching the antigen is selected for fame and fortune. Its division yields a large number of identical B-cells (clones), all with antibodies against the alien antigen molecule which dared intrude into the body.

virgin B-cell B-cell which has an IgM antibody on its surface but which has not yet been stimulated to divide by encountering a matching antigen. Sometimes called a "naive" B-cell by prudes and the politically correct

22.16 PRINCIPLE OF CLONAL SELECTION

B-Cell Development and Somatic Hypermutation

When our waiting virgin B-cell finds an antigen that matches its antibody, it has found its mission in life. Binding of the antigen triggers further development. All kinds of things happen to the B-cells and they happen all at once. One of the most interesting results is an enhanced rate of point mutations known as somatic hypermutation (Fig. 22.17).

How the mutation rate is increased is still unknown. Nonetheless, dividing B-cells undergo hypermutation but only in the DNA coding for the variable regions of both light and heavy chains. Roughly one mutation happens per V-region per cell generation. (Note: hypermutation does not happen to the T-cell receptors - see below.)

Remember that our hypermutating B-cell has already chosen which segments to fuse together into its variable regions and has also found a matching antigen. The original match is good enough for recognition though not likely to be perfect. However, some of the mutations will make the binding of antigen to antibody tighter, and of course, others will make it worse. Those mutated B-cells that bind antigen tighter are stimulated to grow and divide further. Those binding poorly are rewarded with programmed cell death (apoptosis; see Ch. 14). So as time passes, the average affinity of antibodies for their particular antigen increases; this process is called "affinity maturation."

somatic hypermutation deliberately increased rate of mutation in non-reproductive cells - in this case specifically in the antibody encoding genes of antibody producing cells

The mutation rate of variable region DNA in a B-cell is about a million times greater than the mutation rate typical of normal genes

apoptosis programmed suicide of unwanted cells for the good of the whole animal

affinity maturation emergence over time of mutated antibodies with increased affinity for the antigen

22.17 HYPERMUTATION INCREASES AFFINITY

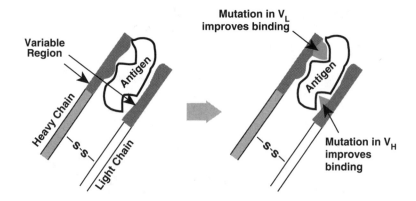

Class Switching

Once a particular line of B-cells has been selected by antigen recognition, other changes also occur. In particular, the antibody is changed by a second genetic rearrangement known as class-switching. The variable region is not affected by this process, but the class of heavy chain is switched. Usually, IgG is produced by splicing out the μ and δ segments from the DNA.

The heavy chain gene still contains the γ, ε, and α segments. As before, RNA processing is involved in making the final IgG heavy chain protein. The ε and α segments are spliced out during conversion of the primary transcript to the messenger RNA (Fig. 22.18).

class-switching genetic rearrangement that replaces one type of heavy chain with another, so changing the class to which an antibody belongs

So what about the other types of heavy chain? If we want to make IgA or IgE then we need a different genetic rearrangement. The principle is the same as for IgG. DNA is deleted between the fused V/D/J variable region and whichever constant region is desired. Any constant regions remaining on the far side of the chosen constant region will be removed during RNA processing, just as the α and ε regions were removed during RNA processing for γ chain mRNA.

Different Roles for Different Classes of Antibody

IgG, with two γ heavy chains plus two light chains (κ or λ), is the most common form of antibody in the blood and accounts for about 80 percent of the total immunoglobulin. It does not remain attached to the B-cell surface (like IgM) but is secreted into the blood. The IgG binds to alien antigen molecules, most of which will be on the outer surface of invading bacteria or viruses. Consequently, these evil germs end up with a coating of IgG on their surfaces (Fig. 22.19).

The constant region of the γ heavy chain is recognized by **Fc receptors** situated on the surface of the **macrophage** and **neutrophil** cells. These cells are the stormtroopers of the immune system. Macrophages stand guard while neutrophils cruise the bloodstream looking for someone to kill.

22.18 RNA PROCESSING IN IgG PRODUCTION

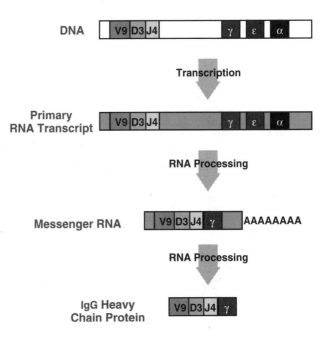

Fc receptor protein on the surface of immune cells that recognizes the Fc region of an antibody once it has bound an antigen

macrophage immune cell, mostly found in solid tissues. It ingests and destroys invading microorganisms

neutrophil immune cell, most often found in blood. It ingests and destroys invading microorganisms

22.19 CELLS ARE COATED WITH IgG

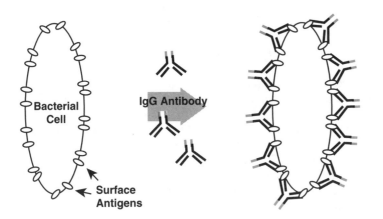

When their Fc receptors recognize IgG that has attached itself to antigen, the macrophage or neutrophil swallows and digests the whole germ with its attached antibodies (Fig. 22.20).

22.20 MACROPHAGE TAKES OUT TAGGED GERM

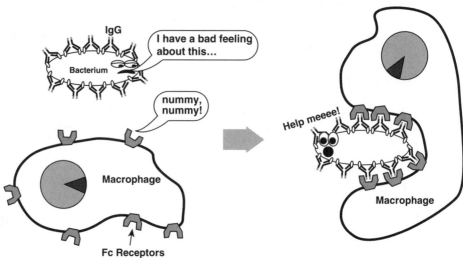

mast cell specialized immune cell responsible for the secretion of histamine

histamine substance made from the amino acid histidine which is released by mast cells and causes expansion of blood vessels and acceleration of the heart rate

leukocyte white blood cell

B-cell type of immune system cell which makes antibodies

T-cell type of immune system cell responsible for cell-mediated immunity and makes T-cell receptors instead of antibodies

killer cell type of T-cell whose job is to kill other cells of the body that have become "foreign" due to cancer or infection

As we have seen, IgM is specialized to bind the antigen when it first appears and this triggers growth of the appropriate B-cell. IgD is made in tiny amounts at the same time as IgM and its role is obscure. IgG is the main antibody circulating in the blood. IgA is found in all secreted body fluids such as saliva, tears, milk, gastric juices, etc. It may actually be made in larger quantities than IgG but only a small amount of IgA is found in blood serum, the fluid usually analyzed for antibodies.

IgE is made only in small amounts and has a special role to play. The Fc region of IgE binds to specific receptors on mast cells and triggers the release of histamine by these cells. This makes blood vessels expand in the area where antigens have been detected, so immune cells can travel more easily to the battleground. Too much histamine release can cause allergic reactions like hay fever and asthma.

Cells of the Immune System

Blood cells are often divided into red cells that possess hemoglobin, the red, oxygen-carrying protein, and the white cells, or leukocytes, that belong to the immune system and include B-cells, T-cells, neutrophils, etc.

All these immune cells stem from ancestral cells known as stem cells in the bone marrow. These divide and give rise to two types of cells smart enough to recognize antigens, B-cells and T-cells. In addition, the stem cells also produce a variety of less intelligent killer cells such as macrophages and neutrophils (Fig. 22.21).

22.21 ORIGIN OF IMMUNE CELLS

Stem Cell

Dumb Immune Cells

Duh!! Heh! Heh! Heh!

Clever Immune Cells

25V x 2D x 12J = 600

Kill!!

Burn!!

Eat!!

Neutrophil

Mast Cell

Macrophage

B-Cells

T-Cells

We are educated! We recognize alien molecules!!

bursa of Fabricius organ of chicken where B-cells develop

thymus organ in the chest where T-cells mature after leaving the bone marrow

cell-mediated immunity immune reactions due to cells of the immune system acting as a whole as opposed to those due to antibody molecules

Both B- and T-cells carry out gene shuffling and then wait until they are triggered to divide further by the binding of an antigen. The B-cells are actually named after the **bursa of Fabricius**. It is one of those strange parts of the chicken whose name you never even wanted to know and which you give to the cat. Since the Bursa is absent in mammals, the B-cells hang out instead in the bone marrow, which luckily also begins with B. The T-cells are named after the **thymus,** where they mature after emigrating from the bone marrow.

The T-Cell Receptor

B-cells make antibody but T-cells do not. Instead T-cells carry out **cell-mediated immunity.** There are two aspects to this and two types of T-cells. Helper T-cells activate and stimulate B-cells. Killer T-cells kill infected cells.

22.22 "HERE KITTY"

Oh kitty, I've got some yummies for you! Tasty chicken innards! Yum-yum!

Oh great! Not another bursa of Fabricius! When will she learn I prefer dark meat!

T-cell receptor protein found on the surface of T-cells that recognizes and binds to foreign protein fragments when they are displayed by the MHC proteins

If a cell of your body is infected by a virus or bacterium, foreign proteins often get into the cell membrane and appear on the cell surface. Killer T-cells detect the foreign proteins and destroy the doomed cells to limit the spread of infection.

To do this, the T-cells must detect foreign antigens. They do not have IgM antibody molecules on the cell surface as do B-cells. Instead they have proteins known as **T-cell receptors** (TCRs) that are similar to antibodies in having a variable region which binds the antigen and a constant region (Fig. 22.23). T-cell receptors differ from antibodies in never being secreted and in having only two protein chains each, the α and β chains.

A great variety of T-cell receptors are generated by gene shuffling, just as with antibodies. The TCR α chain is made by choosing from a large number of V and J segments joined to a single constant region, rather like the antibody light chain. The variable region of the TCR β chain is made from V, D and J segments, as is an antibody heavy chain. In addition, the diversity is increased by sometimes adding or removing a few nucleotides at the V/D/J junctions just as with antibodies. (The hypermutation that occurs in antibody genes in later generations of B-cells does not happen to T-cell receptor genes.)

22.23 STRUCTURE OF THE T-CELL RECEPTOR

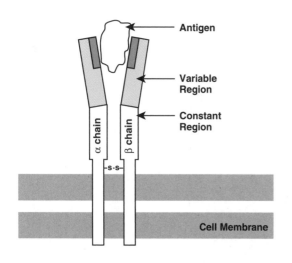

major histo-compatibility complex (MHC) family of genes which encode proteins found on the cell surface where they display fragments of foreign proteins for immune recognition

Antigen Presentation by MHC Proteins

The immune system kills any cells with foreign proteins on their surfaces because it regards them as not self. In real life, of course, such cells are almost always either infected or cancerous. But this can be a snag when performing an organ transplant. The immune system thinks the implanted organ is foreign (which is true enough) and then concludes that it must be an enemy and attacks it.

This brings us to a complicated family of proteins named the **major histocompatibility complex** or **MHC** because they have a big effect on tissue rejection (see below). Let's worry first about their natural role. The MHC proteins mark for destruction cells infiltrated by germs or that have committed treason by going cancerous. Inside cells gone bad, there are foreign proteins. Usually at least some of these will be broken down by digestive enzymes into short fragments. These fragments are bound by the MHC proteins which hold them out on display at the cell surface (Fig. 22.24).

There are two types of MHC proteins, class I and class II, which have similar

22.24 MHC PROTEINS BIND PROTEIN FRAGMENTS

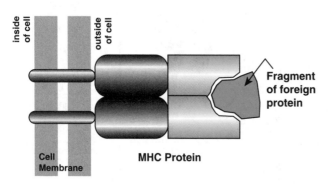

370

structures (Fig. 22.25). The **class II MHC** complex has two chains, α and β, of equal length which share the job of binding the protein fragment. In contrast, the **class I MHC** complex is lopsided, its α chain is longer and folds over to bind the protein fragment on its own. It is paired with a half length β chain, β_2-microglobulin.

class II MHC MHC proteins consisting of two chains of equal length and found only on the surface of certain immune cells. Their role is to display fragments of proteins from digested microorganisms

class I MHC MHC proteins consisting of two chains of unequal length, found on the surface of all cells. Their role is to display fragments of proteins originating inside the cell

22.25 CLASS I AND CLASS II MHC PROTEINS

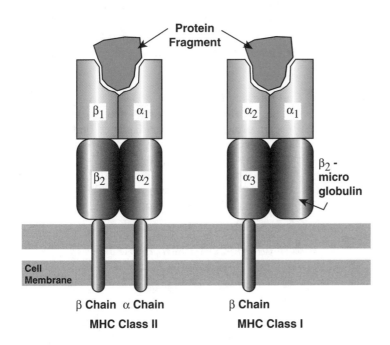

When an alien fragment is held out by the MHC proteins, the immune system goes into attack mode. The fragments displayed by the MHC proteins are detected by the T-cell receptors (Fig. 22.26). Thus, whereas antibodies bind whole foreign proteins, T-cell receptors bind only digested fragments which are bound to the MHC proteins.

22.26 MHC AND FRAGMENT BINDS TO TCR

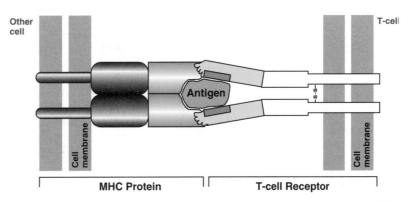

371

The two MHC classes have slightly different roles. Class I MHC are found on all cells and usually carry fragments of proteins that originated from inside the cell, for example, from infecting bacteria or viruses. In contrast, the class II MHC specializes in carrying fragments of alien proteins that originally came from outside the cell. These proteins are swallowed and degraded by macrophages and other immune system cells, and it is only these cells which have class II MHC on their surfaces.

Corresponding to these two situations are the two major types of T-cell. Killer T-cells recognize the class I MHC whose display of alien fragments is a sign that the cell is infected and should be destroyed. Helper T-cells recognize the class II MHC whose display signals enemy proteins in the bloodstream. The helper T-cells then proceed to activate the B-cells that make antibody.

The CD4 and CD8 Cell Surface Proteins

As remarked above, the T-cell receptor actually detects degraded fragments of the original foreign protein. These are displayed on the surface of infected cells by the MHC proteins. The CD4 and CD8 proteins are found in T-cell membranes next to the T-cell receptors. CD4 and CD8 bind to the MHC proteins; CD4 binds to class II MHC proteins, whereas CD8 binds to class I. This helps to stabilize the T-cell receptor/antigen/MHC complex and so CD4 and CD8 are often called co-receptors (Fig. 22.27).

22.27 CD4 BINDS TO MHC PROTEINS

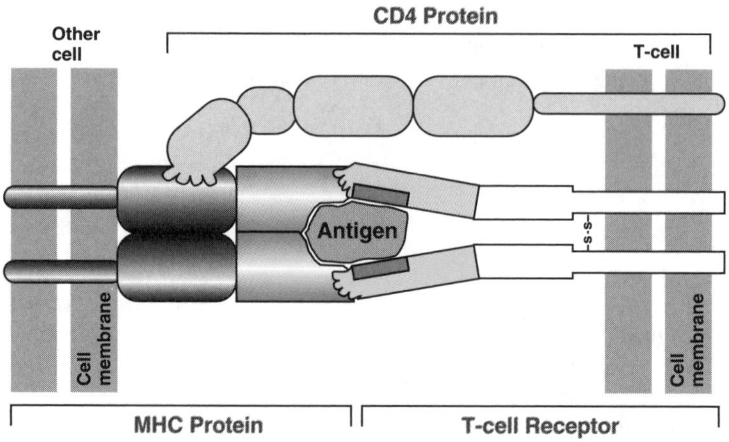

About two-thirds of the T-cells are helper T-cells and have CD4 while the other one-third are killer T-cells and have CD8. The HIV virus that causes AIDS, uses the CD4 protein as a receptor when infecting host cells. The gp120 protein, found on the surface of the virus, binds to CD4. Consequently the AIDS virus preferentially infects and kills helper T-cells, thus undermining the immune response (see Ch. 19).

gp120 ("gene-product" 120) a particular protein found on the surface of the AIDS virus and which binds to the CD4 protein in the T-cell membrane

The Complement Chemical Attack System

The constant region of the heavy chain of IgG has two special functions, both designed to destroy any intruding germ to which the IgG has bound. First, IgG binds to the Fc receptors on macrophages, etc., and prompts them to swallow and munch on the germ. Second, IgG triggers a vicious chemical attack on alien microorganisms. There is no Geneva Convention inside the body!

The **complement** SWAT team consists of 11 proteins, named C1, C2, C3, etc., which are always present in the blood (Fig. 22.28). The C1 protein recognizes and binds to IgG only after it has bound an antigen on the surface of a germ. This causes the successive binding and activation of the other complement proteins. The death blow comes when a cluster of C5, C6, C7, C8 and a dozen C9 land on the bacterial cell surface and drill a hole through its membrane. The cell contents of the bacterium leak out and that's all she wrote.

complement a team of proteins that destroys invading bacteria after being alerted by the binding of antibody

22.28 EXTREME SANCTION BY COMPLEMENT

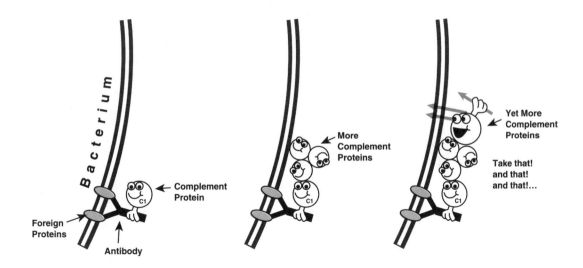

Just for fun, the complement cascade also sends off chemical signals to aggressive white cells (our buddies the neutrophils, etc.), inviting them to come and join the action.

Immune Memory

As we all know, if you survive an infection, you normally become immune to that particular disease, although not to other diseases. In other words, your immune system remembers the bad guys. Next time it sees them, it reacts far more swiftly and aggressively than on the first exposure, and the germs are usually overwhelmed before they can establish a bridgehead and cause a noticeable illness (Fig. 22.29).

Immune memory is due to specialized cells called - yes! - **memory cells**, about the only instance of immunologists actually saying what they mean (see Fig. 22.30)!!

As you already know, if you read this chapter carefully that is, virgin B-cells are triggered to divide when they find the antigen that matches their own personal antibody. Most of the new B-cells devote themselves to mass production of antibodies. This wears them out and they only live a few days.

memory cell a specialized B-cell which waits for possible future infections instead of manufacturing antibodies

A smaller group become memory cells and instead of making antibodies they just hang around waiting, waiting, waiting.... Then one day the antigen they recognize appears again and most of the memory cells switch over very rapidly to antibody production. The invader is snuffed out fast.

22.30 **MEMORY B-CELLS**

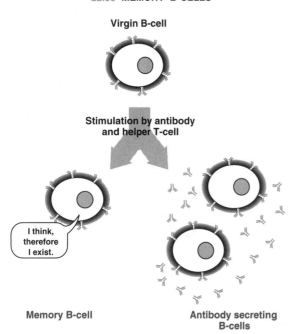

Virgin B-cell

Stimulation by antibody and helper T-cell

I think, therefore I exist.

Memory B-cell

Antibody secreting B-cells

vaccination artificial induction of the immune response by injecting foreign proteins or other antigens

cowpox a disease of cattle caused by a virus closely related to smallpox. Cowpox causes only a very mild disease in humans

Vaccination and Infectious Disease

Vaccination takes advantage of immune memory. Suppose we get hold of a germ and kill it without damaging its surface proteins. We then inject it into humans, usually children of course. The immune system will react against the "invaders" even though they are dead. The B-cells that match the antigens on the dead germs divide and form memory cells. Later, when the vaccinated children are attacked by living germs of this disease they are ready!

In practice, any sample of antigens that doesn't actually cause the disease will work as a vaccine. For example, germs killed by heating are sometimes used. So are harmless natural relatives or deliberately crippled mutants of the disease-causing germ. Purified proteins, including deactivated toxin proteins, have also been used.

The first artificial immunizations were done using **cowpox**, a wimpy relative of **smallpox**. Injecting people with cowpox virus gives a very mild infection. This results in immunity to the related, but deadly, smallpox virus because the virus proteins are extremely similar and the immune system regards them as near enough the same shape.

In medieval times, a substantial proportion of the population caught smallpox. Although only 20 to 30 percent of those infected actually kicked off, the survivors ended up with ugly pock marks on their faces. (Yes, that's why it was called small<u>pox</u>!) Anyway, milkmaids rarely caught smallpox because they all got cowpox first, although back then they didn't even know that there was such a disease, as it caused no disfigurations. Consequently, milkmaids were rarely pockmarked and gained a reputation for beauty due to their smooth unblemished skin. Which, in turn, explains why medieval aristocrats spent so much of their leisure time chasing milkmaids around haystacks (Fig. 22.31).

smallpox a frequently fatal viral disease that only infects humans

22.31 IMMUNOLOGY ON A MEDIEVAL FARM

Outsmarting The Immune System

Although most invaders are terminated relatively quickly and efficiently by the immune system, others are a real problem. You win some and you lose some. There are three main sources of danger.

First, there are macho germs like *Yersinia pestis* (which causes bubonic plague), or Ebola virus that kill you so quickly you never get the chance to develop immunity. Artificial vaccination usually works well with this type of disease as it puts its faith in an all-out frontal assault. These deadly invaders do not interfere with the immune system, they are honorable and manly foes and just try to kill you before your defense forces have time to respond.

22.32 *YERSINIA PESTIS* IS A NOBLE ENEMY

Second, there are diseases that actually damage the immune system. The AIDS virus is named HIV (human immunodeficiency virus) precisely because it wipes out the immune response by killing those T-cells with CD4 proteins on their surfaces (Fig. 22.27).

bubonic plague an often fatal and highly contagious disease caused by bacteria of the species *Yersinia pestis*. It wiped out nearly half the population of Europe in the 1300s and again in the 1600s

Ebola virus a member of the filovirus family that causes a rapid and often fatal disease in humans. It is found in central Africa but its natural host is still unknown

AIDS (acquired immunodeficiency syndrome) a disease that slowly undermines the immune system by destroying helper T-cells

HIV (human immunodeficiency virus) the causative agent of AIDS, is a member of the retrovirus family. Patients do not, strictly speaking, die of AIDS. They die of other infections that would normally have been liquidated by the immune system

amoebic dysentery dis-
ease of intestinal tract
caused by amoebae (singu-
lar: amoeba) that are not
bacteria but single-celled
eukaryotic microorganisms

Third, there are cowardly diseases that adopt molecular camouflage. These sneaky characters constantly change their surface proteins (Fig. 22.33). The idea behind this strategy is really quite simple. The immune memory cells can recognize the proteins on the surface of the original generation of invading germs. However, if each successive wave of invaders changes its surface proteins, they will not be recognized.

22.33 CHANGING SURFACE PROTEINS TO OUTWIT IMMUNITY

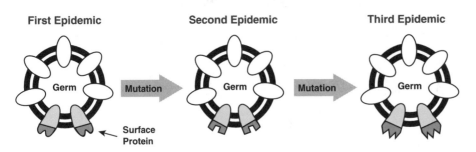

There are several ways to change surface proteins. The simplest is to mutate rapidly. The mutation rate of RNA is much higher than for DNA, so viruses having their genetic information as RNA mutate extremely fast. This is one reason why you never really become immune to RNA virus diseases like influenza (see Ch. 19). After surviving an attack of influenza you are, in fact, immune to that particular strain of virus. However, next time the flu comes around, the virus will have changed its outer protein coat and is not recognized by the immune memory cells.

A more sophisticated approach is for the disease-causing microorganism to shuffle its surface proteins by genetic rearrangement instead of mutation. From the viewpoint of the disease, this is simply poetic justice. The immune system doesn't own genes for each antibody it makes. Instead it generates antibodies to every conceivable antigen by genetic shuffling, a slick maneuver which might be regarded as cheating by a pure-minded geneticist. So why shouldn't an honest germ do the same? Some of them do.

Although most familiar diseases of temperate climates are caused by bacteria or viruses, many tropical or subtropical diseases are due to single-celled eukaryotes. Perhaps the best known are amoebic dysentery, sleeping sickness and malaria.

Sleeping sickness in Africa, and the rarer Chagas' disease of South America, are both caused by trypanosomes. These microorganisms are about 20 microns long (about 20 times longer than average bacteria) and swim around in the blood by a flagellum. They are carried by insects, Tse-tse flies in the case of sleeping sickness.

While growing and dividing in the insect gut, the trypanosomes are covered with a layer of a protein called procyclin that protects them from digestion by the insect (Fig. 22.34). The trypanosomes then move to the salivary glands where they stop dividing and wait for the insect to bite someone. While waiting they change their surface layer to the variant surface glycoprotein, VSG, designed to protect against animal immune systems.

sleeping sickness one
of several tropical diseases
caused by trypanosomes
and which occurs in Africa

malaria tropical disease
caused by *Plasmodium*, a
single-celled eukaryotic
microorganism with a
complicated life cycle

Chagas' disease dis-
ease found in Latin
America and caused by
trypanosomes

trypanosome a single-
celled eukaryotic microorganism that swims by means
of a flagellum and alternates
between insects and humans
during its life cycle

procyclin protein that
covers trypanosomes with a
protective layer while they
are inside the insect

variant surface glyco-
protein (VSG) protein that
covers trypanosomes while
inside humans and which is
constantly altered to mislead
the immune system

22.34 TRYPANOSOME INFECTION CYCLE

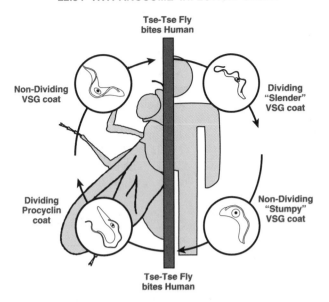

Tse-Tse Fly bites Human

Non-Dividing VSG coat

Dividing "Slender" VSG coat

Dividing Procyclin coat

Non-Dividing "Stumpy" VSG coat

Tse-Tse Fly bites Human

Charles Darwin is thought to have caught Chagas' disease during his voyage on the Beagle to South America. He returned to England with symptoms we now know to be characteristic. But back then, Chagas' disease was unrecognized. Nowadays we could prove this by sampling his corpse and performing a PCR analysis (see Ch. 17) to search for trypanosome-specific DNA sequences.

expression site special location on the chromosomes of a trypanosome from which a few chosen genes of the VSG family can be expressed

After transfer to a human, the trypanosomes grow and divide in the blood. Next, the immune system kills most of them. However, a few of the trypanosomes switch their VSG and escape recognition by the immune system. Eventually, the immune system learns about the new surface protein and kills off most of the second wave of trypanosomes. Meanwhile, some of the invaders have switched their VSG type again. This continues and the infection therefore goes in waves, each spreading the invaders further. The immune system never catches up with the constantly changing outer layer of the trypanosome, and the normal result is death of the victim.

The secret to trypanosome success is that it possesses over 1,000 slightly different copies of the gene for the variant surface glycoprotein or VSG. At any given time only one of these genes is expressed. There are two mechanisms for switching.

Among these 1,000 VSG genes, there are only 10 to 20 that can actually be expressed. These privileged genes are in special locations on the chromosomes known as "expression sites." Only one of them is actually expressed and the others are in standby mode. Every so often, the trypanosome switches from one expression site to another, which results in a different VSG being expressed (Fig. 22.35). That's just the start!

The VSG protein has a variable region which is displayed on the trypanosome surface and a

22.35 SWITCHING BETWEEN EXPRESSION SITES

Transcription

| ON | VSG gene |
expression site 1

| OFF | VSG gene |
expression site 2

| OFF | VSG gene |
expression site.... and so on

Switch Expression Site

| OFF | VSG gene |
expression site 1

Transcription

| ON | VSG gene |
expression site 2

| OFF | VSG gene |
expression site.... and so on

22.36 VSG ON THE SURFACE OF A TRYPANOSOME

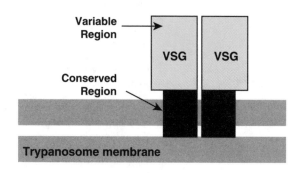

conserved portion that anchors it to the membrane. It is found as a dimer as shown in Figure 22.36.

The 1,000 unexpressed copies of the VSG gene are used to supply sequences for splicing into the privileged VSG genes in the expression sites. Usually, the complete variable region of the VSG gene in the expression site is replaced with the complete variable region from one of the 1,000 extra copies (Fig. 22.37). The constant region stays unchanged, as its name indicates!

22.37 VARIABLE REGION REPLACEMENT: I EARLY

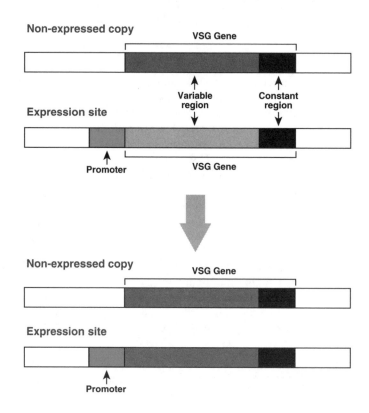

Later in infection, segments of various sizes from the spare VSG genes are used for replacement; anywhere from just a few base pairs to the whole gene may be used (Fig. 22.38). Furthermore, just as with antibodies, point mutations occur in the VSG genes at higher than normal frequency. However, in the case of the VSG genes, the mutations occur during the segment-swapping process, not afterwards.

22.38 VARIABLE REGION REPLACEMENT: II LATE

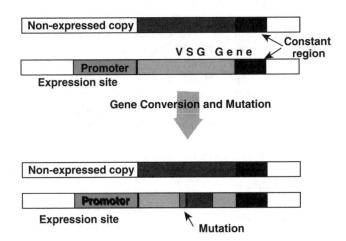

Organ Transplants and Histocompatibility

Although the immune system is vital for resisting infection, it is a major problem when performing blood transfusions or organ transplants. Successful transplantation requires matching the donor and recipient to avoid immune system attack on the transplanted organ. But what exactly do we have to match?

The major histocompatibility complex was so named because it affects the outcome of tissue grafts. Although Class II MHC proteins are found only on cells of the immune system, Class I MHC proteins are found on all cells. There are three separate genes for the α chain of Class I MHC protein and these all have multiple possible allelic variants. (The second chain, the β₂-microglobulin does not vary.) Your immune system monitors the Class I MHC proteins and if it finds a cell with an MHC protein it does not recognize, it kills it.

The Class I MHC genes of humans are still called HLA-A, HLA-B, and HLA-C, for "Human Leucocyte Antigen," an older name. There are about 40 variants of HLA-A, 70 for HLA-B, and 30 for HLA-C, so that the likelihood for two people matching exactly is 1 in 40 x 70 x 30 = 84,000 for each set (Fig. 22.39). But remember that since you are diploid and have two of each chromosome you will have two different sets of MHC genes to complicate matters!

The closer the match in the MHC (HLA) genes the longer the transplanted organ will survive. Transplant patients are also treated with immunosuppressive drugs to help suppress graft rejection. This, of course, increases the risk of cancer or infection.

tissue graft tissue cut out from one animal and inserted into another. It may be a whole organ or a portion of a tissue such as skin or bone marrow

allelic variants different particular versions of the same gene

graft rejection has never been observed in politicians from the inner city

22.39 MULTIPLE ALLELES OF MHC CLASS I

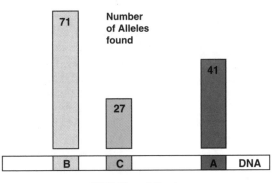

MHC Class I Region

One bizarre facet of transplantation is that organs from male donors are rejected by women due to the presence of a male specific Y-antigen. However, because men have one X-chromosome, they can accept organ transplants from both male and female donors.

monoclonal antibody a pure antibody with a unique sequence which therefore recognizes only a single antigen and which is made by a cell line derived from a single B-cell

myeloma cancer originating from B-cells of the immune system

hybridoma hybrid cell line made by fusion between an antibody secreting B-cell and a cancerous myeloma cell

Monoclonal Antibodies for Clinical Use

There are many possible clinical uses for antibodies. They are used in diagnostic procedures for pregnancy testing and to detect the presence of proteins characteristic of particular disease-causing agents. In the future, they may be used to specifically kill cancer cells or destroy viruses.

For such uses, we need relatively large amounts of a pure antibody that recognizes a single antigen. Blood serum contains a ghastly mixture of antibodies and is of little use for biotechnology. Somehow, we need to pick out a B-cell making the particular antibody we need and grow it in culture. Such a pure antibody made by a single line of cells is known as a monoclonal antibody.

Unfortunately, B-cells live for only a few days and they are unhappy outside the body. The solution to this problem is to use cancer cells. Myelomas are naturally occurring cancers derived from B-cells. Like many tumor cells, myeloma cells will continue to grow and divide in culture forever if given proper nutrients. We take the precious, but delicate, B-cell which is making the antibody we want, and we fuse it to a myeloma cell (Fig. 22.40). (To avoid confusion, we actually use a myeloma that has lost the ability to make its own antibody.) The resulting hybrid is called a hybridoma cell and will make the desired antibody and live forever.

In practice we inject an animal, such as a mouse, with the antigen against which we want antibodies. When antibody production has reached its peak, we remove a sample of antibody-secreting B-cells from the animal. These are fused to immortal myeloma cells to give a mixture of many different hybridoma cells. The tedious part comes next. Many individual hybridoma cell lines must be screened to find the one that recognizes our target antigen.

23.40 MAKING MONOCLONAL ANTIBODIES

Antigen injected

Cancer cells grown *in vitro*

FUSION

Spleen cells

Myeloma cells

Clone 1   Clone 2   Clone 3   Clone 4

TEST FOR ANTIBODIES AGAINST TARGET ANTIGEN

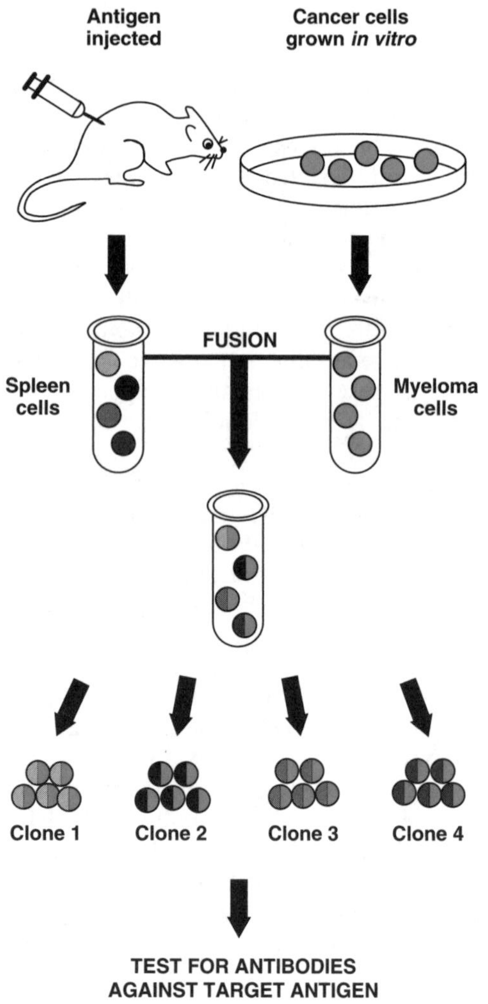

Once found, the hybridoma can be grown indefinitely in culture to give large amounts of the monoclonal antibody.

Monoclonal antibodies could be used as magic bullets to kill human cancer cells by aiming them at specific molecules appearing only on the surface of cancer cells. Ironically, the major snag is that the human immune system regards antibodies from mice or other animals as foreign molecules themselves, and so attempts to destroy them!

One experimental approach which may solve this problem is using genetic engineering to make "humanized" monoclonals. We take our first generation hybridoma and get out the DNA encoding the mouse monoclonal antibody. Then we replace the DNA encoding the constant region of the mouse antibody with the corresponding human DNA sequence. The variable region, which recognizes the target antigen, is left alone. The human/mouse hybrid gene is then put back into a second mouse myeloma cell for production of antibody in culture. Although not totally human, the hybrid is less mouse-like and provokes much less reaction from the human immune system.

A Final Thought

The immune system is a fascinating example of how massive genetic diversity can be generated by shuffling relatively few chunks of genetic information. Animals can make billions of possible antibodies with only a couple of thousand gene segments. The remarkable genetic economy of the immune system stands in bizarre contrast to the phenomenon of junk DNA (see Ch. 20). Sadly, the chromosomes of animals devote more space to junk DNA than to coding sequences. In mammals, typically 95 percent or more of the DNA may be non-coding. Human societies are much the same. Most of the paper used by universities and governments is devoted to futile administration. But here and there a few good men (that's us) write a truly interesting book (like this!)

Additional Reading

Immunobiology: The Immune System in Health and Disease by Janeway CA & Travers P. 2nd edition, 1996. Garland Publishing, Inc., New York & London.

Biology of Microorganisms by Brock TD, Madigan MT, Martinko JM, & Parker J. 8th edition, 1997. Prentice Hall, Englewood Cliffs, New Jersey.

Sequencing DNA

23

How is DNA Sequenced?

We have talked quite a bit about the base sequence of nucleic acids and how it carries the genetic information of the cell. But how do we actually get the base sequence of DNA? When it was first done, DNA sequencing was sophisticated and trendy. Today it is routine and incredibly dull, especially if you have to sequence thousands or millions of base pairs of DNA from higher organisms where most of the DNA doesn't even code for anything but is just intervening sequences!

O.K., so having brought you to the pinnacle of enthusiasm about DNA

DNA sequencing finding out the sequence of bases in a molecule of DNA

23.1 DNA SEQUENCING IS NO LONGER SUCH FUN

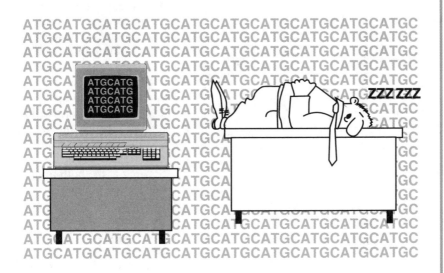

sequencing, let's talk about how it is done. The overall approach involves generating sub-fragments of all possible lengths from the DNA we want to sequence, grouping them according to which base they end in, and

gel electrophoresis
electrophoresis of charged molecules through a gel meshwork in order to sort them out by size (see Ch. 16 for details)

separating them by **gel electrophoresis**. Let's illustrate this using the eight base sequence ACGATTAG as an example.

Our eight fragments are:

ACGATTAG
ACGATTA
ACGATT
ACGAT
ACGA
ACG
AC
A

These are grouped as follows:

Ending in A	Ending in G	Ending in T	Ending in C
	ACGATTAG		
ACGATTA			
		ACGATT	
		ACGAT	
ACGA	ACG		
			AC
A			

23.2 PRINCIPLE OF THE SEQUENCING GEL

We next separate the four groups of fragments by gel electrophoresis. The fragments are all run on the same gel in four parallel lanes. Those fragments ending in A are run in the first lane, those ending in G in the second lane and so on. The fragments will be separated according to their lengths and we will see a separate band for each fragment (Fig. 23.2). Starting at the bottom of the gel and reading upwards, we can read off the sequence directly.

Very clever, you say, but how do we actually make such fragments and, in particular, how do we manage to separate them into four groups depending on the last base?

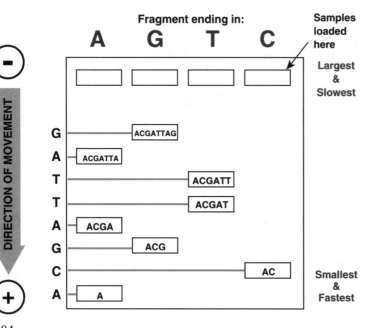

The Chain Termination Method for Sequencing DNA

The method routinely used today is known as the chain termination method or the dideoxy method. Both names refer to the fact that dideoxy analogs of normal DNA precursors cause premature termination of a growing chain of nucleotides being made by DNA polymerase. This allows us to generate the fragments of the stretch of DNA we want to sequence. By using four different dideoxy analogs, one for each of the four bases, we can generate four sets of fragments in four separate tubes.

Let's recall how DNA polymerase works to manufacture DNA. It needs a region of single stranded DNA, the template, and it also needs a primer to which nucleotides are added (see Ch. 5 for details). The DNA polymerase will then elongate the primer and make a new DNA strand complementary to the template strand (Fig. 23.3).

chain termination or dideoxy method a method for sequencing DNA involving the premature termination of growing DNA chains by using dideoxy derivatives of the nucleotides

nucleotide monomer or subunit of a nucleic acid, consisting of sugar + base + phosphate

template strand strand of DNA used as a guide for synthesizing a new strand by complementary base pairing

primer short segment of nucleic acid that binds to the longer template strand and allows DNA synthesis to get started

23.3 DNA POLYMERASE ELONGATING DNA

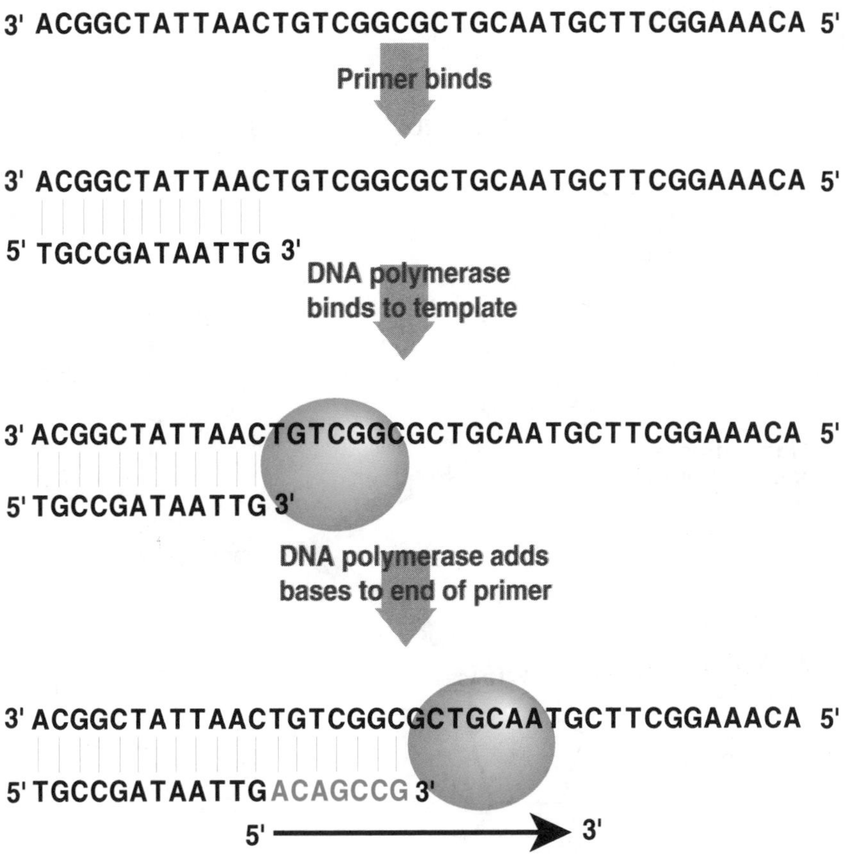

3' ACGGCTATTAACTGTCGGCGCTGCAATGCTTCGGAAACA 5'

Primer binds

3' ACGGCTATTAACTGTCGGCGCTGCAATGCTTCGGAAACA 5'

5' TGCCGATAATTG 3'

DNA polymerase binds to template

3' ACGGCTATTAACTGTCGGCGCTGCAATGCTTCGGAAACA 5'

5' TGCCGATAATTG 3'

DNA polymerase adds bases to end of primer

3' ACGGCTATTAACTGTCGGCGCTGCAATGCTTCGGAAACA 5'

5' TGCCGATAATTGACAGCCG 3'

5' ——————————> 3'

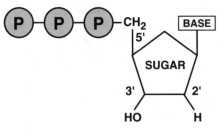

23.4 DEOXYNUCLEOSIDE TRIPHOSPHATE

nucleoside triphosphate (NTP) precursor used in synthesis of a nucleic acid, consisting of sugar + base + three phosphates

dATP, dGTP, dTTP and dCTP deoxyadenosine triphosphate, deoxyguanosine triphosphate, deoxythymidine triphosphate and deoxycytidine triphosphate

Whenever a base is added to a growing strand of nucleic acid it is provided as the nucleoside triphosphate (NTP), a base linked to a sugar and three phosphate groups (Fig. 23.4). The outermost two phosphate groups are chopped off and a nucleotide (sugar + phosphate + base) is added to the end of the growing DNA chain.

Since DNA contains the four bases, adenine, guanine, thymine and cytosine, DNA polymerase must be supplied with a mixture of the four triphosphates, dATP, dGTP, dTT, and dCTP. The small "d" refers to the fact that for DNA (deoxyribonucleic acid) we need deoxynucleoside triphosphates.

When nucleotides are joined to make a nucleic acid, the phosphate group, which is attached to the 5'-carbon atom of the sugar of the incoming nucleotide, is linked to the 3'-hydroxyl group of the sugar belonging to the last nucleotide in the chain (Fig. 23.5). Or, in brief, they are polymerized in the 5' to 3' direction.

23.5 LINKING OF NUCLEOTIDES 5' TO 3'

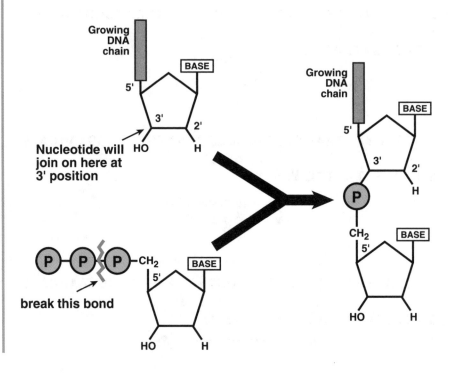

The sugar of RNA is ribose. It has hydroxyl groups on both carbons 2' and 3' of the sugar ring. The sugar of DNA is called *deoxy*ribose because it is missing an oxygen relative to ribose. Deoxyribose has no hydroxyl group on the 2'-carbon atom of the ring. The elongation scheme shown above works for both RNA and DNA since they both have hydroxyl groups at the 3' position on their sugars.

However, suppose we make a sugar with the oxygen missing from both the 2' and the 3' hydroxyl groups, in other words, dideoxyribose (Fig. 23.6).

Nucleotides containing dideoxyribose can be incorporated into a growing nucleic acid chain. But that is literally the end of the chain. Incoming nucleotides are added to the 3'-hydroxyl group of the previous nucleotide. But since dideoxyribose has no 3'-hydroxyl group, no further nucleotides can be added, and the chain is terminated.

O.K., so we are capable of blocking the elongation of a growing DNA chain. Just as we use dG to refer to a normal deoxynucleotide with the base guanine we will use ddG to refer to the dideoxy version. Consider what happens if we add ddGTP, dideoxy guanosine triphosphate, the dideoxyribose analog of dGTP, to a growing DNA chain. When the polymerase reaches the next G, it puts in ddG instead of dG. Then the chain is terminated (Fig. 23.7). If we use a mixture of dGTP and ddGTP, then sometimes we will put in dG and sometimes ddG. We will get a mixture of chains that were terminated by ddG at all positions where there is a G (Fig. 23.8).

23.6 "HEY SWEETIE"

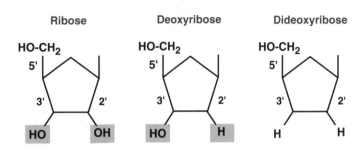

dideoxyribose derivative of ribose missing both the 2' and 3' hydroxyl groups

23.7 DIDEOXYRIBOSE BLOCKS ELONGATION

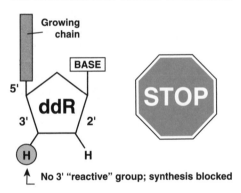

ddGTP dideoxy guanosine triphosphate

23.8 RANDOM TERMINATION AT "G" POSITIONS

Original sequence:
T C G G A C C G C T G G T A G C A

Mixture of chains terminated at G using mixtures of dGTP and ddGTP:

1. T C G
2. T C G G
3. T C G G A C C G
4. T C G G A C C G C T G
5. T C G G A C C G C T G G
6. T C G G A C C G C T G G T A G

Gee-whiz!

Next we load our mixture onto a gel and electrophorese it (Fig. 23.9). The chains are separated according to their sizes. We get a series of bands, each corresponding to a piece of DNA of a particular length. The lengths reveal the positions of the G bases in the original DNA. The shortest pieces are closer to the bottom as they move fastest during electrophoresis. To completely sequence the DNA we do the same thing simultaneously for all four bases, A, G, T, and C. So there are four reaction mixtures, one for each of the four bases. Each contains a series of artificially terminated chains. Next we load these four samples side by side onto a gel and separate the chains by electrophoresis. We get four ladders, side by side (Fig. 23.10).

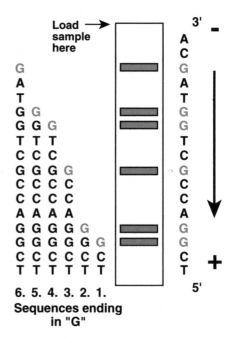

23.9 DNA SEQUENCING GELL FOR "G" ONLY

6. 5. 4. 3. 2. 1.
Sequences ending in "G"

As before, the position of each band corresponds to a chain of DNA of a particular length and reveals the position of one base. We read off the sequence from the bottom, combining results from all four bases. Several hundred bases can usually be obtained from one gel.

Since DNA is colorless, odorless and tasteless, we need some way to detect it. In practice, a radioactively labeled molecule is included in the reaction so that all the DNA chains are radioactive. After running, the gel is dried and a sheet of photographic film is laid on top of the gel. Darkened bands are formed where the radioactive DNA is located (Fig. 23.11). This procedure is known as **autoradiography**. Part of a real sequencing gel is shown in Figure 23.12.

23.10 DNA SEQUENCING GEL ALL FOUR BASES

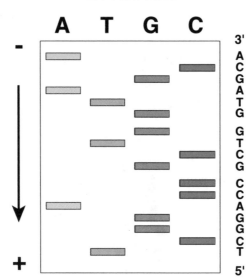

A T G C

autoradiography
allowing radioactive materials to take pictures of themselves by laying them flat on photographic film

23.11 AUTORADIOGRAPHY

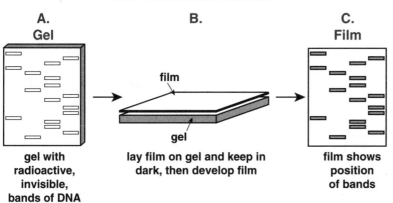

A. Gel

B.

C. Film

film

gel

gel with radioactive, invisible, bands of DNA

lay film on gel and keep in dark, then develop film

film shows position of bands

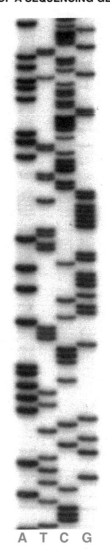

23.12 A SMALL PORTION OF A SEQUENCING GEL

A T C G

Using M13 To Make Single Stranded DNA

For high quality DNA sequencing we first need to obtain purified DNA as single strands to which the sequencing primer can bind.

The original method uses the bacterial virus **M13** (Fig. 23.13). This virus is rod-shaped and contains a circle of single stranded DNA (**ssDNA**). Upon infecting an *E. coli* cell, the single stranded viral DNA is converted to a double stranded form, the **replicative form,** or **RF**. After replicating itself for a while, the RF then turns its efforts to manufacturing large numbers of single stranded circles of DNA to pack into newly made virus particles.

23.13 LIFE CYCLE OF M13

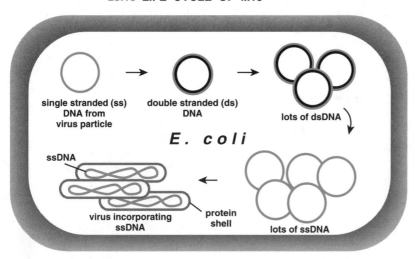

single stranded (ss) DNA from virus particle

double stranded (ds) DNA

lots of dsDNA

E. coli

ssDNA

virus incorporating ssDNA

protein shell

lots of ssDNA

Not only does M13 give us single stranded DNA but it purifies it for us, too! Unlike most viruses, M13 doesn't destroy the bacterial cells. Instead, the cells continuously secrete virus particles containing ssDNA into the surrounding medium. The DNA making up the bacterial

M13 a bacterial virus containing single stranded circular DNA

ssDNA single stranded DNA

replicative form (RF) double stranded circular DNA found as an intermediate form during the replication of viruses that contains single stranded DNA

389

chromosome is left behind inside the cells. So we just collect the virus particles and extract the DNA they contain.

23.14 INSERTION INTO M13 OF DNA TO SEQUENCE

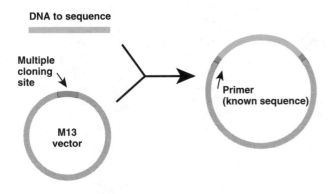

23.15 SEQUENCING BY PRIMER WALKING

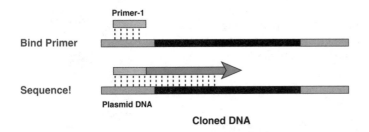

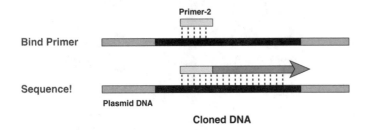

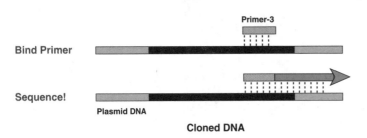

The DNA that we wish to sequence is cloned into the double-stranded replicative form of M13. The engineered virus is used to infect *E. coli*, and virus particles containing single strands are manufactured in large quantities.

Normally we use an M13 vector that has already been engineered and contains a convenient **multiple cloning site** into which the DNA to be sequenced is inserted (Fig. 23.14). This multiple cloning site is contained within the N-terminal fragment of the *lacZ* gene of *E. coli*. This allows us to use blue/white screening to decide whether or not we were successful in inserting our piece of DNA into the M13 vector (see Ch. 9 for details). Furthermore, the sequence to the side of the inserted DNA is already known and gives us a good starting point. This is essential, as the primer for sequencing must be complementary to a known sequence on the template strand in order to hybridize in the correct position.

Nowadays, bacterial plasmids containing the M13 origin of replication are used to manufacture single-stranded DNA. The use of intact virus is avoided and improved yields of DNA can be obtained more conveniently.

Primer Walking Along a Strand of DNA

Sequencing long pieces of DNA was originally done by cutting the DNA into smaller segments with restriction enzymes and then sub-cloning each fragment separately into M13. Nowadays, we usually do "**primer walking**" (Fig. 23.15). This involves first sequencing the cloned DNA as far as we can get using the primer belonging to the M13 or

390

plasmid vector. Next, we use the newly-obtained sequence to make another primer and continue sequencing as far as this allows. Then we make another primer, and another, and so on until we reach the end of the cloned DNA.

"Double Stranded" DNA Sequencing and Using PCR Products

So much for the classic methods. A variety of technical improvements have made DNA sequencing a little less tedious. Using double stranded DNA (dsDNA) directly for sequencing is more convenient than is generating single strands. In reality, "double stranded" DNA sequencing involves a preliminary step, either heat or alkali treatment, to denature the dsDNA into single strands. The actual sequencing reactions therefore use single stranded DNA just as described above.

In fact, it is now possible to completely avoid cloning of DNA into either M13 or a plasmid vector by using PCR to generate segments of DNA (see Ch. 17). PCR products are linear double stranded lengths of DNA and they can be directly sequenced after separation into single strands. (As noted in Chapter 17, one drawback of PCR is that we need to know enough sequence on each side of the target DNA to construct primers for PCR. Hence, we cannot always avoid cloning.)

The Cat's Meow: Automated Sequencing

Best of all the technical improvements is to let a machine do the work. Today, the majority of sequencing is done this way. The main modification here is that we use dyes to label the DNA instead of radioactivity. Each of the four sequencing reactions is done just as before, but the DNA is labeled by attaching a fluorescent dye to the primer before running the reactions.

Although we use the same DNA primer for each of the four reactions, we use four different fluorescent dyes (how about red, green, blue and yellow, for example!). The first color is used when carrying out the A-reaction, another one for G, another for T, and the fourth for C. When we run our sequencing gel we get bands of four different colors, a separate color for each base. In fact, since the bases are color coded, we can load all four completed reactions in the same track on the sequencing gel as shown in Figure 23.16.

While we go out for lunch, the gel is run and the bands are scanned with a laser beam and the four different dyes fluoresce in different colors. The color of each band is recorded by a computer that analyzes the data and prints out the sequence.

The Human Genome Project

One of those famous people whose names no one can ever remember, once remarked, "The proper study of mankind is man." We may not be able to read each others' minds but at least we can read our own genetic code. The objective of the human genome project is to sequence all the DNA making up the human genome by the year 2005.

The human genome contains 3×10^9 (or in simple language – three billion) base pairs of DNA. A typical page of text (in a proper book for literate

multiple cloning site stretch of DNA with recognition sites for several restriction enzymes

primer walking sequencing a long stretch of DNA by using a series of primers

polymerase chain reaction (PCR) artificial amplification of a DNA sequence by repeated cycles of replication and strand separation

human genome project program to completely sequence human DNA

23.16 AUTOMATED FLUORESCENT SEQUENCING

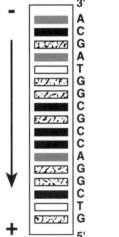

P.S. We are not as ignorant as you think!

"Know then thyself, presume not God to scan;
The proper study of mankind is man."

is by Alexander Pope, modeled on Chaucer's

"Full wise is he that can himselven knowe."

391

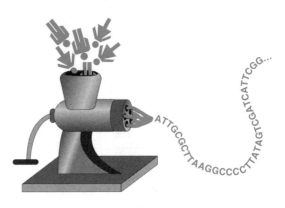

readers, not one like this which is mostly cartoons!) contains about 3,000 letters. So the human genome would fill about a million pages, without any pictures! I can read at nearly 50 pages an hour (if the plot is interesting!). That's 20,000 hours or, at eight hours a day, five days a week, 50 weeks a year, roughly 10 years to read the human genome.

intergenic regions DNA sequence between genes

intron segment of a gene that does not code for protein

Unfortunately, for those looking forward to a good gripping plot, most DNA from higher organisms is non-coding DNA, intergenic regions, introns and so forth.

If English Had Introns It Would Look Like This (sans color):

Onlyaksjcbakcnamcnabhjbkakjcncknmncsmallcbbhbhjcgjhcgchcg

pergtehfamaasecnsbcbidcentagegcueyfgshdnclpwsjof

humandcgschjgchDNAisbvchdjcsvchvajshcahdsnactuallymksiw

sanqmnskjmeanjhsfbbddjbcbnsmsmndingful

Will the total human DNA sequence be of much real value? Does it contain any thigh-slapping humor? The often heard claim that once we have sequenced ourselves we will know how to fix all diseases, is doubtful, to put it mildly. We have known for years the complete gene sequences of several viruses, including HIV, yet no cure is in sight. Deducing the function of a gene given only the DNA sequence is hazardous at best. Although DNA sequences are useful, in practice a great deal of experimental work must be done before inherited defects are understood (see Ch. 13 for examples).

Another, rather squalid angle of sequencing human genes is the rush by scientists and biotechnology companies to patent human DNA sequences. If you invent the camera it's fair enough for you to own the patent to it. But what if you now go on to take a picture of the moon? There is no doubt that you own the picture, but do you also own the moon?? The greed of some of those involved in sequencing DNA for financial rewards reminds us that some scientists and televangelists evolved from the same common stock not too long ago.

The Oligonucleotide Array Detector

Despite its name, the oligonucleotide array detector is not a chain of radar installations deployed across the arctic to detect incoming nuclear missiles. Its purpose is to detect and simultaneously identify lots of short DNA fragments (*i.e.*, oligonucleotides). It can be used both for diagnostic purposes and for large scale DNA sequencing.

The key principle is DNA-DNA hybridization (see Ch. 16). Suppose that we have a piece of DNA of unknown sequence. We denature this to give single strands and test one of these to see if it hybridizes to a known sequence of eight bases (an octonucleotide), say CGCGC-CCG. If the answer is "yes" then we know that this eight base sequence occurs somewhere in the other, complementary, strand of unknown DNA.

We then hybridize the unknown DNA to all other possible stretches of eight bases, one at a time, to see which are found.

Suppose our unknown sequence is: **TCCAACGATTAGTCG**
Its complementary strand will be: **AGGTTGCTAATCAGC**

Consequently of all possible eight base sequences, only the following can hybridize with the original sequence:

**AGGTTGCT   TAATCAGC   TGCTAATC   GCTAATCA
GTTGCTAA   GGTTGCTA   TTGCTAAT   CTAATCAG**

Given this information, a computer program will test all possible overlaps for these eight base sequences and generate the solution as shown in Figure 23.18.

In practice the hybridizations are all carried out at once. There are actually 65,536 possible eight base sequences, trust me! Samples of each of the eight base sequences to be used as probes are arranged in a square array and anchored to the surface of a glass chip. The glass chip can then be dipped in a solution of the target DNA which will hybridize simultaneously to all those eight base sequences to which it can base pair. This has provided a more user-friendly name for the nucleotide array detector, the

The first free-living organism to have its genome completely sequenced was the bacterium *Haemophilus influenzae* whose genome is 1.8 megabases (Mb) long - 1,830,121 base pairs to be precise - and contains 1749 genes. Some of the model organisms which are being sequenced at present are as follows:

Escherichia coli (4.6 Mb). The most intensively investigated bacterium. Most of its 3,000 genes have by now been identified and the functions of around half are known.

Saccharomyces cerevisiae (12 Mb). Yeast is single celled and is the eukaryote about which we know the most at the level of molecular genetics. It was completed in the spring of 1997 and found to contain 12,057,500 bp and roughly 6,000 genes.

Arabidopsis thaliana (70 Mb). The wall cress has the smallest genome of any flowering plant. It is now used as the model organism in plant molecular biology.

Caenorhabditis elegans (100 Mb). This roundworm is 1 mm long and is the model for development in multicelled animals. It has 959 cells and the lineage of each has been completely traced.

Drosophila melanogaster (165 Mb). The humble fruit fly has been intensively analyzed at the genetic level.

Mus musculus (3000 Mb). The mouse is the mammal about which most is known genetically. Its genome is the same size as ours, giving rise to the saying "Of mice and men."

oligonucleotide array detector   square array of short DNA probes fixed on a glass chip and used for carrying out multiple hybridizations all at once

oligonucleotide   short stretch of DNA (or RNA) consisting of just a few nucleotides

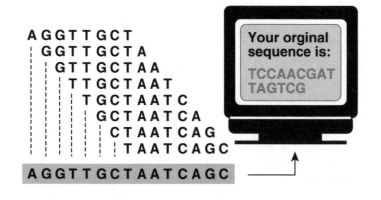

```
A G G T T G C T
  G G T T G C T A
    G T T G C T A A
      T T G C T A A T
        T G C T A A T C
          G C T A A T C A
            C T A A T C A G
              T A A T C A G C
A G G T T G C T A A T C A G C
```

Your orginal sequence is:

TCCAACGAT TAGTCG

GeneChip® another name for the oligonucleotide array detector

GeneChip®. In practice a GeneChip® about 1 cm square can carry 100,000 or more nucleotide probe sequences.

To sequence a large piece of DNA it is first broken into relatively small pieces and tagged with a fluorescent dye. The unknown DNA will bind to just a few of the many eight base probes on the GeneChip®. The GeneChip® is then scanned by a laser which locates the fluorescent tagged DNA. The positions to which it has bound are recorded and therefore we, or rather our computer, has a list of all the possible eight base sequences that occur in this fragment of unknown DNA. It is then simple (for the computer, not for us!) to calculate the complete sequence of the unknown DNA.

For simplicity, the GeneChip® illustrated in Figure 23.19 is designed to detect all possible four base sequences. The "unknown" sequence, ACTGGC, contains three overlapping four-base sequences: ACTG (No. 1), CTGG (No. 2), and TGGC (No. 3). Their positions are shown on the array.

The GeneChip® will run into trouble if the target DNA contains repeated sequences. Therefore, completely new DNA sequences must be checked by conventional sequencing. However for diagnostic tests to check for hereditary defects (*i.e.*, mutations in known genes) and for forensic analysis, the GeneChip® should be faster and simpler. According to Affymetrix Corporation, the first GeneChip® is designed to detect mutations in the reverse transcriptase gene of the AIDS virus. It is presently in pre-market testing.

23.19 GENECHIP® IN ACTION

PUZZLE

SOLUTION

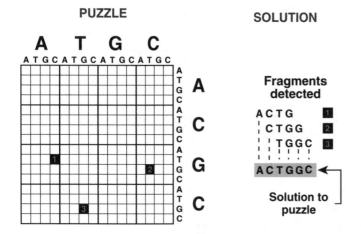

Fragments detected

```
A C T G      [1]
  C T G G    [2]
    T G G C  [3]
A C T G G C
```

Solution to puzzle

Data Banks and Computer Analysis

Assuming we have sequenced a stretch of DNA in which we are interested, what can we do with the information?

codon bias tendency to use some of the codons more frequently than others

1) Perform a codon bias analysis to locate coding regions. Due to third base redundancy and the preferential use of some codons over others (in coding regions but not in random, intergenic DNA), a smart

computer can often tell you if a stretch of DNA codes for a protein. If so, have the computer translate the DNA into protein (Fig. 23.20).

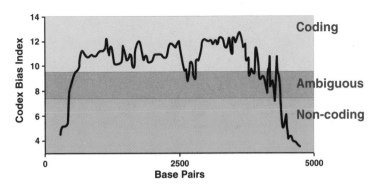

23.20 CODON BIAS PLOT TO FIND READING FRAMES

2) Search for known consensus sequences, indicating the presence of promoters, ribosome binding sites (in prokaryotes only), terminators and other regulatory regions. Inverted repeats in the DNA may imply stem and loop structures which are often sites for the binding of regulatory proteins, etc.

3) Search for related sequences. Since vast amounts of sequence information are now stored in data banks, have your computer compare the sequence of a stretch of DNA you have found with other available sequences. You can run DNA searches or protein sequence searches (if you had coding DNA and have translated it). If you find another protein with a sequence similar to yours, it may give you some idea of the function of your protein. Of course, this assumes that the function of the other protein has already been deciphered!

To do this we need to connect to a data bank. Nowadays scientists usually submit their own sequences for comparison or download others from databanks via the World Wide Web.

One of the major uses of sequence comparisons is to trace the evolution both of individual genes and of the organisms that carry them. This topic is discussed in the following chapter.

The two most prominent molecular biology databanks are:

1 **GenBank** is run by the National Center for Biotechnology Information (NCIB) which can be reached at: http://www.ncbi.nlm.nih.gov

2 **EMBL** is run by the European Bioinformatics Institute: http://www.ebi.ac.uk

Additional Reading

Molecular Biology of the Cell by Alberts B, Bray D, Lewis J, Raff M, Roberts K, & Watson JD. 3rd edition, 1994. Garland Publishing, Inc., New York & London.

Molecular Biotechnology: Principles and Applications of Recombinant DNA by Glick BR & Pasternak JJ. 1994. American Society for Microbiology, Washington, DC.

23.21 FINDING A RELATED SEQUENCE

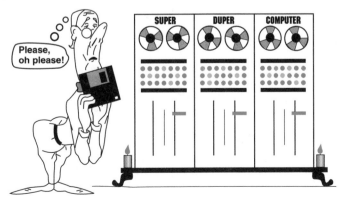

Molecular Evolution: "The Way We Were"

24

Getting Started - The Big Bang

The story goes like this. Several zillion years ago there was a big bang. A really big,

BIG, BIG, **BIG**, **BIG**, **BANG !!**

About 15,000,000,000 years later a cloud of interstellar dust and gas condensed due to gravity into a large ball of gas (the sun) surrounded by smaller lumps of miscellaneous crud (the planets). The universe consists mostly of the light molecular weight gases hydrogen and helium.
The heavier elements together comprise only about 0.1 percent of the total and form the planets.

24.1 THE SOLAR SYSTEM FORMS

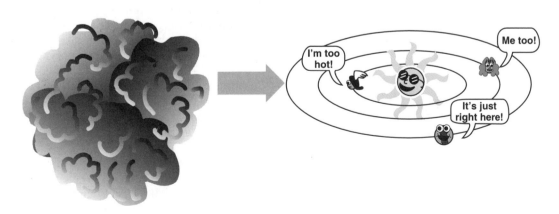

Don't ask me how, but heat was released by the collapse due to gravity and also by the radioactivity of elements present in the original dust. During the first few hundred million years of the Earth's existence, water was in the form of steam, condensing later as the Earth cooled off, to form oceans and lakes. Life is thought to have originated by means of chemical reactions occurring in the atmosphere followed by reactions in the primeval oceans and lakes.

Table 24.1 Approximate Evolutionary Time Scale

Millions of Years Ago	Major Events
20,000	Big Bang
5,000	Origin of planets and sun
3,500	Origin of life
3,000	Primitive bacteria start using solar energy
2,500	Advanced photosynthesis releases oxygen
1,500	First eukaryotic cells
1,000	Multicellular organisms
600	First skeletons give nice fossils
1.8	First true humans - *Homo erectus*
0.2	Modern man
0.01	Beginnings of civilization
0.002	Roman Empire civilizes assorted savages
0.00005	Molecular biology gets going
0.000001	We started writing this book

The Early Atmosphere

The story continues. The Earth's original atmosphere consisted mostly of hydrogen and helium but the Earth is too small a planet to hold such light gases and they floated away into space. The Earth then accumulated a secondary atmosphere, mostly by volcanic gases that yielded steam, carbon dioxide, nitrogen and lesser amounts of other molecules (Fig. 24.2). In addition, water vapor reacted with primeval minerals such as nitrides to give ammonia and with carbides to give methane. There was no free oxygen.

Our present atmosphere is of biological origin. The methane and ammonia, etc., have been consumed and the inert components (nitrogen, traces of argon, etc.) have remained unchanged. Large amounts of oxygen have been produced by photosynthesis. This could not occur until the first true photosynthetic organisms had evolved about 2.5 thousand million years ago. The oxygen content of the atmosphere reached 1 percent about 800 million years ago and 10 percent about 400 million years ago. Today it is about 20 percent.

24.2 ATMOSPHERE FROM VOLCANIC GASES

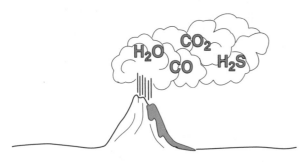

Oparin's Theory of the Origin of Life

Ultraviolet radiation from the sun, together with lightning discharges caused these gases to react forming simple **organic compounds**. These dissolved in the primeval oceans and continued to react, forming what is often referred to by storytellers as the "primitive soup" (Fig. 24.3). (Today, this is only produced in the kitchens of the military and in student cafeterias.)

organic compounds chemical compounds characteristic of living creatures and containing carbon and hydrogen

The primitive soup contained amino acids, sugars, and nucleic acid bases among other random molecules (Fig. 24.4). Further reactions formed polymers and these associated, eventually forming globules. Ultimately, these evolved into the first primitive cells.

This theory of the origin of life was put forward by the Russian biochemist, Alexander Oparin, in the 1920s. Charles Darwin himself had actually proposed that life might have started in a warm little pond provided with ammonia and other goodies. However, it was Oparin who outlined all the necessary steps and realized the critical point, that life evolved before there was any oxygen in the air.

A year or so after Oparin put his theory forward, a British biochemist called Haldane, invented the same theory "independently." Turns out that Haldane just happened to be a member of the communist party and took an interest in the goings on in Russia at the time. Not to be outdone, the Americans claim that a scientist named Bernal also invented the same theory - also "independently" - a year or so after Haldane!!!

24.3 PRIMITIVE SOUP

24.4 PRIMITIVE SOUP FORMS

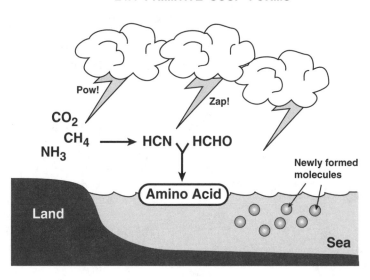

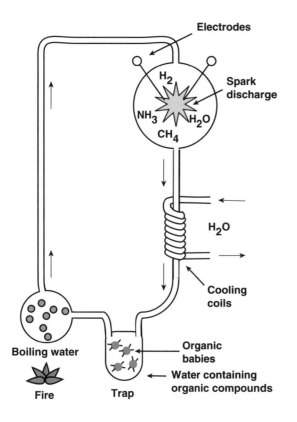

Electrodes

Spark discharge

H_2

NH_3 H_2O

CH_4

H_2O

Cooling coils

Boiling water

Fire

Organic babies

Water containing organic compounds

Trap

The Miller Experiment

In the 1950s, the biochemist Miller mimicked the primitive Earth atmospheric reactions. An imitation atmosphere containing methane, ammonia and water vapor was subjected to a high voltage discharge (to simulate lightning) or to ultraviolet light (Fig. 24.5). Organic compounds formed by reactions in the atmosphere would then dissolve in a flask of water, intended to represent the primeval ocean. Further reactions between the compounds formed could, of course, take place in the water.

There are many variants of this experiment (different gas mixtures, different energy sources, etc.). As long as oxygen is excluded, the results are similar. About 10 to 20 percent of the gas mixture is converted to soluble organic molecules and quite a lot more is converted to the sort of un-analyzable organic tar you might expect to find in a smoker's lung. First, aldehydes and cyanides are formed, and then a large variety of organic compounds. Most of the naturally occurring amino acids, hydroxyacids, purines, pyrimidines, and sugars have been produced in variants of the Miller experiment.

The same energy sources which produce organic molecules are also very effective at destroying them. The long term build-up of organic material requires its protection from the energy sources which created it. This is the function of the imitation primeval ocean in the Miller experiment. Water shields molecules from ultraviolet radiation and from electric discharges. The survival of organic molecules on the primitive earth would have depended on their escape from UV radiation and lightning either by dissolving in seas or lakes or by sticking to minerals. Most organic molecules formed too far up in the sky would have been destroyed again very quickly (Fig. 24.6), while those who made it to the sea would have survived (Fig. 24.7).

Note that organic acids, and in particular amino acids, are water soluble and involatile. Once they are safely dissolved in water there is little tendency for such molecules to return to the atmosphere. Their precursors, the aldehydes and cyanides are not only reactive but also

24.6 CREATION AND DESTRUCTION OF MOLECULES

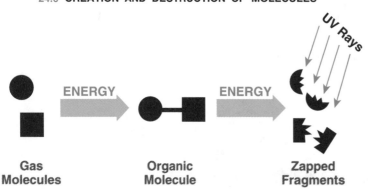

ENERGY

ENERGY

UV Rays

Gas Molecules

Organic Molecule

Zapped Fragments

volatile. Consequently, these molecules do not survive for long. Thus even at this early stage there was natural selection, though between molecules, not organisms.

Evolving Macromolecules

Assembly of macromolecules such as proteins and nucleic acids needs energy to polymerize the small molecules which make them up. Before the high energy phosphates used in modern cells were available, some other form of energy was needed.

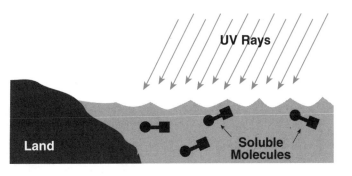

Imitation protein polymers, containing randomly linked amino acids, are known as "proteinoids." They can be formed by heating dry amino acid mixtures at around 150°C for a few hours (Fig. 24.8). They contain up to 250 amino acids and can sometimes perform primitive enzymatic activities. Such dry heat could have occurred near volcanoes or when pools left behind by a changing coastline evaporated.

proteinoids random proteins made by artificial chemical assembly of amino acids as opposed to genuine proteins made on ribosomes by cells

24.8 HEAT FORMS RANDOM PROTEINOIDS

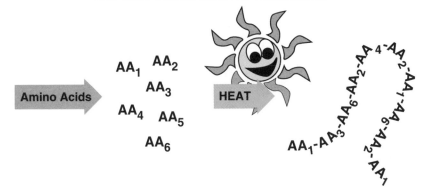

montmorillonite clay type of clay that promotes polymer formation by vigorously sucking out water

Another way to get amino acid polymers is by using clay minerals with special binding properties (Fig. 24.9). Binding of small molecules to the surface of catalytic minerals can promote many reactions. Montmorillonite clay will condense amino acids to form polypeptides up to 200 residues long.

The amino acids of genuine biological proteins are linked using only their α-amino and

24.9 CATALYSIS BY BINDING TO CLAY YIELDS POLYMERS

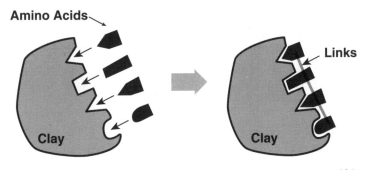

α-carboxyl groups, the ones attached to the central or α-carbon atom (see Fig. 24.10). However, these "primeval polypeptides" or "proteinoids" contain lots of incorrect bonds involving side-chain groups.

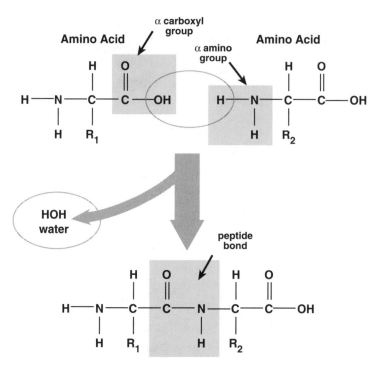

24.10 **ALPHA LINKAGES IN PROTEINS**

Enzyme Activities of Random Proteinoids

Interestingly, random proteinoids stewed up in modern laboratories under fake primeval Earth conditions will carry out some simple enzyme reactions (Fig. 24.11). They are far slower and less accurate than enzymes made by real cells, but nonetheless they can perform recognizable enzymatic reactions.

For example, random proteinoids can often remove carbon dioxide from biochemicals like pyruvate or oxaloacetate and split organic esters. The presence of traces of the metal copper allows them to move amino groups around and zinc allows the breakdown of ATP, which is used by modern cells both as a precursor of nucleic acids and as an energy carrier.

24.11 **PRIMITIVE VERSUS MODERN**

Primitive Enzyme

Modern Enzyme

So although no one was present to see what happened, scientist generally believe that the essential ingredients for life could form spontaneously under the conditions that prevailed on the primeval Earth.

Did Nucleic Acids or Proteins Come First?

Which came first, the chicken or the egg? - protein or nucleic acid? There are two main theories for the origin of the first cells:

1) Proteins first theory. Primitive proteins coalesced into blobs. Nucleic acids were incorporated to carry the genetic information at a later date.
2) Naked genes theory. The primitive nucleic acid replicated alone, later a protective protein coat was added.

Nowadays most molecular biologists opt for the second theory, for two reasons. First, it is possible for random RNA molecules in solution to duplicate themselves under certain conditions. Second, although most modern day enzymes are indeed proteins, examples of RNA acting as an enzyme and catalyzing reactions on its own without help from proteins have been found.

When a strand of RNA is stewed up with suitable precursors, a complementary piece of RNA is synthesized (Fig. 24.12). This reaction can be catalyzed by zinc even in the absence of enzymes, and lengths of up to 40 bases are produced with an error rate of about 1 in 200. Almost all modern day enzymes that manufacture RNA or DNA contain zinc atoms.

24.12 PRIMEVAL RNA DUPLICATION

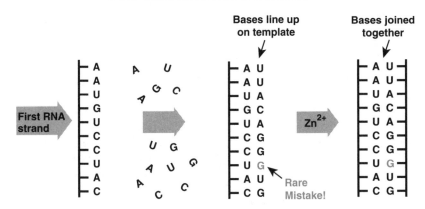

Ribozymes and the RNA World

One rather extreme viewpoint is the idea that the earliest organisms had both genes and enzymes made of RNA and formed a so-called "RNA World." This seeks to avoid the paradoxical problem that nucleic acids are needed to encode proteins, but that enzymes made of protein are needed to replicate nucleic acids. During the RNA world stage, RNA supposedly carried out both functions. Later, proteins infiltrated and took over the role of

RNA world hypothetical stage of evolution when RNA played the role of both gene and enzyme

enzymes and DNA appeared to store the genetic information, leaving RNA as a mere intermediate between genes and enzymes. Several lines of evidence favor the primacy of RNA.

1) Self-Splicing Introns. We have seen that the genes of higher, eukaryotic, cells are often interrupted by non-coding regions (the introns) which must be removed from the messenger RNA before translation into protein (Ch. 11). Normally, this is done by a **spliceosome** made up of several proteins and small RNA molecules. Occasionally, however, splicing out of an intron is catalyzed by the RNA itself without help from any protein! This **self-splicing** is found in a few nuclear genes of some protozoans, in the mitochondria of fungal cells and the chloroplasts of plant cells (Fig. 24.13).

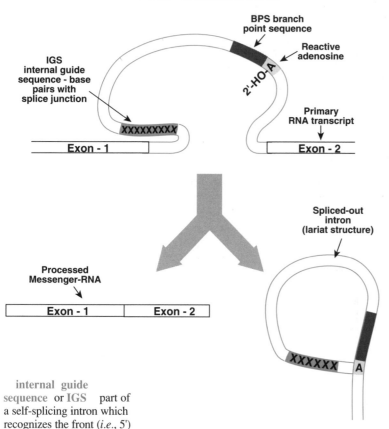

24.13 SELF SPLICING INTRON

For self-splicing, the RNA forming the intron is twisted in such a way that one of its constituent nucleotides is more chemically reactive than usual. The actual reactive group is the 2'-hydroxyl of the ribose sugar of this special nucleotide. Since 2'-hydroxyl groups are only present in RNA, not in DNA, only RNA can perform such reactions.

In addition, part of the self-splicing intron, known as the **internal guide sequence (IGS)** must recognize the correct site for the splice junction, that is, it plays the same role as the **U1 snRNA** of the spliceosome (see Ch. 11).

2) Ribozymes. An RNA molecule that is enzymatically active is called a **ribozyme**. A genuine enzyme processes large numbers of other molecules but is not altered itself. Therefore, self-splicing RNA is not truly enzymatic because it works only once.

There is a growing list of suspected ribozyme candidates, including the possibility that ribosomal RNA may be directly involved in the reactions of protein synthesis.

However, the only absolutely proven ribozyme is **ribonuclease P**. This enzyme has both RNA and protein components and its job is processing certain transfer RNA molecules. It is the RNA part of ribonuclease P which carries out the reaction. The protein serves only to hold the ribozyme and the transfer RNA it operates on, together. In fact, in concentrated solution, the protein is not even necessary, and the ribozyme will work on its own.

3) Other Suspicious Roles of RNA. As mentioned in Chapter 4, during replication of DNA, primers made of RNA are used whenever new strands of DNA are started. Further, small guide molecules of RNA are used not only by spliceosomes, but also by telomerase when extending the ends of eukaryotic chromosomes.

Although the "RNA world" concept was fashionable for a while, it is no longer so trendy. One problem is that RNA is more reactive than DNA. Although RNA would form more easily than DNA under primeval conditions, it would also be less stable. Thus DNA, though slower to form initially, might tend to accumulate under such conditions. Moreover, the primeval soup would contain a mixture of the sub-components of both types of nucleic acid as well as proteins, lipids and carbohydrates. So it seems more likely that our proto-cell contained an ill-defined mixture, perhaps even hybrid nucleic acid molecules with both RNA and DNA components.

The First Cells

So perhaps random proteins and grease molecules collected around the primeval RNA (or DNA), forming a microscopic membrane-covered organic blob. Eventually this proto-cell learned how to use RNA to code for its protein sequences. The grease formed a membrane around the outside to keep the other components together (Fig. 24.14).

ribozyme RNA molecule which acts as an enzyme

ribonuclease P an enzyme which processes the precursors to some transfer RNA molecules

primer short segment of nucleic acid which binds to the longer template strand and allows DNA synthesis to get started

telomerase enzyme which adds DNA to the end, or telomere, of a chromosome

24.14 **RNA - BASED PROTOCELL**

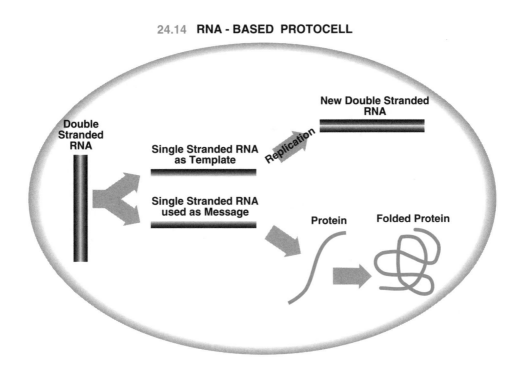

Early on, enzymatic functions were shared by protein and enzymatic RNA (ribozymes). Later, RNA lost most of its enzymatic roles as the more versatile proteins took these over. Today the vast majority of enzymes are proteins. It is generally thought that RNA was the first information carrying molecule and that DNA was a later invention. Because DNA is more stable than RNA, it would store and transmit information with fewer errors.

We now have something vaguely resembling primitive bacteria and which lived off the organic compounds in the primitive soup. Eventually this primeval free lunch ran out. The proto-cell was forced to find a new source of energy and it turned to the sun (Fig. 24.15).

24.15 EARLY CELL SUNBATHING

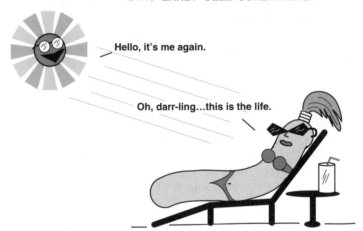

photosynthesis process of trapping light energy to form food molecules

respiration process of releasing energy from food molecules by reacting them with oxygen

The earliest forms of **photosynthesis** probably used solar energy coupled to the use of sulfur compounds as reducing power. Later, more advanced photosynthesis used water instead of sulfur compounds (Fig. 24.16). The water was split, releasing oxygen into the atmosphere.

Up to this point there had been no oxygen in the atmosphere. From our modern-day perspective, the atmosphere before the invention of photosynthesis was poisonous. Once oxygen became available, **respiration** was invented. Cells reorganized components from the photosynthetic machinery to release energy by oxidizing food molecules with oxygen

24.16 PHOTOSYNTHETIC APPARATUS

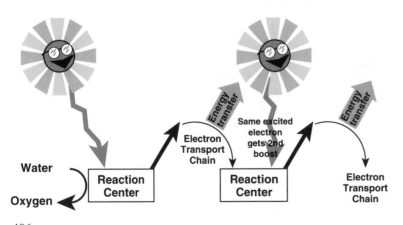

406

(Fig. 24.17). Photosynthesis emits oxygen and consumes carbon dioxide, whereas respiration does the reverse. The overall result is an ecology where plants and animals complement each other biochemically.

Evolution of DNA, RNA and Protein Sequences

Let's consider the genes of an ancient ancestral organism. Over millions of years mutations will occur in the DNA sequences of its genes at a slow but steady rate (see Ch. 12.) Most mutations will be selected against because they are detrimental, but some will survive. Most mutations that are incorporated permanently into the genes will be neutral mutations with no harmful or beneficial effects on the organism. Occasionally muta-

24.17 RESPIRATION

Food molecule

Hydrogen — Carbon

Carbon Dioxide

Oxygen

ELECTRON TRANSPORT CHAIN

Water

ENERGY

tions that improve the function of a gene and/or the protein encoded by it will occur. These are very rare. Sometimes a mutation that was originally harmful may turn out to be beneficial under new environmental conditions.

What matters in most cases is how well the protein encoded by a gene functions. If the protein can still operate normally, a mutation may be acceptable (Fig. 24.18). In practice, many of the amino acids making up a protein chain can be varied, within reasonable limits, without damaging the function of the protein. For example, glutamic acid and aspartic acid are both acidic and water soluble and not very different in size. Hence, replacement of either one by the other in a protein is usually tolerable and often has no noticeable effect at all.

24.18 SIMILAR AMINO ACIDS ARE O.K.

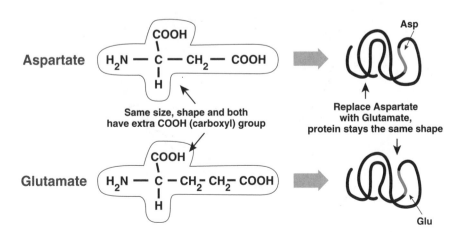

Aspartate

$H_2N — C — CH_2 — COOH$
COOH
H

Same size, shape and both have extra COOH (carboxyl) group

Replace Aspartate with Glutamate, protein stays the same shape

Asp

Glutamate

$H_2N — C — CH_2 — CH_2 — COOH$
COOH
H

Glu

hemoglobin red protein that carries oxygen in blood

cytochrome c protein that is part of the respiratory chain

Suppose we compare the sequences of the same protein taken from many different modern day organisms. We will find that the sequences can be lined up and are very similar. For example the α chain of hemoglobin is identical in humans and chimpanzees. The same protein from pigs has 13 percent of its amino acids different, that from chickens 25 percent, and that from fish 50 percent. This divergence in sequence is pretty much what you would expect from other estimates of relatedness based on an evolutionary perspective.

It is possible, then, to construct an evolutionary tree using a set of sequences for a protein as long as it is found in all the creatures being compared. The α chain of hemoglobin is only found in our blood relatives. In contrast, **cytochrome c** is a protein involved in energy generation in all higher organisms, including plants and fungi. It even has recognizable relatives in many bacteria. A cytochrome c tree is shown in Figure 24.19. Humans and fish differ in amino acid sequence by only 18 percent for cytochrome c. From humans to either plants or fungi gives about 45 percent divergence. However plants differ from fungi also by 45 percent, which tells us that, by

Divergence of Hemoglobin of Animals % Difference in Amino Acid Sequence From Humans				
	Rhesus Monkey	Cow	Pig	Rabbit	Chicken
α-chain	3	12	13	18	25
β-chain	5	17	16	10	26

24.19 CYTOCHROME C TREE
(numbers indicate base pair changes)

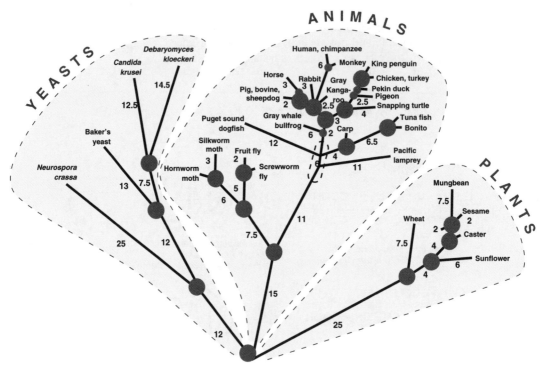

408

this measure, plants have diverged as far from fungi as animals have from plants.

Why Don't We Evolve Backwards?

Genes almost never mutate backwards to resemble the ancestors they diverged from many mutations ago. Why not? It's really a matter of probability. There is nothing forbidding any particular mutation to revert to the original sequence, but the likelihood of reversing exactly each of dozens of mutations is infinitesimally small.

Molecular Clocks to Track Evolution

Obviously we should not rely on just one protein. If we make trees for several proteins we get pretty much the same evolutionary relationships. However, the interesting part is that different proteins evolve at different speeds. As already noted, humans and fish differ by 50 percent in the α chain of hemoglobin but by less than 20 percent in their cytochrome c. If we plot the amino acid changes versus the evolutionary time scale (Fig. 24.20), we can see this easily for cytochrome c (slow), hemoglobin (both α and β chains alter at medium speed), and fibrinopeptides A and B (rapid evolution).

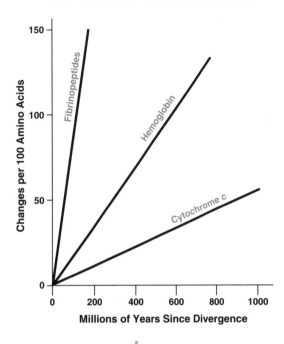

24.20 RATES OF PROTEIN EVOLUTION

Table 24.2

The rate of evolution is the number of mutations per 100 amino acids per 100 million years.

Protein	Rate of Evolution
Neurotoxins	110-125
Immunoglobulins	100-140
Fibrinopeptide B	91
Fibrinopeptide A	59
Insulin C peptide	53
Lysozyme	40
Hemoglobin α chain	27
Hemoglobin β chain	30
Growth hormone	25
Insulin	7.1
Cytochrome c	6.7
Histone H2	1.7
Histone H4	0.25

Rates of Evolution for Assorted Proteins (Table 24.2)

Fibrinopeptides are involved in the blood clotting process. They need an arginine at the end and must be mildly acidic overall. They can vary a lot because there are so few constraints on what is needed. In contrast histones bind to DNA and are responsible for its correct folding. Almost all changes to a histone would be lethal for the cell, so they evolve very slowly.

Cytochrome c is an enzyme whose function depends most critically on a few amino acid residues at the active site which bind to its heme cofactor. It should come as no surprise to find that these active site residues of cytochrome c rarely vary, even though amino acids around them change. Of 104 residues, only Cys-17, His-18 and Met-80 are totally invariant.

fibrinopeptides small protein fragments involved in blood clotting

24.21 IT'S THE 3-D STRUCTURE OF A PROTEIN THAT MATTERS

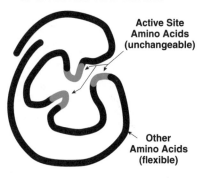

Active Site Amino Acids (unchangeable)

Other Amino Acids (flexible)

insulin a protein hormone that controls the sugar level in the blood

In other places variation is low; positions 35 and 36 are always filled by large and nonpolar amino acid residues. Several cytochrome c molecules have been examined by X-ray crystallography and they all have the same 3-D structure. Some pairs of cytochrome c can vary by as many as 88 percent of their residues, but as long as the protein assumes the correct 3-D shape around the active site, such variation is acceptable (Fig. 24.21).

Insulin is a hormone that evolves at much the same rate as cytochrome c. Insulin consists of two protein chains (A and B). Nonetheless there is only one insulin gene. During protein synthesis a long pro-insulin molecule is made. This has the middle, the C-peptide, cut out and discarded. The A and B chains are then held together by disulfide bonds (see Ch. 10 for insulin processing). Since the C chain is not part of the final hormone, it is free to evolve much faster, in fact it changes at almost 10 times the rate of the A and B chains.

Eventually, a rapidly evolving protein will become so altered in sequence between diverging organisms that the relationship will no longer be recognizable. Conversely, a protein that evolves very slowly will show little or no difference between two different organisms. Therefore we need to use slowly changing sequences to work out distant evolutionary relationships and fast evolving sequences for closely related organisms. Even if we examine the rapidly evolving fibrinopeptides, let alone hemoglobin or cytochrome c, humans and chimpanzees end up on the same branch of the evolutionary tree.

So, how can we tell people apart from chimps? As I sit here, looking at my fellow author, I realize that it's not always easy. If we look at most proteins they are identical in us and our primate brothers and sisters.

24.22 CLOSE RELATIVES ON THE SAME BRANCH

third codon position since many amino acids have several codons, the base in the third position of a codon can often be changed without changing the amino acid encoded

non-coding sequence region of a DNA sequence that does not code for any gene product, either protein or RNA

However, if we look at the DNA sequence of very closely related organisms we find many more differences. These are found in two main places, a) in the **third codon position,** and, b) in **non-coding sequences.** As discussed in Chapter 7, changing the third base of most codons does not alter the amino acid for which they code. So we can change the DNA sequence at the third base of most codons while leaving the protein the gene encodes unaltered (Fig. 24.23).

Introns are intervening (non-coding) sequences that are spliced out and do not appear in the messenger RNA and are therefore not represented in the final protein (see Ch. 11). Apart from the intron boundaries and splice

24.23 THIRD BASE POSITION MUTATIONS

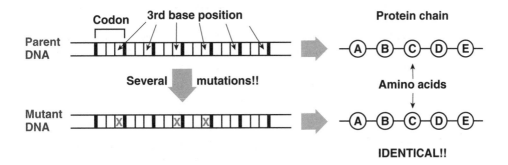

recognition sites, the DNA sequence of an intron is free to mutate extensively (Fig. 24.24). Other non-coding sequences exist between genes and if not involved in regulation, are also relatively free to mutate.

24.24 NON-CODING DNA EVOLVES FASTER

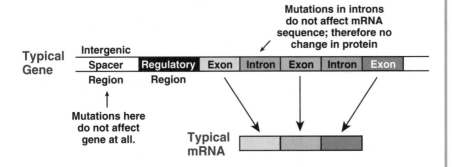

The early data on cytochrome c, hemoglobin, etc., were obtained by direct sequencing of proteins. But nowadays, nobody who is anybody sequences proteins any more. In fact most of the protein sequences available today were obtained by sequencing the DNA and then calculating the protein sequence using the genetic code. Hence, we have a lot of information about the detailed differences at the DNA level between closely related animals. Yes, chimps are slightly different after all but I still don't know where my coauthor fits in.

Creating New Genes By Duplication

Bacteria have a couple of thousand genes and higher eukaryotic cells have 50,000 to 100,000. How do we create new genes? The standard way is via gene duplication (Fig. 24.25). A mutation may cause the duplication of a segment of DNA that carries a whole gene. The mutant now has an extra copy. The original copy must be kept for its original function but the copy is free to mutate like crazy.

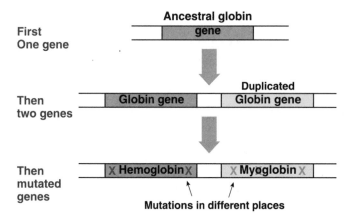

| | Mutation → | Original gene | Duplicate gene |

One of the best examples of gene duplication is in the globin family of genes. Hemoglobin carries oxygen in the blood whereas myoglobin carries it in muscle. These two proteins have much the same function, similar 3-D shapes and their sequences are related. After the ancestral globin gene duplicated, the two genes for hemoglobin and myoglobin slowly diverged as they specialized to operate in different tissues (Fig. 24.26).

But that's not all. The ancestral hemoglobin gene itself split to give alpha (α)-globin and ancient beta (β)-globin. Then the ancestral β-globin gene split again, twice, to give modern β-globin and the gamma (γ)-, delta (δ)- and epsilon (ε)-globins (Fig. 24.27).

The actual hemoglobin floating around in your blood has two α-globin and two β-globin chains forming a tetramer, unlike myoglobin which works on its own.

myoglobin oxygen carrier protein found in muscle tissue

24.26 DIVERGENCE OF HEMOGLOBIN AND MYOGLOBIN

First
One gene

Ancestral globin gene

Then
two genes

Globin gene   **Duplicated Globin gene**

Then
mutated
genes

X **Hemoglobin** X   X **Myoglobin** X

Mutations in different places

Withdrawing oxygen from the mother's blood by using the hemoglobin γ-chain is the earliest known case of "Fetal Attraction."

The globin variants are used during different stages of development. The ε-globin chain appears in early embryos, it is replaced by the γ-chain in the fetus and then by the β-chain in adults. A fetus needs to attract oxygen away from the mother's blood so it has an α2/γ2 hemoglobin that binds oxygen better than the adult α2/β2 hemoglobin (Fig. 24.28).

A group of closely related genes that arose by successive duplication is a gene family. The individual members are obviously related and carry out similar duties. When continued, gene duplication gives rise to a whole slew of new genes with differing functions; this is called a gene superfamily. The genes of the immune system provide good examples of gene families and superfamilies (see Ch. 22).

gene family group of closely related genes with a common ancestor and which carry out similar functions

gene superfamily group of more distantly related genes that have often diverged to carry out different functions

24.27 GLOBIN FAMILY TREE

α   γ   ε δ   β   Myoglobin

Ancestral α-globin

Ancestral β-globin

Ancestral hemoglobin

Ancestral globin

24.28 FETAL HEMOGLOBIN IS BETTER

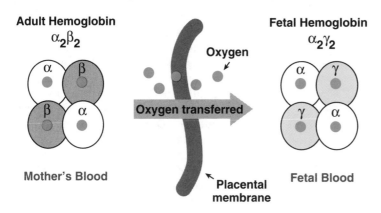

Adult Hemoglobin
$\alpha_2\beta_2$

Fetal Hemoglobin
$\alpha_2\gamma_2$

Oxygen

Oxygen transferred

Mother's Blood

Placental membrane

Fetal Blood

Another way to create new genes is by using premade parts. In other words, we start with two or more genes and take useful chunks from each. We then string these together to make a new gene (Fig. 24.29).

24.29 MODULAR GENE EVOLUTION

gene 1

gene 2

recombination

new gene

An example of the formation of a new gene from several diverse components is the **LDL receptor** (Fig. 24.30). LDL is low density lipoprotein that carries cholesterol around in the blood. The LDL receptor is on the cell surface and does just what its name suggests. The gene for the LDL receptor consists of several regions. Two of these have been plagiarized from other genes. Towards the front are seven repeats of a sequence also appearing in the C9 factor of complement, an immune system protein (see Ch. 22). Farther along is a chunk related to part of epidermal growth factor (a hormone). When such a gene mosaic is transcribed and translated, we get a patchwork protein consisting of several different regions.

LDL receptor receptor on cell surface for low density lipoprotein

24.30 LDL RECEPTOR – A GENETIC MOSAIC

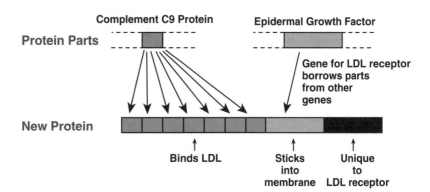

Complement C9 Protein

Epidermal Growth Factor

Protein Parts

Gene for LDL receptor borrows parts from other genes

New Protein

Binds LDL

Sticks into membrane

Unique to LDL receptor

Ribosomal RNA - A Slowly Ticking Clock

A major problem is how to construct an evolutionary tree with everybody on it. What is important here is the big picture. We want a tree showing the relationships between all the main groups of organisms. To achieve this, first we need a molecule that is present in all organisms. Second, our chosen molecule must evolve slowly so as to still be recognizable in all major groups of living things.

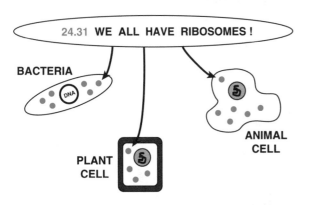

24.31 **WE ALL HAVE RIBOSOMES !**

BACTERIA

PLANT CELL

ANIMAL CELL

Although, as we saw above, histones evolve very slowly, only eukaryotic cells possess them; they are missing from bacteria. The solution is to use ribosomal RNA (Fig. 24.31), or rather sequence the DNA of the genes that code for rRNA. All living organisms have to make proteins and they all have ribosomes. (Viruses have no ribosomes but then, many biologists do not regard them as being truly alive; see Ch. 19.) Furthermore, since protein synthesis is so vital, ribosomal components are highly constrained and evolve slowly.

fungi these higher organisms have mitochondria but no chloroplasts. Although fungi do not move, their cell walls are made of chitin which is also found in the shells of insects, not cellulose as in plants. They are, if anything, immobile animals rather than colorless plants

Once upon a time things were divided into animal, vegetable and mineral. Even before molecular biology came to the rescue, biologists had realized that **fungi** were of equal rank to the plant and animal kingdoms and that most microorganisms didn't really fit in anywhere neatly. Use of relationships based on ribosomal RNA has allowed the creation of large scale evolutionary trees encompassing all the major groups of organisms. Higher organisms do consist of three main groups - plants, animals and fungi (Fig. 24.32). A variety of single celled organisms sprout off the tree near the bottom.

24.32 **THREE KINGDOMS AND MINOR PRINCIPALITIES**

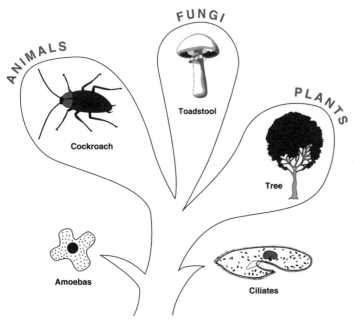

ANIMALS

Cockroach

FUNGI

Toadstool

PLANTS

Tree

Amoebas

Ciliates

However, there was a major surprise when the bacteria were incorporated into this scheme. It turns out that life on Earth consists of three lineages: the **eubacteria** ("true" bacteria), the **archebacteria** ("ancient" bacteria) and the **eukaryotic cells** (higher organisms, including animals, plants and fungi). But hold on, the surprise is yet to come.

The Surprise: The Symbiotic Theory of Evolution

As you know, all eukaryotic cells (animal, plant or fungus) contain mitochondria, and plant cells also contain chloroplasts. These organelles contain their own ribosomes (Fig. 24.33). So when we use ribosomal RNA from a eukaryotic cell for making an evolutionary tree, which ribosomes do we use? The trees shown above were made by using the ribosomes found in the cytoplasm of eukaryotic cells. These ribosomes have their ribosomal RNA coded for by genes in the cell nucleus.

eubacteria prokaryotic cells of the more common type, including most bacteria, mitochondria and chloroplasts

archebacteria prokaryotic cells of the less usual type, including many bacteria living in bizarre or extreme environments. They differ greatly from both eukaryotes and eubacteria in their rRNA sequences

eukaryotic cell cell type found in higher organisms and which has several chromosomes within a compartment called the nucleus

24.33 DIFFERENT RIBOSOMES IN EUKARYOTIC CELLS

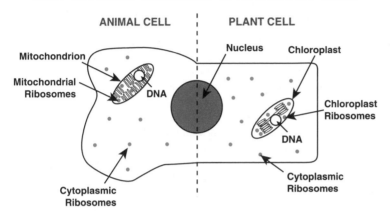

But what about the ribosomes from the mitochondria and chloroplasts? These organelles contain not only their own ribosomes, but also their own DNA! Mitochondrial ribosomes resemble those of bacteria rather than those found in the cytoplasm of eukaryotic cells. Mitochondrial ribosomes contain ribosomal RNA (rRNA) encoded by genes which are part of the mitochondrial DNA. Amazingly, organelle gene sequences are not related to the DNA sequences in their own cell nucleus! Sequences from these organelles indicate that mitochondria, and chloroplasts too, belong to the eubacterial lineage. Figure 24.34 shows the relationships of the three main lineages and where the organelles fit in.

Both mitochondria and chloroplasts are roughly the same size and shape as bacterial cells and each have their own circular DNA molecules. Finally, both mitochondria and chloroplasts grow and divide much as

24.34 ORGANELLES BELONG TO THE EUBACTERIA

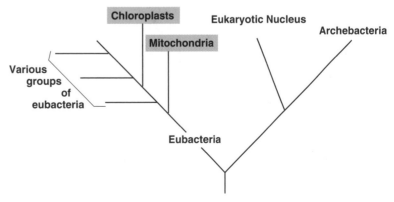

24.35 MITOCHONDRIA ARISE BY DIVISION

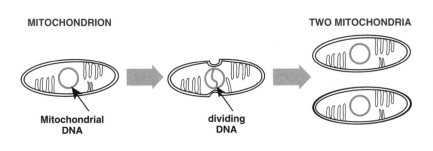

MITOCHONDRION   TWO MITOCHONDRIA

Mitochondrial
DNA

dividing
DNA

bacteria do (Fig. 24.35). When a eukaryotic cell divides, each daughter cell inherits some of its parents mitochondria and chloroplasts. If either organelle is lost, it cannot be reconstructed because the nucleus does not have all of the genetic information needed.

Taken together, these findings lead to the belief that mitochondria are enslaved bacteria which were trapped long ago by the ancestors of modern eukaryotic cells. The trapped bacteria received shelter and nutrients, and in return, devoted themselves to generating energy by respiration. Over the millions of years following their capture, these bacteria became narrowly specialized for energy production, lost the ability to survive on their own and evolved into mitochondria. Most of their original genes have now been lost and only a few remain on the mitochondrial DNA, which is much smaller than the typical circular chromosome of bacteria. But, you say, "Mitochondria are surrounded by two membranes, so where did they get the second one?" The outer "mitochondrial" membrane is actually derived from the host cell membrane as shown in Figure 24.36.

In other words, a modern eukaryotic cell is actually an association of two quite distinct genetic lineages that have shacked up in the same cell (Fig. 24.37). This is referred to as the symbiotic theory (from the Greek sym, meaning together, and bios, meaning life). The nuclear genes of a

Mitochondria and chloroplasts may be slaves, but what an easy life they have!

symbiotic theory the idea that the organelles of eukaryotic cells are derived from bacteria that took up residence as symbionts inside the ancestral eukaryotic cell

24.36 THEORY: BACTERIUM BECOMES MITOCHONDRION

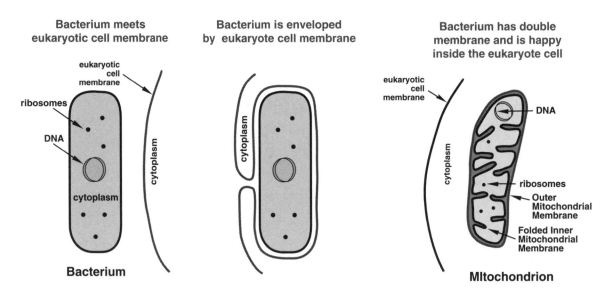

Bacterium meets
eukaryotic cell membrane

Bacterium is enveloped
by eukaryote cell membrane

Bacterium has double
membrane and is happy
inside the eukaryote cell

eukaryotic
cell
membrane

ribosomes

DNA

cytoplasm

cytoplasm

Bacterium

cytoplasm

eukaryotic
cell
membrane

cytoplasm

DNA

ribosomes

Outer
Mitochondrial
Membrane

Folded Inner
Mitochondrial
Membrane

MItochondrion

eukaryotic cell are sometimes referred to as coming from the "urkaryote." The urkaryote is the hypothetical ancestor that provided the genetic information found in the present day eukaryotic nucleus.

Plant cells also contain chloroplasts that perform photosynthesis. The rRNA from chloroplasts matches rRNA from photosynthetic bacteria better than rRNA from the plant cell nucleus. Thus, chloroplasts are strongly believed to have descended from photosynthetic bacteria trapped

urkaryote the ancestral eukaryotic cell before it gained its mitochondria and chloroplasts

A nucleus without its tamed organelles is like a cowboy without his horse.

24.37 SYMBIOTIC ORIGIN OF EUKARYOTIC CELL

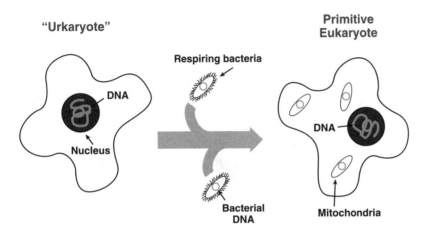

by the ancestors of modern day plants. Although chloroplasts are no longer independent, they have kept much more of their own DNA over the millions of years than have mitochondria.

It used to be thought that fungi were degenerate plants that had lost their chlorophyll, the green, light-absorbing pigment, and therefore could no longer photosynthesize. Today we believe, based on rRNA analysis, that the ancestral fungus never was photosynthetic but split off from the plant ancestor before the capture of the chloroplast (Fig. 24.38).

Thus, mitochondria and chloroplasts may be viewed as domestic animals at the cellular level. Symbiosis means "living together" and the cells of higher organisms are not individuals but associations.

24.38 SYMBIOTIC ORIGIN OF PLANTS

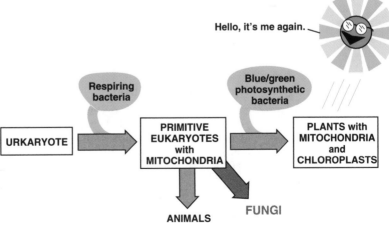

21.39 SYMBIONT THEORY

The Archebacteria Versus the Eubacteria

Although most common bacteria are eubacteria, there is another group, the archebacteria. Both types of bacteria have microscopic cells without a nucleus. They both have single circular chromosomes and divide in two by simple binary fission. In short, they both conform to the definition of a prokaryotic cell and there was no reason to suspect from their structure that there was anything but one type of bacteria. However, sequence analysis of ribosomal RNA indicates that there is about as much genetic difference between the eubacteria and archebacteria as between either of these two groups and eukaryotic cells.

Of the two groups of prokaryotes, the archebacteria are actually slightly more closely related to the urkaryote, the primeval ancestor of the eukaryotic nucleus.

Archebacteria tend to be found in bizarre environments and many of them are adapted to extreme conditions. They are found in hot sulfur springs, thermal vents in the ocean floor, in the super salty Dead Sea and Great Salt Lake and also in the guts of cows where they make methane (Fig. 21.40).

24.40 ARCHEBACTERIA LIVE IN WEIRD PLACES

Wanted: Dead or Alive!

One bizarre aspect of classifying life forms by ribosomal RNA is that you do not even need the organism itself. A sample of DNA containing the genes for 16S rRNA will do just fine. Although many microorganisms present in the sea or in soil have never been successfully grown in captivity, DNA can be extracted from the soil or seawater directly. Using PCR (see Ch. 17) it is possible to amplify the DNA from a single cell and get enough of the 16S rRNA gene to obtain a sequence (Fig. 24.41). Several new groups of bacteria that branched off very early from the archebacterial lineage have been discovered by this method, although none have ever been cultured alive.

Mitochondrial DNA - A Rapidly Ticking Clock

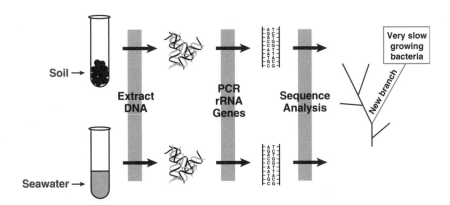

24.41 **STRANGE OCEAN AND SOIL LIFE FORMS**

Although mitochondria contain circular molecules of DNA reminiscent of bacterial chromosomes, the DNA is much smaller. The mitochondrial DNA codes for some of the proteins and ribosomal RNA of the mitochondrion, but most components are now encoded by the eukaryotic nucleus. The more advanced an organism, the shorter its mitochondrial DNA. Primitive eukaryotes still have quite a lot of mitochondrial genes, whereas humans and other higher animals have very few left (Fig. 24.42).

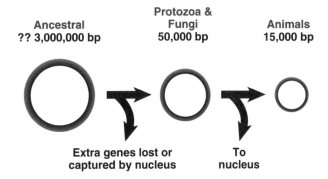

24.42 **EVOLUTION OF MITOCHONDRIAL GENOME**

The mutation rates for much of the mitochondrial DNA are higher than for nuclear genes. In particular, mutations accumulate rapidly in the third codon position of structural genes and even faster in the control region. This means that mitochondrial DNA can be used to study the relationships of closely related organisms in the relatively recent past. We can study the evolutionary history of closely related species or of races within the same species. Most of the variability in mitochondrial DNA occurs within the **D-loop** segment of the control region. Sequencing this segment allows us to distinguish between people of different racial groups.

One snag with using mitochondrial DNA is that you inherit all your mitochondria from your mother. (Though sperm cells do contain mitochondria, these are not released during fertilization of the egg cell and are not passed on to the descendants.) On the other hand, a eukaryotic cell contains only one nucleus but has many mitochondria so there are often thousands of copies of the mitochondrial DNA.

D-loop segment
region of about 500 bases of mitochondrial DNA, next to the origin of replication and replicated first, so forming a loop

DNA from Extinct Animals

Mitochondrial DNA can even be obtained from museum samples and extinct animals. Mitochondrial DNA extracted from frozen mammoths found in Siberia differed in four to five bases out of 350 from both Indian elephants and African elephants. This confirms the three way split proposed on

24.43 THREE - WAY SPLIT

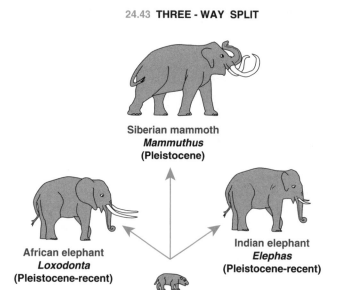

Siberian mammoth
Mammuthus
(Pleistocene)

African elephant
Loxodonta
(Pleistocene-recent)

Indian elephant
Elephas
(Pleistocene-recent)

Elephant Ancestor
Moeritherium
(Eocene-Oligocene)

anatomical grounds before DNA technology was used (Fig. 24.43).

The quagga is an extinct animal, similar to the zebra. It grazed the plains of Southern Africa only a little over a hundred years ago. A pelt preserved in a German museum has yielded muscle fragments from which DNA has been extracted and sequenced. The two gene fragments used were from the quagga mitochondrial DNA. The DNA from the quagga differed in about 5 percent of its bases from the modern zebra. The quagga and mountain zebra are estimated from this to have had a common ancestor about three million years ago.

The African Eve Hypothesis

Attempts to sort out human evolution from skulls and other bones led to two alternative schemes. The multiregional model proposes that *Homo erectus* evolved gradually into *Homo sapiens* simultaneously throughout Africa, Asia and Europe. The Noah's Ark model proposes that most branches of the human family became extinct and were replaced, relatively recently, by descendants from only one local sub-group (Fig. 24.44). Although both are taken seriously by anthropologists, few geneticists regard the multiregional model as plausible.

multiregional model
theory that multiple lines of humans, spread over the various continents, evolved simultaneously into modern man

Noah's Ark model
theory that only a single ancestral line of humans, localized in a relatively restricted area, evolved into modern man

Not surprisingly, recent molecular analysis has tended to support the Noah's Ark model.

Although mitochondria evolve fast, the overall variation among people of different races is surprisingly small. Calculations based on the observed divergence and the estimated rates suggest that our common ancestor lived in Africa between 100,000 and 200,000 years ago. Since mitochondria

24.44 MULTIREGIONAL AND NOAH'S ARK MODELS

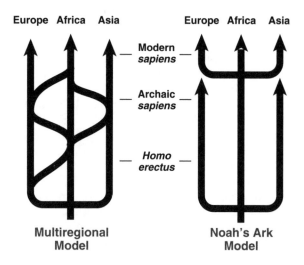

Europe Africa Asia

Europe Africa Asia

Modern
sapiens

Archaic
sapiens

*Homo
erectus*

Multiregional
Model

Noah's Ark
Model

are inherited only from your mother, this ancestor has been named African Eve.

African Eve female human ancestor derived by comparing maternally inherited sequences of mitochondrial DNA

The African origin is supported by the deeper "genetic roots" of modern day African populations. In other words, different sub-groups of Africans branched off from each other before everybody else branched off from the Africans as a whole (Fig. 24.45). The ancestors of today's Europeans split off from their Euro-Asian forebears and wandered into Europe via the Middle East around 40,000 to 50,000 years ago (Fig. 24.46). American Indians appear to derive from two major migrations originating from mainland Asian populations. The earlier, Paleo-Indians, (around 30,000 years ago) populated the whole American continent, while the more recent migration (less than 10,000 years ago) produced the Na-Dene peoples who are mostly North American Indians.

Recent racial comparisons of the non-coding sequences of repeated DNA in the microsatellite regions of chromosomes from the human cell nucleus tell much the same story. They also give a primary African - non-African split, and if anything, suggests an even more recent date for a common ancestor, nearer 100,000 than 200,000 years ago.

24.45 **MITOCHONDRIAL TREE FOR HUMANS**

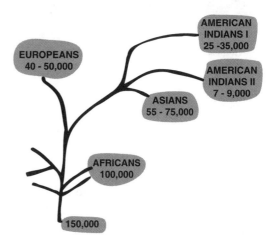

24.46 **OUT OF AFRICA**

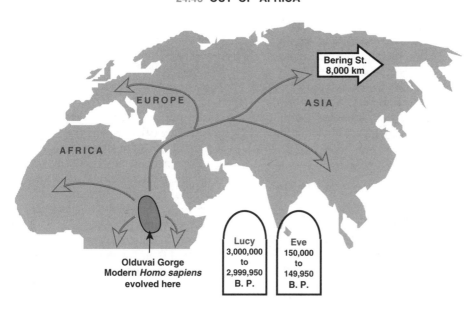

As you might have guessed, this relatively recent African origin has resulted in a lot of sentimentalist mush about the "unity of humankind." However, things might well be viewed rather differently from the perspective

Mitochondrial Eve's Boyfriend

Neanderthals

Mitochondrial Eve

of Neanderthal Man (and Woman) and other less successful lines who were "replaced."

To ensure approval by both male chauvinists and liberal feminists we have considered the ladies first. All the same, what about Adam? The ZFY gene on the Y chromosome is handed on from father to son and is involved in sperm maturation. The sequence data for ZFY suggest a split between humans and chimps about 5 million years ago and a common ancestor for modern mankind (no, not humankind, MANkind, this is the Y chromosome, remember) about 250,000 years ago.

With Eve at 125,000 years ago and Adam at 250,000 it would seem that the tendency for older men to marry younger wives goes back a long way!

Divine DNA from Mummies

While on the topic of female ancestors we should not forget that DNA has been successfully extracted from Egyptian mummies. Although the amounts of DNA obtained are only 5 percent or so of those from fresh, modern, human tissue, DNA sequences have been obtained from a mummy 2,400 years old. Although several thousand base pairs were sequenced, no actual human genes were identified. Since most higher animal DNA consists of intervening sequences, this is not unexpected. Nonetheless, the mummy DNA did contain **Alu elements** (see Ch. 11) which are characteristic of human DNA.

The Ancient Egyptians considered their rulers to be gods and practiced incest (brother/sister marriages) in order to preserve the divine bloodline (Fig. 24.49). We now have the opportunity to sneak a look at the genes that confer divinity!!!

24.48 **OLD MEN MARRY YOUNGER WIVES**

I love her for her mitochondria.

Alu elements a particular DNA sequence found in many copies on the chromosomes of animals, but which is of no use to the animal whose chromosomes it inhabits

24.49 **DIVINE DNA**

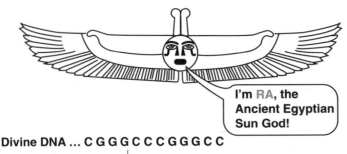

I'm RA, the Ancient Egyptian Sun God!

Divine DNA ... C G G G C C C G G G C C

Protein ... Arg Ala Arg Ala   or in one letter code RA RA RA!

Resurrecting Extinct Life

Apart from the odd mummy or mammoth, we have been using DNA sequences from still living creatures to construct evolutionary schemes. But wouldn't it be nice to get hold of DNA from somebody who lived long enough ago to provide a check on our estimated evolutionary rates?

The oldest available DNA so far successfully analyzed comes from amber. Amber is polymerized, hardened resin from extinct trees, a sort of primeval maple syrup that has turned from gooey to glassy over millions of years. Sometimes creepy crawlies were stuck in the resin when it oozed out of the trees and have been preserved there ever since (Fig. 24.50). Most of the trapped animals are insects, but occasionally worms, snails, and even small lizards have also been found.

Amber acts as a preservative and the internal structure of individual cells from trapped insects can still be seen with an electron microscope. It has proven possible to recover DNA that is 25 to 125 million years old from some insects. Some of this DNA has been amplified by PCR and sequenced.

Forget the insects, what about the dinosaurs? The largest chunks of amber are no more than 6 inches across, so there is no way you are going to find a whole embedded dinosaur. Nonetheless, a few blood cells preserved in the gut of a blood-sucking insect could, in theory, provide the complete DNA sequence of a large animal. This was the scenario for Michael Crichton's high-tech thriller, Jurassic Park, where dinosaurs were resurrected by having their DNA inserted into amphibian eggs (Fig. 24.51). In real life, the dino-DNA would probably be severely damaged and only short segments would be readable. However, the likelihood of someday obtaining at least a few fragments of some *Tyrannosaurus rex* genes is no longer a fantasy.

While not quite on the same scale, some 30 million-year-old bacteria have actually been revived in real life. Bacterial spores were found inside a bee trapped in a piece of amber. When provided with nutrients, the spores grew into bacterial colonies (Fig. 24.52). (Spores, covered by a protective coat are formed by some bacteria to survive bad conditions.)

24.50 INSECT TRAPPED IN AMBER

I feel like life has passed me by!

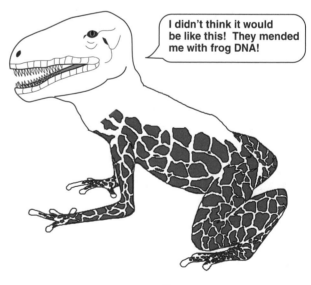

24.51 HALF FROG, HALF DINOSAUR

I didn't think it would be like this! They mended me with frog DNA!

24.52 REVIVING ANCIENT BACTERIAL SPORES

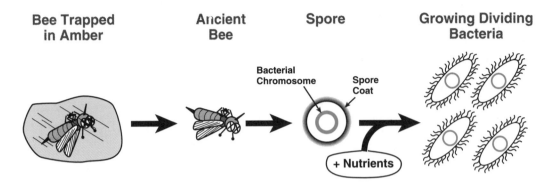

Bee Trapped in Amber   **Ancient Bee**   **Spore**   **Growing Dividing Bacteria**

Bacterial Chromosome   Spore Coat   + Nutrients

These reawakened bacteria were identified as *Bacillus sphaericus*, which is found today in association with bees. DNA from the ancient *Bacillus* was similar in sequence to its modern relative, but not identical as it would have been if the old-timer was just a contaminant!!

Evolving Sideways

No, this is not about crabs. Standard evolution involves slight changes in genetic information passed from one generation to its descendants. In general, worms breed worms, vegetables breed vegetables and bureaucrats breed paperwork.

However genetic information can be passed sideways. When antibiotic resistance genes are carried on plasmids they can be passed between unrelated types of bacteria (see Ch. 8). Since genes carried on plasmids are sometimes incorporated into the chromosome, a gene can move from one organism to an unrelated one in a couple of steps. This is known as horizontal gene transfer.

horizontal gene transfer transfer of genes between unrelated organisms by a vector such as a virus or plasmid

24.53 HORIZONTAL TRANSFER OF GENES

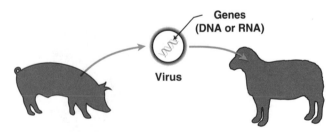

Genes (DNA or RNA)

Virus

Horizontal gene transfer depends on carriers that cross the boundaries from one species to another. Viruses, plasmids and transposons are all involved in such sideways movement of genes (Fig. 24.53). These genetic elements are discussed in their own chapters (see Chs. 19 and 20). Retroviruses in particular, are capable of inserting themselves into the chromosomes of animals, picking up genes and moving them into another animal species.

One well-described example of horizontal transfer stars the type-C virogene shared by baboons and all other Old World monkeys. The type-C virogene was present in the common ancestor of these monkeys, about 30 million years ago, and since then has diverged in sequence just like any other normal monkey gene. Related sequences are also found in a few species of cats. Only the smaller cats of North Africa and Europe possess the baboon type-C virogene. American, Asian and Sub-Saharan African cats all lack this sequence. Therefore the original cat ancestor did not have this type-C virogene. Furthermore, the sequence found in North African cats resembles that of baboons more closely than the sequences in monkeys closer to the ancestral stem (see Fig. 24.54).

24.54 HORIZONTAL TRANSFER BY RETROVIRUS

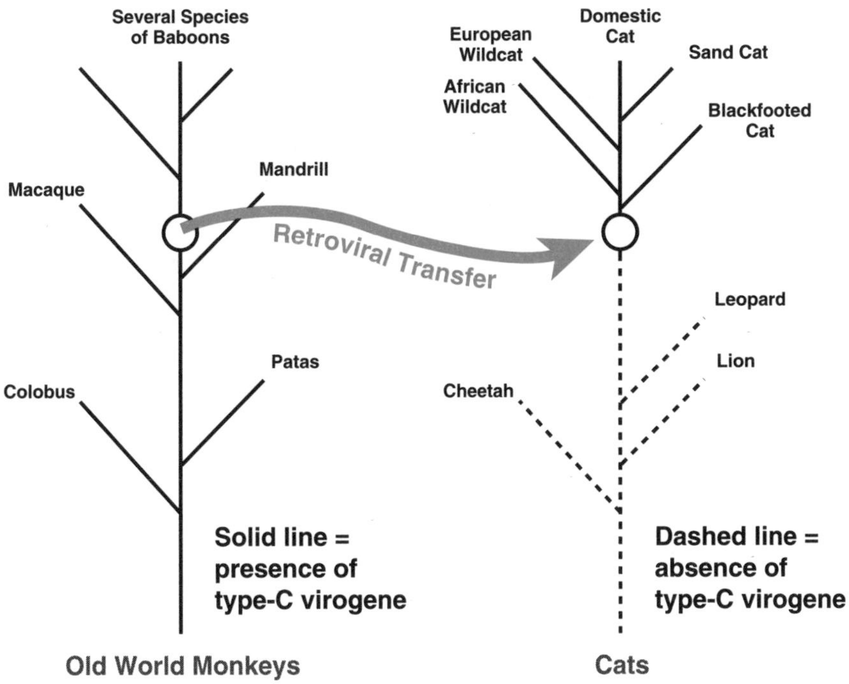

This all suggests that about 5 to 10 million years ago a retrovirus carried the type-C virogene horizontally from the ancestor of modern baboons to the ancestor of small North African cats. (The domestic European pussy-cat originally came from Egypt.) Hence, other cats that diverged more than 10 million years ago missed out.

Additional Reading

Fundamentals of Molecular Evolution by Li W-H & Grauer D. 1991. Sinauer Associates, Inc., Sunderland, MA.

The Origin of Life by Oparin, AI. 1929 [Translated from Russian and reprinted in JD Bernal, The Origin of Life, Cleveland: World, 1967].

The Quest for Life in Amber by Poinar G & Poinar R. 1994. Addison-Wesley Publishing Company, Reading, MA.

Classification: Biology For the Neurotic and the Obsessive-Compulsive

25

How do we classify the living world? For that matter, why? Why do we divide people into neurotics, obsessive-compulsives, manic depressives, schizophrenics and psychotics? What more characteristic sign of modern civilization is there than the ubiquitous filing cabinet? Is man driven to classify by some deep primal urge? Perhaps it's genetic, maybe those messy slobs who leave their stuff all over the place are classification-defective mutants.

Even other species classify. When tame chimpanzees were trained to sort a stack of photos into people versus animals, guess which pile they put their own photos onto? Yes, the human pile. Of course, they were quite correct, human and chimp DNA is just over 99 percent identical in base sequence.

Come to that, what use is classification? Is it really worth the bother? If you discover some new creature and classify it, you can be pretty sure that it has a lot in common with other creatures in the same group. If these have already been studied, this could save you the trouble of analyzing everything about your new organism from scratch.

Before we can start classifying and arranging we must do what Adam did in the Garden of Eden, give every living creature a distinct name. Unfortunately, at the Tower of Babel, mankind was divided into many groups, speaking different languages. So identical creatures may have different names in different countries. Still worse, confusion may result if different creatures have the same name:

25.1 THE URGE TO CLASSIFY

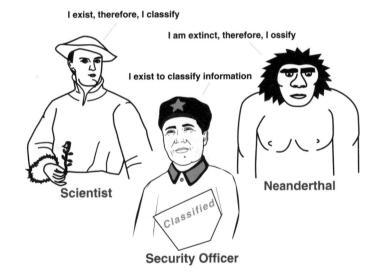

I exist, therefore, I classify

I am extinct, therefore, I ossify

I exist to classify information

Scientist

Classified

Security Officer

Neanderthal

did i ever tell you that mrs mccave
had twenty three sons and she named them all dave?
well, she did. and that wasnt a smart thing to do.
you see, when she wants one and calls out 'yoo-hoo!
come into the house, dave!' she doesnt get ONE.
all twenty three daves of hers come on the run!
this makes things quite difficult at the mccaves'
as you can imagine, with so many daves.

(from "Too Many Daves" by Dr. Seuss)

Even when people try to speak the same language, there may still be confusion. Though they are both small birds with red breasts, what an American calls a robin is not the same as an English robin and what an American calls corn is known as maize across the Atlantic.

So we need someone to do Adam's job all over again. Enter Carl von Linné, usually known by the Latinized form of his name, Carolus Linnaeus. With a characteristic lack of false modesty he announced, "God creates, Linnaeus arranges." In between writing five, yes five, autobiographies, Linnaeus managed to find time to publish his Systema Naturae in 1735. Looking down on the chaotic mess made by lesser mortals, Linnaeus proclaimed himself God's Registrar and suggested that all life forms should be given official two-word names in Latin, which at that time was universally used by educated Europeans. Though totally arrogant, Linnaeus was also a workaholic and set a good example by systematically naming and classifying 12,000 or so organisms. Since his scheme was sensible and no one else, however irritated, had the energy to re-name that many animals and plants, Linnaeus' system prevailed.

25.2 LINNAEUS WROTE A LOT OF BOOKS

428

A Question with No Answer: What is a Species?

Biological classification or **taxonomy**, consists of dividing life forms into smaller and smaller groups. The most specific is the **species**, a group of closely related organisms such as *Homo sapiens* (that's us!), *Esherichia coli* (the colon bacterium), or *Tyrannosaurus rex*. Each species has a double name, printed in italics. Pussy cats and tigers are clearly different yet clearly related. They are two species in the same **genus**, the next level up of classification. The pussy cat, *Felis domesticus* and the tiger, *Felis tigris* share the same first name, the generic name, and have a different second name, the specific name.

But how exactly do we define a species? Unfortunately, this is not as easy as you might think. The classic definition, really intended for animals, is that a species is a group of individuals who breed among themselves but do not interbreed with individuals of other species. But what about *Tyrannosaurus rex*? In such cases we only have skeletons and we simply have to guess who mated with whom.

But even living organisms can be awkward. Most bacteria do not mate with each other, they simply divide in two. There are also many plants and fungi with no sexual reproduction. Such organisms have common ancestors but do not interbreed with each other. Here species are groups of individuals whose shared characteristics imply a common ancestry. But even if we can trace their descent accurately, how far back do we go for the common ancestor to a species as opposed to a genus or a family? In other words, where do we draw the divisions in Figure 25.3 to separate the species? In practice, for bacteria and fungi, the divisions between species and even between genera are somewhat arbitrary.

taxonomy the science of naming and classification

There are two things that no living creature can avoid - death and taxonomy

species a group of closely related organisms with a relatively recent common ancestor

genus next level of organization up from the species; a group of related species

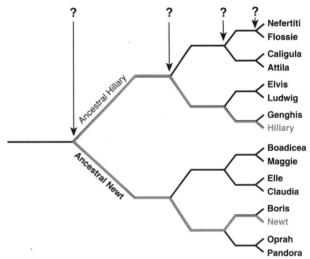

25.3 DIAGRAM OF ANCESTORS: WHERE DO WE DIVIDE ?
(ARE NEWT AND HILLARY RELATED ?)

Natural and Artificial Classifications

There is no totally satisfactory way to classify living things. Worse, there are two conflicting approaches, artificial and natural. We may design an **artificial classification** scheme that gives every creature a unique identity and position based on characters that are unambiguous and easy to detect. This is like allocating social security numbers. All the government cares about is making sure it knows where you live and how much tax you owe!

Natural classification attempts to place genuine relatives next to each other even if their relationships are not immediately obvious. If you have children, they are not likely to get the next social security number after yours! Yet a natural classification must obviously put them in your family, whether they look much like you or not. Natural classification means grouping together life forms that have common ancestors. The more recent the common ancestor, the more closely related are the two organisms.

artificial classification classification based on arbitrary characters selected for convenience

natural classification classification based on genuine biological ancestry

Molecules and Classification

Before molecular biology arrived on the scene, animals and plants were classified reasonably well, fungi and other primitive eukaryotes were classified poorly, and bacterial classification was pretty much a lost cause. The reason is obvious enough, most animals and plants have plenty of parts like legs, wings, leaves and flowers that can be examined and compared. In contrast, lower organisms, especially bacteria, really do not have much in the way of anatomy to be observed (Fig. 25.4). Using gene sequences for classification was developed for bacteria and has since spread to other types of organism.

Even higher organisms can pose problems. If we just look at superficial likenesses, we might group whales and sharks together because they both swim in the sea and have no legs. If we probe deeper by slicing them open and poking around inside, we find that whales are not fish but unusual mammals that have adapted to a marine existence. When members of different ancestral lines evolve similar adaptations to the same environment (like whales and fishes) this is called convergent evolution and is sometimes confusing.

25.4 BACTERIA HAVE NO ANATOMY

Squawk! I belong right here!!

I'm trying to "file'm," but I can't make heads nor tails of these things!

Heads - 1
Limbs - 4
Wings - 2
Tails - 1

Heads - 1
Limbs - 2
Wings - 2
Tails - 1

Heads - 1
Limbs - 6
Wings - 0
Tails - 1

Heads - 1
Limbs - 4
Wings - 0
Tails - 0

Heads - 1
Limbs - 8
Wings - 2
Tails - 1

Heads - 2
Limbs - 8
Wings - 4
Tails - 1

PHYLUM FILING SYSTEM

Bacteria

convergent evolution when two organisms of different ancestry evolve similar adaptations and so come to resemble each other superficially

What can molecular biology do for classification? First, we can trace ancestries by comparing the sequences of DNA, RNA or proteins which are more representative of fundamental genetic relationships than are many superficial characters. (This has been discussed in the chapter on molecular evolution.)

Second, in situations where division into species, genera, families, etc., is arbitrary, sequence data can provide quantitative measurements of genetic relatedness. Even if we cannot define a species unambiguously, at least we could be consistent in practice by how many bases two organisms must differ in their gene sequences before we allocate them to different species or families.

How are molecular family trees constructed? First we must agree on a molecule to compare. Most often we use ribosomal RNA sequences, since all living cells must make proteins and so possess ribosomes. So from each organism we get the sequence of the rRNA found in the small subunit of the ribosome. This is known as 16S rRNA in prokaryotes but 18S rRNA in eukaryotes where it is a little longer. This information is normally obtained by sequencing the DNA (see Ch. 23) of the genes encoding the rRNA.

Next the sequences are aligned and compared. Computer programs exist for calculating the relative divergence of the sequences and can generate trees such as that in Figure 25.5. Here we have four bacteria, all in different genera but belonging to the same family, the Enterobacteria. To root such a tree correctly we also need the sequence from an organism in an "out-group," in this case we have used the bacterium, *Pseudomonas*, which is known to be only distantly related to enteric bacteria. The nodes in Figure 25.5 represent the deduced common ancestors. The numbers indicate how many base changes are needed to convert the sequence at each branch point into the next. (The total length of the 16s rRNA of Enteric bacteria is 1542 bases.) The branch lengths are often scaled, as in Figure 25.5, to represent the number of mutations needed.

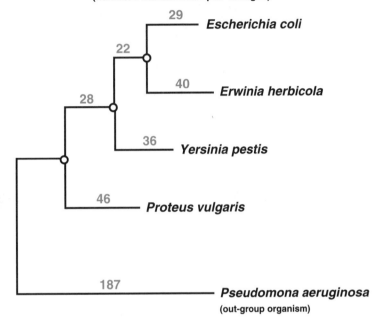

25.5 TREE SHOWING SEQUENCE DIVERGENCE
(numbers indicate base pair changes)

29 — *Escherichia coli*

22

40 — *Erwinia herbicola*

28

36 — *Yersinia pestis*

46 — *Proteus vulgaris*

187 — *Pseudomona aeruginosa*
(out-group organism)

Righting Old Wrongs

If an organism evolves into a parasite, many of its organs tend to atrophy and eventually disappear. Thus, parasitic plants often lose their leaves and roots and come to rely on the host plant for food and water (Fig. 25.6). Free living plants take up water with their roots and get energy by trapping sunlight with chlorophyll, the green pigment in their leaves. In the first stage of parasitism, the parasite connects its own roots to those of its victim and steals water. The modified roots that invade the host plant are called **haustoria**. Next, the leaves get smaller and less chlorophyll is made as the parasite steals pre-made nutrients from the victim. Finally, the parasite become completely dependent. Its leaves have faded away and the only roots left are haustorial.

haustorium modified root used by parasite to connect itself to host plant to get water and/or food

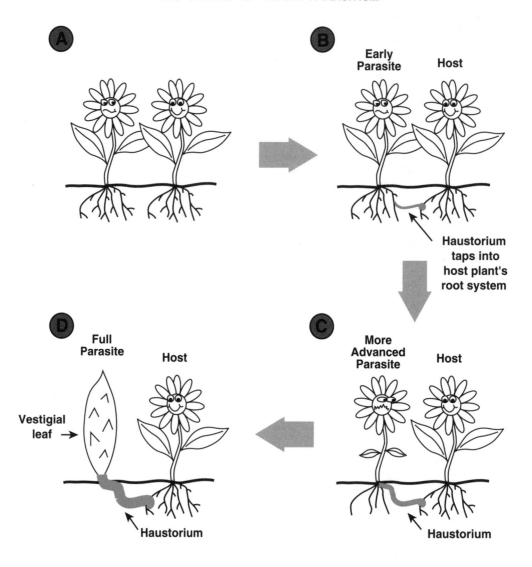

When an organism has been degenerating into a parasite for a long time, it may have lost so many of its original features that we cannot trace its relatives accurately. Gene sequences can often be used to trace the ancestry of parasitic or aberrant life forms. Dwarf mistletoes are a group of parasites which grow on the branches of coniferous trees. Externally, they only have a stem carrying berries and a few shrunken scale-like leaves. The stem is connected to a haustorial strand that penetrates the host plant. Nonetheless, some dwarf mistletoes are more degenerate than others and two of the smallest, *Arceuthobium pusillum* (infects spruce) and *A. douglasii* (infects Douglas fir), used to be classified close together. However, sequencing of DNA from rRNA genes has resulted in re-classification

(see Fig. 25.7). These two plants have converged by both losing everything except a tiny stem and a single berry totalling less than 1 cm in size.

Not all organ losses are due to parasitism. Moles and other animals adapted to living underground or in caves have lost their eyes. And what about those whales we mentioned above? From their internal structures, including the atrophied remains of hindlimbs, we deduce that whales are not fish, but mammals that have lost their legs and become fish-like in general form as they have adapted to life in the ocean. But which other mammals are the whale's closest relatives? Until gene sequencing came along, no one was sure. Whales were simply put into their own **order** and left to sink or swim. It now appears that whales are related to the artiodactyls, hoofed mammals such as hippos, giraffes, pigs and camels!

25.7 RECLASSIFICATION OF DWARF MISTLETOES

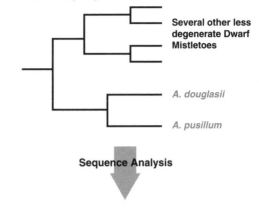

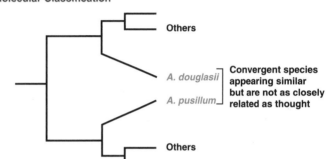

25.8 WHERE DO WHALES BELONG ?

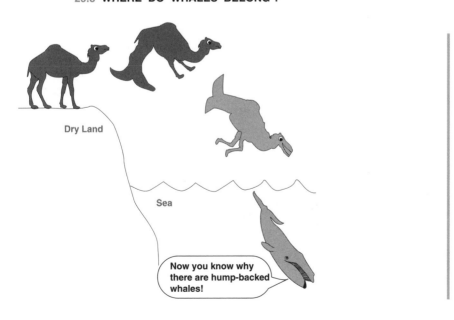

order group intermediate in rank between a class and a family

Are We All Upside Down??

In 1822, long before Darwin proposed the theory of evolution, Geoffrey Saint-Hilaire noticed that the overall body plan of vertebrates was upside down relative to that of other animal groups, such as arthropods (insects and crustaceans), annelid worms, mollusks, etc. (Fig. 25.9).

Saint-Hilaire's original proposal, that vertebrates were designed by inverting the fundamental animal plan, got a belly laugh from his contemporaries. Nonetheless, molecular analyzes of genes involved in embryo development suggest he may have been correct after all.

The very early embryos of most animals are rather similar and worm-like. During development the **homeobox genes** oversee the production of a head at the front, a tail at the rear, and the correct number of segments and limbs in between. In addition, there are regulatory genes that make sure that structures belonging to the **dorsal** (back) and **ventral** (underside) surfaces are correctly positioned.

homeobox genes family of genes which share a sequence of 180 bases and which encode transcription factors that regulate development

dorsal surface the back or upper surface

ventral surface the belly or under surface

In the fly, the *dpp* gene encodes a regulator protein that turns on those genes needed for the development of dorsal structures. It also turns off genes for ventral structures, including the nervous system, in those regions of the embryo that will develop into the back. The *sog* gene does the opposite; it promotes ventral structures and suppresses dorsal ones.

In frogs, whose embryos are used to study vertebrate development, there are also two master control genes for dorsal and ventral development. These are the *bmp-4* and the *chordin* genes. In terms of sequence homology, *bmp-4* is related to *dpp* and *chordin* to *sog*. But they act upside down relative to the insect versions!! The *bmp-4* gene promotes ventral structures and the *chordin* gene promotes dorsal structures!! What's more, if you transplant the *sog* gene from flies into frog embryos it is functional, but instead of promoting ventral structures, as it would in an insect, it works upside down and promotes dorsal development; that is, it mimics the action of *chordin*. (The opposite experiment, putting *chordin* into flies, also gives an upside down result.)

So perhaps it's not just the Australians who are upside down, perhaps we all are (Fig. 25.10). When the chordate ancestor

25.9 BODY PLAN OF ARTHROPODS VERSUS VERTEBRATES

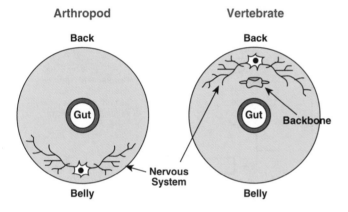

25.10 CORRECTED MAP OF THE WORLD

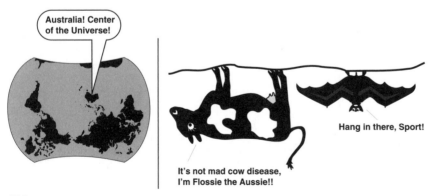

434

diverged from the rest of the animals, it may have suffered an inversion in its general body plan. If you think such a change would have caused too much confusion, consider that from the perspective of a primitive worm-like ancestor, being the right way up may not have been so terribly important.

A Brief Classification of All Life Forms

For reference, let's outline the various different life forms and their relationships. The first great split is between living cells and those lesser life forms that possess genes but not cells, the gene creatures (see Chs. 19 and 20). Each of these in turn is subdivided into many other groups as shown in Figure 25.11.

25.11 TREE OF LIFE

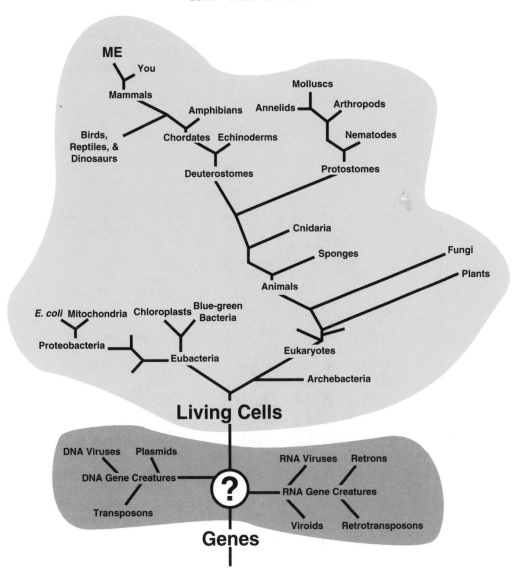

Gene Creatures are subcellular life forms that possess genetic information coded by nucleic acids, but do not consist of cells, do not possess ribosomes and cannot generate energy themselves.

DNA-based	RNA-based
DNA Viruses	RNA Viruses
Plasmids (most)	RNA Plasmids (rare)
Transposons	Retrotransposons
	Retrons
	Viroids

prions distorted disease-causing form of a normal brain protein that can transmit an infection

O.K., what about **prions** you demand, eager to confuse the issue? True, they possess genetic information in protein form, but the information is not even their own! The prion protein is encoded by a gene in the DNA of the cell it infects (see Ch. 21 for details). Even a virus uses its *own* genetic information. Whereas viruses steal a cell's physical resources, prions violate a cell's intellectual property rights. Perhaps prions should be called Gene Plagiarist Creatures!

25.12 THE PEOPLE VERSUS PRION

domain highest ranking group into which living creatures are divided, based on the most fundamental genetic properties

Cellular organisms are living cells surrounded by a cell membrane and which have genes made of DNA, possess ribosomes and use RNA as a genetic messenger. They are divided into three **domains**:

436

1) Eubacteria are traditional bacteria. These are prokaryotic cells without a nucleus and with only a single chromosome. This group includes the genomes of mitochondria and chloroplasts. Divided into several divisions, some still without proper names.

2) Archebacteria: From a structural viewpoint they are prokaryotes like eubacteria because they lack a nucleus. However, their gene sequences and other biochemical features indicate they are, if anything, slightly more closely related genetically to eukaryotes than to eubacteria.

3) Eukaryotes: Higher organisms whose DNA is carried on several chromosomes which are found inside the nucleus. Their cells are divided into separate compartments and usually contain other organelles in addition to the nucleus. They are divided into four **kingdoms**, listed below.

Protoctista - A rather bogus grouping of primitive, mostly single-celled eukaryotes that don't belong to the three main kingdoms. In fact there are several groups that are sufficiently distinct to merit status as miniature kingdoms.

Plants - Possess both mitochondria and chloroplasts and are photosynthetic. Typically they are non-mobile and have rigid cell walls made of cellulose.

Fungi - Possess mitochondria but lack chloroplasts. Once thought to be plants that had lost their chloroplasts, it is now thought they never had them. Although fungi are non-mobile, they lack cellulose and their cell walls are made of chitin, a polymer also found in the shells of insects. They are, if anything, more closely related to animals than plants.

Animals - Lack chloroplasts but possess mitochondria. Differ from fungi and plants in lacking a rigid cell wall. Typically mobile. They are divided into 20 to 30 **phyla** (singular, phylum), depending somewhat on personal taste. Some of these include:

> **Porifera** - sponges
> **Cnidaria** - sea anemones and jellyfish
> **Platyhelminthes** - flatworms
> **Nematoda** - roundworms
> **Arthropoda** - insects, crustaceans, etc.
> **Annelida** - segmented worms such as earthworms
> **Mollusca** - snails, squids, etc.
> **Echinodermata** - starfish, sea urchins
> **Chordata** - vertebrates and their relatives.

Phyla are divided into **classes**, such as mammals.
Classes are divided into **orders**, such as primates.
Orders are divided into **families**, such as hominids.
Families are divided into **genera**, such as *Homo*.
Genera are divided into **species**, such as *Homo sapiens* - that's our own name for ourselves. It's Latin for "wise guy." Linnaeus wasn't the only one lacking in humility!

Additional Reading

Biology by Campbell NA. 4th edition, 1996. The Benjamin/Cummings Publishing Company Inc., Menlo Park, California.

Five Kingdoms: An Illustrated Guide to the Phyla of Life on Earth by Margulis L & Schwartz KV. 1982. WH Freeman & Co., San Francisco, California.

---

prokaryotes lower organisms with a primitive type of cell that contains a single chromosome and has no nucleus

divisions major groups into which bacteria or plants are divided, roughly equivalent in rank to the phyla of animals

kingdoms major sub-divisions of eukaryotic organisms, especially the plant, fungus and animal kingdoms

phylum (plural, phyla) major groups into which animals are divided, roughly equivalent in rank to the divisions of plants or bacteria

class a subdivision of a Phylum

order a subdivision of a class

family a subdivision of an order

genus a subdivision of a family

species a subdivision of a genus whose precise definition is currently in question

Whodunnit First:
A Brief History of Molecular Biology

26

Not so long ago, it was thought that the night air caused disease and maggots were spawned spontaneously from dirt. On the heredity front, it was widely believed that if a woman became pregnant as a result of activities carried out indoors in the dark, her child would have brown eyes. In contrast, sex in the open air under a blue sky would produce children with blue eyes. Don't mock the middle ages. To this day, many people who claim a modern outlook persist in the medieval superstition that playing with toy guns or dolls determines adult behavior patterns.

Misery and Vice

To understand the foundations of modern genetics we must go back to the eighteenth century. The most important contribution of the 1700s to the understanding of both wild populations and human society (always assuming there is any difference) was that of Thomas Robert Malthus. Born in 1766, he published his infamous *An Essay on the Principle of Population,* in 1798.

Malthus realized that the supply of food, or of any other natural resource for that matter, is limited. Furthermore, any population of animals or humans will tend to increase rapidly in numbers as long as surplus food is available. Eventually the population will reach a point where there is not enough food to go around. (Although Malthus talked mostly about food, much the same applies to other resources such as water, shelter, space, etc.) Inevitably then, sooner or later, starvation will ensue, unless there is some other check on population growth.

The true significance of Malthus' discovery emerged when Charles Darwin asked the next question. In a Malthusian situation, where there is not enough food to go round, who will eat and who will starve? This led to Darwin's idea of **natural selection** - *the survival of the fittest*. Darwin's basic proposal was that there would be competition for the resources available. The fittest animals would get the resources and the losers would either fail to survive or rely on the welfare state.

Nature Red in Tooth and Claw

The idea that only the fit survive and that the inferior members of each species died or failed to reproduce, would result in the gradual improvement

Malthus said: "In short, it is difficult to conceive of any check to population, which does not come under the description of some species of misery or vice."

natural selection the evolutionary process in which the unfit fail to reproduce

of each species. The more romantic side of Darwinian selection gave rise to the dictum "nature red in tooth and claw." This refers to the fate of the slower zebras, who became lunch for the lions (Fig. 26.1).

26.1 LUNCH AT THE LION'S CLUB

Hey Leo, pass the salt please!

I'll take mine rare

The faster zebras survived to breed and pass on their genes to the next generation.

Of course, natural selection also occurs, at an altogether more boring level, when two plants compete for a place in the soil or for water in dry conditions. Darwin realized that if the inferior members of each species were weeded out by natural selection each generation, then only the superior individuals would survive to breed.

This was the critical point - the descendants represent only the successful individuals of the previous generation. Since like breeds like, succeeding generations would gradually improve. Over successive generations, zebras would get faster.

In order not to miss lunch, the lions would be forced to get faster too. So, over long periods of time, there would be gradual but steady changes in all living things. This process, Darwin referred to as evolution by natural selection.

Although, at first, Darwin took it for granted that "like breeds like" this became the burning issue of future work on evolution. There were really several questions.

1) How do living organisms inherit the characteristics of their parents? In other words what is the mechanism of inheritance?

2) What accounts for the differences between closely related individuals? Although we resemble our parents we are not identical to them, nor to our brothers and sisters.

3) How does novelty arise in biology? Are new characters and "improvements" merely due to re-shuffling of previously existing genetic information? Or does genuinely new genetic information come into existence, and if so how?

How Long is a Giraffes Neck?

Lamarck was a contemporary of Darwin who wondered about giraffes reaching for the last few remaining leaves high up on the tree. Obviously, the giraffes with the longest necks would get the food and those with shorter necks would not. Lamarck believed that a giraffe who tried very hard to get the highest leaves off the tree would stretch its neck as a result of continuous exercise. Lamarck proposed that such **acquired characteristics** could be inherited. The belief that changes due to conscientious efforts in

acquired characteristics properties which develop as a result of exercise or effort during your own lifetime

440

self-improvement by the parents can be passed on to their offspring has been a cherished myth of wishful thinkers ever since.

Darwin argued that even if a giraffe succeeded in stretching its own neck by its own heroic efforts, this extra long neck would not be inherited. Darwinian evolution has provided the giraffe with a long neck as the result of natural selection. Short-necked animals were less successful in feeding whereas their longer-necked competitors survived and passed this characteristic on to their descendants (Fig. 26.2).

26.2 SELECTION OF GIRAFFES WITH LONGER NECKS

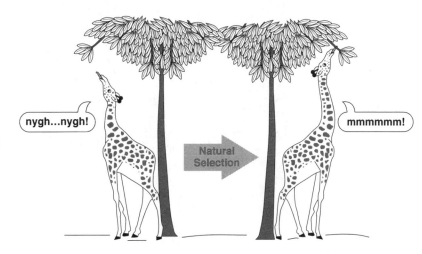

nygh...nygh!

Natural Selection

mmmmmm!

aim high! if you are a giraffe it would be nice to inherit a longer neck. The only question is how?

The Farmer's Wife Experiment

If acquired characteristics are inherited, then we should be able to demonstrate this in the laboratory. Let's take some mice (not necessarily three that are visually impaired) and cut off their tails. Their babies should be born tail-less. This procedure was carried out on a large number of mice for many generations. But not a single tail-less mouse was born. Many other less melodramatic experiments have been performed and all gave negative results.

Although the inheritance of acquired characteristics has been disproven, the question remains, - where do new characteristics come from? Ultimately, all our giraffes will have necks as long as the tallest original giraffe, and what then?

26.3 THE FARMER'S WIFE DISPROVES LAMARCK

OUCH!! Thwack!!

Like Peas in a Pod

A major obstacle to understanding heredity is that many characteristics of living organisms are due to the complex interaction of multiple factors. Although we can all see that a mother and her child look like each other and are therefore related, we are mostly unable to say exactly what about them is the same. Perhaps we are not looking in fine enough detail. Look closely at a newspaper picture of a face, better still use a magnifying glass, and you will see it consists of many thousand dots of ink. Each point on the paper may be either black or white - thus a complex image actually consists of many simple characters (Fig. 26.4). In a similar way, many human characteristics are built

26.4

THE COMPONENTS

OF A PHOTOGRAPH

ARE LESS COMPLEX

THAN THE GENETIC

MAKEUP OF AN

INDIVIDUAL

up of subcomponents so numerous as to defy analysis upon gross examination.

The great breakthrough in genetics came when Gregor Mendel decided to study the inheritance of simple, clear-cut, characters. The question whether the offspring inherit such a character from their parents can be answered by a simple "yes" or "no." In honor of Mendel's insight these are now referred to as Mendelian characters. Although twins, triplets or even very similar brothers and sisters have been referred to for eons as resembling peas in a pod, Mendel was the first to dispense with the children and study the peas.

Mendel was a monk and grew the pea plants in his monastery garden where he followed the inheritance of such characteristics as whether the seeds were smooth or wrinkled, whether the flowers were red or white and whether the pods were yellow or green, etc. Another advantage of using peas was that large numbers could be grown and scored in a relatively short time. Mendels results were published in 1865 but were not appreciated until much later.

In the early 1900s T. H. Morgan confirmed and greatly extended the science of genetics using an animal species, *Drosophila melanogaster,* the humble fruit fly. Flies are handy as they give a new generation every couple of weeks, unlike peas which take a whole year.

Mendel was a stickler for detail. So he got the monastery gardener to grow his plants and to count the peas of each type!

"Time flies like an arrow; fruit flies like a banana."
Groucho Marx

Molecules Enter Biology

Until early in the nineteenth century it used to be thought that there was a special vital force which energized living creatures, - from earthworms to Darth Vader. It was believed that living matter was quite different in nature from inanimate matter and was not subject to the normal laws of chemistry. This was overthrown in 1828 when Friedrich Wohler demonstrated the conversion, in a test tube, of a laboratory chemical, ammonium cyanate, to urea, a "living" molecule generated by animals.

This demonstrated that there was nothing magical about living matter. Many further experiments showed that the molecules found in living organisms were often very large and complex but could be understood and even synthesized by the normal laws of chemistry.

The de-mysticism of life reached its peak in the 1930s when the Russian biochemist Alexander Oparin

26.5 LIFE EMERGES

wrote a whole book outlining the chemical origin of life. Although the nature of the genetic material had not then been nailed down, Oparin put forward the idea that life evolved from small molecules in the primeval ocean as a result of standard physical and chemical forces (see Ch. 24).

Genetic Information is a Chemical Compound

Until the time of the Second World War, the nature of the inherited genetic information remained very vague and elusive. Then in 1944 Oswald Avery showed that the virulent nature of some strains of pneumonia could be transmitted to harmless strains by a chemical extract. Avery purified the essential molecule and demonstrated that it was DNA. When DNA from virulent strains was added to the harmless bacteria some of them took up the DNA and were "transformed" into virulent strains.

Avery concluded that the genes were made of DNA and that somehow genetic information was encoded in this molecule. Since DNA was known to have only half a dozen components it had not been a leading competitor for the role of genetic material; it was regarded as too simple!

The Post War Biology Boom

The question of how DNA with only half a dozen components could act as the genetic information was solved by James Watson and Francis Crick in 1953. Their now famous double helix provided a chemical basis for the genetic code and a mechanism for DNA replication and so for accurate inheritance. Today this central finding underlies our whole understanding of how living cells operate and what life means.

There are a bunch of other whodunnit firsts that led up to and followed Watson and Crick's Nobel prize-winning discovery (see Table 26.1).

Molecular Biology and the Future

So much for the history of molecular biology. But what about the future? Every advance in technology is sure to bring the cry of the

---

circa 10,000 BC wolves were tamed and mankind's first animal breeding experiments eventually produced the dog

---

Table 26.1 Brief History of DNA and Genetic Engineering

Year	Investigators and their Discoveries
1869	**Frederich Miescher** discovers DNA (in pus from infected wounds!)
1941	**Beadle and Tatum** Each gene codes for a single protein
1944	**Avery** DNA is the genetic material
1953	**Watson and Crick** Double helical structure for DNA
1958	**Meselson and Stahl** DNA replicates semiconservatively
1961	**Jacob and Monod** Operon model for genetic regulation
1970	**Temin and Baltimore** Reverse transcriptase in retroviruses
1977	Sequencing of DNA; intervening sequences in genes
1981	Catalytic RNA; transgenic animals are made
1987	**Kary Mullis** Polymerase Chain Reaction
1995	The bacterium **Haemophilus influenzae** is the first living cell to have its DNA completely sequenced

---

1993 *Jurassic Park* - the movie

1995 **Kary Mullis** serves on O. J. Simpson trial defense team

2345 Terminator cyborgs made and sent back through time

paranoid that Big Brother is coming (or is it Big Sister/Brother in political-ly correct America). In fact, a casual glance at history shows the reverse may be true. Take for example, computers, which many have feared allow the government to maintain data banks on their subject populations. To some degree this has happened, but in practice the greatest effect of personal computers has been to allow private individuals to distribute their own opinions, either by desk-top publishing or over the internet, without needing approval from big organizations, whether governments, universities or major publishing houses.

But what BIG changes will genetic engineering make to the lives of ordinary people? One of us thinks that major health care improvements will result from the Human Genome Project. The other (that's me, just in case you haven't figured it out yet) feels that improved agriculture due to transgenic animals and plants will have the biggest effect. In particular, adaptation of crop plants to growth in poor soils (too salty or too dry) or to yield new products, will have a colossal impact.

Actually, the unknown area most ripe for future advance is not so much preserving our bodies but probing our minds. Molecular biology is just now beginning to relate mental abilities to actual genes.

Those of you determined to wallow in gloom and doom may feel that manipulating the human mind by genetic engineering is frightening and therefore immoral. But imagine the possible future benefits - smart football players, honest politicians, rational women and sensitive, caring, men (like us)!!!

Another, more subtle, effect of genetic engineering will be on political beliefs. When genetics was still unknown to the general public, it was pos-sible for governments to base social policies on pre-scientific egalitarian mythology. The advent of genetic engineering has brought genetics to the public eye and what's more, the availability of actual products has con-vinced most ordinary people that genetics is not only real but applies to people. In spite of our equal rights, genetics is incompatible with the myth that all of us are created with the same mental and physical capabilities. In other words we are not all physically and mentally equal. As realization of this spreads, there will no doubt be a re-examination of many current societal practices.

Enough idle philosophizing! We admit that we really don't really know what the future will bring. One thing we do know is that to be an intelligent player in the future molecular revolution you should be literate in the language that helps you understand what you are. Chances are that if you have made it this far in the book you are well on your way to accomplishing that goal.

Additional Reading

The Origin of Species 1872 by Charles Darwin. John Murray, London. (reprinted by numerous publishers)

The Double Helix 1968 by James D. Watson, New York, Atheneum.

Glossary of Terms

-10 region	Region of promoter 10 bases back from the start of transcription and which is recognized by RNA polymerase
-35 region	Region of promoter 35 bases back from the start of transcription and which is recognized by RNA polymerase
2,3-butanediol	A particular chemical substance with two alcohol groups, made by certain bacteria, but a rare by-product in cells of higher organisms
A (acceptor) site	Binding site on the ribosome for the tRNA that brings in the next amino acid
ABO blood group system	Classic blood typing system for specific antigens found on red blood cells
Ac/Ds family	A family of transposons found in multiple copies, some of which are damaged, in cells of the maize plant
acetaldehyde dehydrogenase	The enzyme that converts acetaldehyde to acetate
acetosyringone	One of several related chemicals released by wounded plants
acquired characteristics	Properties that develop as a result of exercise or effort during your own lifetime
active site	Special site or pocket on a protein where the chemical reaction occurs
adenine (A)	One of the purine bases that pairs with thymine, found in DNA or RNA
adenosine deaminase	An enzyme that converts adenosine to inosine; its absence indirectly causes a lack of immune T- and B-cells
adenoviruses	Class of virus that infects many animals and contains DNA
adenylate cyclase	The enzyme that uses ATP to make cyclic AMP
African Eve	Female human ancestor derived by comparing maternally inherited sequences of mitochondrial DNA
agarose	A gel-forming polysaccharide found in some types of seaweed
Agrobacterium	Type of bacterium that infects plants and lives inside the tumors it causes
AIDS	Acquired Immuno-Deficiency Syndrome - a disease caused by the HIV retrovirus which slowly undermines the immune system by destroying helper T-cells
alcohol dehydrogenase	The enzyme that converts alcohol to acetaldehyde
alkaline phosphatase	An enzyme that chops phosphate groups off from a wide range of different molecules
allele	One particular version of a gene
allelic variants	Different versions of the same gene
allosteric proteins	Proteins that change shape when they bind a small molecule
alternative splicing	Variations in processing mRNA allowing more than one possible protein to be made from a single gene
alu element	A particular DNA sequence found in many copies on the chromosomes of animals, but which is of no use to the animal whose chromosomes it inhabits
amino acid	Monomer from which proteins are built
amino- or N-terminus	The end of a polypeptide chain that is made first and that has a free amino group
amino-acyl tRNA synthetase	Enzyme that attaches an amino acid to tRNA
amoebic dysentery	Disease of intestinal tract caused by amoebae (singular, amoeba) which are not bacteria but single-celled, eukaryotic microorganisms
ampicillin	A commonly used antibiotic of the penicillin family
amp^R or *bla* gene	Gene encoding β-lactamase

analog	One chemical substance that mimics another well enough to be mistaken for it by a living cell
anaphase	The step of mitosis during which separate halves of each chromosome are drawn apart by the spindle fibers towards the poles of the cell
annealing	The rejoining of separated single strands of DNA to form a double helix
anthrax	A cattle disease caused by the spore-forming bacterium, *Bacillus anthracis*
anti-oncogene/tumor-suppressor gene	Gene that acts to prevent unwanted cell division
antibiotic	Chemical substance that kills bacteria selectively, that is, without killing the patient, too
antibiotic resistance gene	Gene conferring resistance to an antibiotic
antibody	Protein of the immune system that recognizes and binds to foreign molecules (antigens)
antibody classes	Five classes of antibody that correspond to the five types of heavy chain that may be present
anticodon	Group of three complementary bases on tRNA that recognize and bind to a codon on the mRNA
antigen	A molecule that causes an immune response and which is recognized and bound by an antibody
antiparallel	Parallel, but running in opposite directions
antisense RNA	RNA complementary in sequence to messenger RNA and which therefore, base pairs with it
Apc and Dcc anti-oncogenes	Two particular anti-oncogenes often involved in colon cancer
apoprotein	Protein without any extra cofactors or prosthetic groups; that is, just the polypeptide chain(s)
apoptosis	Programmed suicide of unwanted cells for the good of the whole animal
Arabidopsis	A small plant used for genetics because of its genetic simplicity; it has only about 10 times as much DNA as bacteria
archebacteria	Prokaryotic cells of the less usual type, including many bacteria living in bizarre or extreme environments; they differ greatly from both eukaryotes and eubacteria in their rRNA sequences
aroA gene	The gene that codes for EPSPS
artificial classification	Classification based on arbitrary characters selected for convenience
auto-inducer	Signal molecule that can freely exit and enter cells and is involved in quorum sensing
autoradiography	Allowing radioactive materials to take pictures of themselves by laying them flat on photographic film
auxin	Plant hormone that typically promotes cell growth
azidothymidine (AZT)	A base analog that terminates a growing DNA chain and is used to treat AIDS
B-cell	The type of immune system cell that produces antibody
β-galactosidase	Enzyme that splits lactose and related compounds
β-lactamase	Enzyme that destroys antibiotics of the penicillin class
Bacillus thuringiensis	Type of bacterium that makes toxins for killing insects
bacteria	Primitive, relatively simple, single-celled organisms often used by geneticists and molecular biologists
bacterial ligase	The DNA ligase found in bacterial cells
bacterial plasmid	Circular molecule of double helical DNA that replicates independently of the bacterial cell's chromosome
bacteriocin	A toxic protein made by bacteria to kill other, closely related, bacteria
bacteriophage	Any virus infecting bacteria
bacteriophage ΦX174	A small spherical bacterial virus containing single-stranded DNA that infects *E. coli*
bacteriophage Lambda	A bacterial virus that infects *E. coli* and which is used both as a vector for cloning and for transfering bacterial genes by specialized transduction
bacteriophage M13	A small, filamentous bacterial virus containing single stranded DNA which only infects male *E. coli* and is used to make single stranded DNA for sequencing

bacteriophage Mu	A virus that infects *E. coli* and which causes <u>mutations</u> by inserting its DNA into the bacterial chromosome
bacteriophage P1	A bacterial virus that infects *E. coli* and which is used by geneticists to transfer bacterial genes by a mechanism known as transduction
bacteriophage Qβ	A small spherical bacterial virus containing single stranded RNA and which only infects male *E. coli*
base	Alkaline chemical substance, in particular the cyclic nitrogen compounds found in DNA and RNA
base analog	Chemical that resembles a base of a nucleic acid well enough to fool a cell into using it instead
base pair	Two bases held together by hydrogen bonds
base pairing	When two complementary bases (A with T or G with C) recognize each other and are held together by hydrogen bonds
Bcl-2 gene	A gene that probably controls apoptosis in humans and may contribute to cancer when mutated
benign	When a tumor stays in one place
beta- (or β-) galactosidase	Enzyme that splits lactose and related compounds
binary fission	Simple form of cell division, by splitting down the middle, found among bacteria
biosafety levels	A series of levels of increasingly strict laboratory containment intended for dealing with dangerous agents; from 1 (mild) to 4 (bad!!)
biotin	A vitamin
blunt ends	Ends of a DNA molecule that are fully base paired
bubonic plague	An often fatal and highly contagious disease caused by bacteria of the species *Yersinia pestis*
bursa of Fabricius	Organ of chicken where B-cells develop
c-onc	Oncogene carried on a cell's chromosome
C-peptide	Connecting peptide that originally links the A and B-chains of insulin but which is absent from the final hormone
Caenorhabditis elegans	A species of nematode or roundworm, widely used in studying animal development because of its relative simplicity
cancer	Disease due to unplanned growth and division of mutant somatic cells
capsid	Protein shell surrounding the DNA or RNA of a virus particle
carboxy- or C-terminus	The end of a polypeptide chain that is made last and has a free carboxy-group
carcinogen	Any agent that causes cancer
carriers	Individuals who have a single defective copy of a gene but show no clinical symptoms
cauliflower mosaic virus	A small virus with double stranded circular DNA that attacks cauliflower and related plants
C-terminus	The end of a polypeptide chain that is made last and has a free carboxy-group. Also known as carboxy-terminus
CD4 protein	A protein found on the surface of many cells of the body's immune system
cDNA library	A collection of cloned genes present as their cDNA versions and carried on an appropriate plasmid or virus vector. cDNA is generated by the reverse transcription of messenger RNA
Ced genes	Genes which control apoptosis in nematode worms
cell	A cell is the basic unit of life. Each cell is surrounded by a membrane and has a full set of genes that provide it with the genetic information necessary to operate
cell cycle	The sequence of events required for growth and division of a cell
cell-mediated immunity	Immune reactions due to cells of the immune system acting as a whole as opposed to those due to antibody molecules
cellulose	Major carbohydrate of plant cell walls; a polymer of glucose
central dogma	Basic plan of genetic information flow in living cells which relates genes (DNA), message (RNA) and proteins

centromere	Structure found on a chromosome and used to build and organize microtubules during mitosis
cephalosporin C	Original member of a family of antibiotics closely related to the penicillins
Chagas' disease	Disease found in Latin America, caused by trypanosomes
chain termination or dideoxy method	A method for sequencing DNA that involves the premature termination of growing DNA chains by using dideoxy derivatives of the nucleotides
charged tRNA	tRNA with an amino acid attached
chemiluminescence	Emission of light as a side product of a chemical reaction
chloride	A dissolved inorganic ion found in major levels in body fluids
chlorinated hydrocarbons	Class of organic chemical compounds containing chlorine which are often toxic and difficult to break down
chloroplast	Organelle found in plant cells which traps energy from sunlight to make sugars by photosynthesis
choleratoxin	A toxic protein made by cholera bacteria and which causes the symptoms of acute diarrhea
chromatid	Single double-helical DNA molecule making up whole or part of a chromosome
chromatin	Complex of DNA plus protein which constitutes eukaryotic chromosomes
chromosome walking	Method for cloning neighboring regions of a chromosome by successive cycles of hybridization using overlapping probes
chromosome	Structure bearing the genes of a cell and made of DNA
cistron	Segment of DNA (or RNA) which encodes a single polypeptide chain
Class I MHC	MHC proteins consisting of two chains of unequal length and found on the surface of all cells. Their role is to display fragments of proteins originating inside the cell
Class II MHC	MHC proteins consisting of two chains of equal length and found only on the surface of certain immune cells. Their role is to display fragments of proteins from digested microorganisms
class switching	Genetic rearrangement that replaces one type of heavy chain with another and so changes the class to which an antibody belongs
clone a gene	To obtain the DNA making up a particular gene and put it on a suitable vector such as a plasmid
co-dominance	When two functional alleles both contribute to the observed properties
coconut cadang-cadang viroid	A specific viroid that infects coconut trees
coding strand	The strand of DNA equivalent in sequence to the messenger RNA
codon bias	Tendency to use some of the codons for a particular amino acid more frequently than others
codon	Group of three RNA or DNA bases that encodes a single amino acid
cofactor or prosthetic group	Extra chemical group attached to a protein and which is not part of the polypeptide chain
"cold"	Slang for non-radioactive
ColE-plasmid	A small, high copy plasmid that carries genes for a toxin known as colicin E and whose derivatives have been widely used as vectors
colicin	A bacteriocin made by some strains of *Escherichia coli* to kill other strains of *E. coli*
competent	Capable of taking up pure DNA from the external medium
competitive inhibitor	Chemical substance that inhibits the action of an enzyme by mimicking the true substrate well enough to be mistaken for it
complement	A team of proteins that destroy invading bacteria after being alerted to their presence by the binding of antibody
complementary DNA (cDNA)	The DNA sequence complementary to an RNA sequence, usually to mRNA
complementary sequences	Two base sequences whose bases pair off with each other because A, T, G, C in one sequence correspond to T, A, C, G, respectively, in the other
complex transposon	Type of transposon that leaves behind one copy in its original location when it moves, and inserts a second copy into the new location

448

composite transposon	Segment of DNA flanked by two insertion sequences which can jump as a complete unit
conjugation	Transfer of genes between bacteria involving cell to cell contact
conjugation bridge	Channel that forms where the cell envelopes of mating bacteria touch and fuse together and through which DNA is transferred
consensus sequence	Ideal sequence to which real life sequences approximate
conservative substitution	Replacement of an amino acid in a protein with another that is similar in its chemical properties
conservative transposition	The version of transposition in which the transposon is removed from its original location, so leaving a gap in the DNA, and is inserted unaltered into a new site
constant region	Region of an antibody protein chain that remains constant
contact inhibition	When normal cells prevent their neighbors from dividing by touching them
convergent evolution	When two organisms of different ancestry evolve similar adaptations and so come to resemble each other superficially
copy number	Number of copies of a plasmid in each host cell
core enzyme	Bacterial RNA polymerase without the sigma (recognition) subunit
coupled transcription-translation	When ribosomes of bacteria start translating an mRNA molecule that is still being transcribed from the genes
cowpox	A disease of cattle caused by a virus closely related to smallpox. Cowpox only causes a very mild disease in humans
Creutzfeldt-Jacob disease	Inherited brain degeneration disease of humans caused by prion
crossing over	When two strands of DNA are broken and are then spliced to each other
cyclic AMP	Cyclic adenosine monophosphate, a signal molecule used in global regulation
cyclin dependent kinases (CDK)	Subordinate proteins that transmit orders from a cyclin by adding phosphate groups to the enzymes they control
cyclin	Protein that controls the cell cycle
cystic fibrosis	A disease in which the major symptom is the accumulation of fibrous tissue in the lungs
cytochrome C	Protein that is part of the respiratory chain
cytokinin	Plant hormone that typically promotes cell division
cytosine (C)	One of the pyrimidine bases found in DNA or RNA and which pairs with guanine
D-loop segment	Region of about 500 bases of mitochondrial DNA, next to the origin of replication and replicated first, thus forming a loop
Daf-2 gene	Gene involved in life length determination in nematode worms
dATP, dGTP, dTTP and dCTP	Deoxyadenosine triphosphate, deoxyguanosine triphosphate, deoxythymidine triphosphate, and deoxycytidine triphosphate
ddGTP	Dideoxyguanosine triphosphate
degenerate primer	Primer with several alternative bases at certain positions
degenerate	In speaking of genes, it refers to the use of more than one codon to encode the same amino acid
deletion	Removal of one or many nucleotides from DNA
denaturation	When used of proteins or other biological polymers, refers to the loss of correct 3-D structure
denature	To destroy the 3-D folding of a protein or other polymeric molecule
deoxyribonuclease	An enzyme that cuts DNA into shorter pieces
deoxyribonuclease I (DNAse I)	An enzyme that degrades DNA by cutting between individual nucleotides in a non-specific manner
deoxyribose	The sugar with five carbon atoms that is found in DNA
detergent	Molecule that binds grease at one end and uses its other end to dissolve in water
diabetes	Disease causing inability to control level of blood sugar due to defect in insulin production
dideoxyribose	Derivative of ribose missing both the 2' and 3' hydroxyl groups

diffraction pattern	Array of spots formed by X-rays after traveling through a crystal
digoxigenin	Steroid molecule from foxglove, commonly used as a linker for attaching a fluorescent tag to DNA
diploid	Possessing two copies of each gene
disulfide bond	Chemical linkage between two cysteine residues that binds together two protein chains
divisions	Major groups into which bacteria or plants are divided, roughly equivalent in rank to the phyla of animals
Dmd gene	Gene which, when defective, causes Duchenne muscular dystrophy
DNA	Deoxyribonucleic acid, nucleic acid polymer of which the genes are made
DNA fingerprint	Individually unique pattern due to multiple bands of DNA produced using restriction enzymes, separated by electrophoresis and usually visualized by Southern blotting
DNA gyrase	The enzyme that unwinds supercoiled DNA
DNA helicase	The enzyme that unwinds double helical DNA
DNA ligase	An enzyme that joins up DNA fragments end to end
DNA polymerase I	Enzyme that makes small stretches of DNA to fill in gaps between Okazaki fragments or during repair of damaged DNA
DNA polymerase III	Enzyme that makes most of the DNA when chromosomes are replicated
DNA sequencing	Finding out the sequence of bases in a molecule of DNA
domain	Highest ranking group into which living creatures are divided, based on the most fundamental genetic properties
dominant allele	The allele whose properties are expressed as the phenotype
dorsal surface	The back or upper surface
dot blot	A method used to recognize specific sequences of DNA with a tagged DNA probe
double helix	Structure in which two strands of DNA are twisted spirally around each other
Down syndrome	Defective development, including mental retardation, resulting from an extra copy of chromosome No. 21
Duchenne muscular dystrophy	One particular form of muscular dystrophy, a group of degenerative muscle diseases
dystrophin	Protein encoded by the *dmd* gene whose malfunction causes muscular dystrophy
Ebola virus	New and very virulent member of filovirus family which appeared in Africa in the 1980s
EcoRI	A restriction enzyme found in *E. coli*
eicosahedron	Solid shape with 20 faces
electrophoresis	Movement of charged molecules towards an electrode of the opposite charge; used to separate nucleic acids and proteins
electroporator	Device used to apply high voltage to cells in order to make them permeable to DNA
elongation factors	Proteins that oversee the elongation of a growing polypeptide chain
enhancers	Regulatory sequences outside the actual promoter that bind transcription factors
enteric bacteria	A family of related bacteria often found in the intestines of animals
enzyme	A protein which carries out a chemical reaction
epistasis	When a mutation in one gene masks the effect of alterations in another gene
epitope	Localized region of an antigen to which the antibody binds
EPSPS or 5-enolpyruvoyl shikimate-3-phosphate synthase	An enzyme essential for making amino acids of the aromatic family
ethidium bromide	A dye that stains DNA and RNA orange if viewed under UV light
eubacteria	Prokaryotic cells of the more common type, including most bacteria, mitochondria and chloroplasts

450

eugenics	Artificially improving a species by genetics
eukaryote	Organism that shelters its genes inside a nucleus and has several linear chromosomes
eukaryotic cell	Advanced cell of higher organisms which has several chromosomes within a compartment called the nucleus
eukaryotic promoter	This has three subcomponents: a) initiator box where transcription starts, b) TATA box that first binds RNA polymerase, and, c) upstream elements needed for specific control by transcription factors
excision repair (cut and patch repair)	Cutting out a stretch of damaged DNA and replacing it with new DNA
exon	Segment of a gene that codes for protein
expression site	Special location on chromosome of a trypanosome from which a few chosen genes of the VSG family can be expressed
F-plasmid	A particular plasmid which confers ability to mate on its bacterial host, *Escherichia coli*
Fab fragment	Antigen binding fragment of an antibody
FACS (fluorescence activated cell sorter)	Machine which sorts particles, such as cells or chromosomes according to their fluorescence
Fc fragment	The stem region of an antibody, *i.e.*, the fragment which does not bind the antigen
Fc receptor	Protein on the surface of immune cells which recognizes the Fc region of an antibody once it has bound an antigen
fertility plasmid	Type of plasmid that confers ability to mate on its bacterial host
ferric uptake regulator	A global regulator protein that detects and binds iron
fibrinopeptides	Small protein fragments involved in blood clotting
filial generations	Successive generations of descendants from a genetic cross which are numbered F_1, F_2, F_3, etc., to keep track of them
filoviruses	Family of negative-stranded RNA viruses with a filamentous structure
FISH (fluorescence *in situ* hybridization)	Using a fluorescent tagged probe to see a molecule of DNA or RNA in its natural location
fluorescence	When a molecule absorbs light of one wavelength and then emits light of another, longer, lower energy wavelength
fMet or formyl-methionine	Chemically tagged version of methionine used to start the polypeptide chain in prokaryotic (bacterial) cells
footprint	When a protein binds to DNA at a specific site, it can protect the DNA from being cut in this region. The result may be visualized as a missing group of bands when the DNA is run on a gel
founder mice	Original engineered mice that receive a single copy of a transgene and are bred together to found a line of transgenic animals
frameshift mutation	A mutation that changes the reading frame of the protein encoded by that gene
fructose	A type of simple sugar commonly found in fruits
fungi	These higher organisms have mitochondria but no chloroplasts. Although fungi do not move, their cell walls are made of chitin which is also found in the shells of insects, not cellulose, as in plants. They are, if anything, immobile animals rather than colorless plants
G_1 phase	First stage in the cell cycle, cell growth
G_2 phase	Third stage in the cell cycle, preparation for division
gametes	Cells specialized for sexual reproduction which are haploid, that is, they have one set of genes
gel	A semi-solid made by a polymer that forms a cross-linked meshwork in water
gel electrophoresis	Electrophoresis of charged molecules through a gel meshwork in order to sort them by size
gel retardation (bandshift)	When a protein binds to a segment of DNA, the DNA will move slower through a gel and the DNA band will be shifted to a new position
gelatin	A natural protein extracted by boiling from the connective tissue of animals

gene	A unit of genetic information
gene activator proteins	Proteins that switch on genes by binding to DNA and helping RNA polymerase to bind
gene cassette	Artificially constructed segment of DNA containing a genetic marker, usually an antibiotic resistance gene, and convenient restriction sites at each end
gene creature	Life form that consists primarily of genetic information, sometimes with a protective covering, but without its own machinery to generate energy or replicate macromolecules
gene family	Group of closely related genes with a common ancestor and which carry out similar functions
gene fusion	Hybrid in which the regulatory sequences from one gene are joined to the coding region of another gene
gene knock out	Technically, knocking out a gene means disrupting it by inserting foreign DNA
gene superfamily	Group of more distantly related genes which have often diverged to carry out different functions
Gene Chip®	Another name for the oligonucleotide array detector
generalized transduction	Transduction in which the transported genes are picked at random
genetic engineering	Alteration of an organism by deliberately changing its DNA
genotype	The description of an organism at the genetic level
genus	Next level of organization up from the species, a group of related species
germ line cells	Reproductive cells producing eggs or sperm that take part in forming the next generation
global regulation	Regulation of a large group of genes in response to the same environmental stimulus
global regulator protein	Protein that controls expression of many genes in response to the same signal
glucose	A type of simple sugar usually found making up polymers such as starch and cellulose
glyphosate	A weedkiller that inhibits synthesis of amino acids of the aromatic family in plants
gp120 ("gene-product" 120)	A particular protein found on the surface of the AIDS virus and which binds to the CD4 protein in T-cell membranes
growth hormone or somatotropin	The hormone made by the pituitary gland whose major effect is on skeletal growth in young animals
guanine (G)	One of the purine bases found in DNA or RNA and which pairs with cytosine
haploid	Possessing only a single copy of each gene
haustorium	Modified root used by parasite to connect itself to a host plant
heavy chain	The longer of the two pairs of chains comprising an antibody molecule
hemoglobin	Oxygen carrier protein that carries oxygen in blood
herpesvirus	Family of DNA-containing viruses that cause a variety of diseases and sometimes cause tumors. They contain dsDNA and an outer envelope surrounding the nucleocapsid
heterozygous	Having two different copies, or alleles, of the same gene
Hfr-strain	Bacterial strain with F-plasmid integrated into the chromosome, so allowing high frequency transfer of chromosomal genes
histamine	Substance made from the amino acid histidine that is released by mast cells and which causes expansion of blood vessels and acceleration of the heart rate
histones	Special positively charged proteins that bind to DNA
HIV	Human immunodeficiency virus, the causative agent of AIDS, is a member of the retrovirus family
HLA	Human leukocyte antigens are proteins found on the cell surface that allow cells to be recognized by the immune system. Same as MHC proteins
homozygous	Having two identical copies, or alleles, of the same gene
horizontal gene transfer	Transfer of genes between unrelated organisms by a vector such as a virus or plasmid
hormone	Regulatory molecule that carries commands from one tissue to another in the body fluids
host range	The range of different types of host organism which a plasmid, or a virus, can infect
"hot"	Slang for radioactive

housekeeping genes	Genes that are switched on all the time because they are needed for essential life functions
human genome project	Program to completely sequence human DNA
hybrid DNA	Artificial double stranded molecule of DNA formed by two single strands from two different sources
hybridoma	Hybrid cell line made by fusiing an antibody secreting B-cell with a cancerous myeloma cell
hydrocarbon	Molecule made only of carbon and hydrogen
hydrogen bond(ing)	Bond resulting from the attraction of a positive hydrogen atom to both of two other atoms with a negative charge
hydrophilic	Water loving
hydrophobic	Water hating
ice nucleation factor	Protein found on the surface of some bacteria which acts as a seed for ice crystal formation
immunization	Process of preparing the immune system for future infection by treating the patient with weak or killed versions of an infectious agent
immunoglobulin	Another name for an antibody protein
incompatibility group	A family of related plasmids. Two members of the same family cannot inhabit the same cell simultaneously
incompatibility	The inability of two related plasmids to live together in the same bacterial cell
induced mutations	Mutations caused by chemical damage or radiation
influenza virus	Member of the orthomyxovirus family with eight separate ssRNA molecules
initiator tRNA	The tRNA which brings the first amino acid to the ribosome when starting a new protein
inosine	An unusual nucleoside derived from guanosine
insertion sequence (IS)	Special mobile chunks of DNA. The simplest type of transposable element, consisting only of two terminal inverted repeats and a gene for the transposase enzyme
insulin	A protein hormone that controls the sugar level in the blood
intercalating agent	Chemical agent that inserts itself into DNA between two base pairs
intergenic regions	DNA sequence between genes
internal guide sequence	Part of a self-splicing intron which recognizes the front (5') splice site by base pairing
internal resolution site	DNA sequence inside a transposon that is recognized by resolvase
intron	Segment of a gene that does not code for protein
inversion	When a segment of DNA is removed, flipped and reinserted facing in the opposite direction
inverted repeat or "palindrome"	Length of DNA which has the same sequence if read in the opposite direction, but on the other strand
ion	Any molecule that carries an electrical charge
isoenzyme	Variant forms of the same enzyme found within a single species, often showing tissue specific distribution
junctional flexibility	Increasing antibody diversity by alterations introduced when joining together the separate segments that make up the variable region
junk DNA	Defective selfish DNA which is of no use to the host cell it inhabits and which is no longer capable of either moving or expressing its genes
κ (kappa) chain	One of the two alternative types of antibody light chain
Kappa particle	Symbiotic bacteria found inside killer *Paramecium* and needed for killing its sensitive relatives
killer cell	Type of T-cell whose job is to kill other cells of the body that have become "foreign" due to cancer or infection
kilobase ladder	A set of standard DNA fragments with lengths differing by one kilobase
kingdoms	Major subdivisions of eukaryotic organisms, especially the plant, fungus and animal kingdoms
knockout mice	Mice containing genes which have been inactivated by genetic engineering, usually by insertion of a DNA cassette to disrupt the coding sequence

kuru	Brain degeneration disease of cannibals caused by prion
lambda attachment site	Specific site on the chromosome of the bacterium *Escherichia coli* at which λ phage inserts its DNA
λ (lambda) chain	One of the two alternative types of antibody light chain
λ (lambda) phage	A virus which infects the bacterium *Escherichia coli* and is used as a cloning vector or in transduction
lactose	A type of sugar found in milk and made of glucose plus galactose
lacZ gene	The gene which codes for beta-galactosidase
lagging strand	The new strand of DNA which is synthesized in short pieces during replication and then joined later
Lassa Fever	New and very virulent virus disease which appeared in Africa in the 1960s
latency	Another name for lysogeny, normally applied to animal viruses
LDL receptor	Receptor for low density lipoprotein on cell surface
leading strand	The new strand of DNA that is synthesized continuously during replication
leptin	A protein hormone that controls the appetite and the burning of fat by the body
leukocyte	White blood cell
light chain	The shorter of the two pairs of chains comprising an antibody molecule
LINES	The long interspersed sequences which make up much of the highly repetitive DNA of mammals
linkage	Two genes are linked when they are on the same DNA molecule, *e.g.*, on the same chromosome
living cell	An independent unit of life, surrounded by a membrane, that contains genes made of DNA, and which possesses its own machinery to generate energy and synthesize proteins, DNA and RNA
locus (plural, loci)	A place or location on a chromosome, it may be a genuine gene or just any site with variations which can be measured, like RFLPs or VNTRs
long terminal repeat (LTR)	Repeats at end of retrovirus DNA which are necessary for integrating it into the host cell chromosome
luc genes	Genes which encode luciferase from animals
luciferase	The enzyme that generates light
luciferin	Molecule which is decomposed by luciferase to generate light
lux genes	Genes which encode luciferase from bacteria
lysogen	Cell containing a virus in harmless, lysogenic mode
lysogeny	Mode of virus growth in which duplication of the virus genome keeps in step with division of the host cell which is therefore not destroyed
lysozyme	An enzyme which breaks down the tough, structural layer of bacterial cell walls
lytic growth	Mode of virus growth resulting in destruction of the infected host cell
M13	A bacterial virus which contains single stranded circular DNA
M phase	Fourth stage in the cell cycle, cell division or mitosis
macrophage	Immune cell, mostly found in solid tissues, that ingests and destroys invading microorganisms
mad cow disease	Brain degeneration disease of cattle caused by prion that originated in sheep with scrapie
major histocompatibility complex (MHC)	Family of genes that encode proteins found on the cell surface where they display fragments of foreign proteins for immune recognition
malaria	Tropical disease caused by *Plasmodium*, a single celled eukaryotic microorganism with a complicated life cycle
male-specific bacteriophage	Bacterial virus such as Qβ or M13, which only infects male bacteria; in the particular case of *E. coli*, those bacteria containing the F-plasmid which confers the ability to conjugate
malignant	When cancer cells from a tumor disperse throughout the body
maltose	A type of sugar consisting of two glucose units and found in malt where it is derived from starch breakdown

mast cell	Specialized immune cell responsible for the secretion of histamine
meiosis	Formation of haploid gametes (*i.e.*, eggs and sperm) from diploid parental cells
melting	When used of DNA, refers to its separation into two strands as a result of heating
melting temperature	The temperature at which the two strands of a DNA molecule are half-way unpaired
memory cell	A specialized B-cell which waits for possible future infections instead of manufacturing antibodies
Mendel, Gregor	Discovered the basic laws of genetics by genetically crossing pea plants
Mendelian ratios	Whole number ratios found as the result of a genetic cross
messenger RNA (mRNA)	The molecule that carries genetic information from the genes to the rest of the cell
metaphase	Step of mitosis during which chromosomes move to the cell equator where they align themselves in pairs
metastasis	Spreading of cancer cells from their original site to form new secondary cancers
methionine	One of the 20 amino acids found in proteins. It is always used to start the polypeptide chain
micron	A millionth part of a meter
microtubule	Tubular structure made of protein used for support and guidance when internal components are moved around in the large cells of higher organisms
minus strand	The strand of DNA or RNA complementary to the coding sequence
mismatch repair	DNA repair system which recognizes and corrects wrongly paired bases
missense mutation	When a sequence change in DNA results in the replacement of one amino acid by another in the encoded protein
mitochondrion	Organelle found in eukaryotic cells that produces energy
mitosis	Division of eukaryotic cell into two daughter cells with identical sets of chromosomes
mobilizability	Ability of a small plasmid to move into another bacterial cell if assisted by a self-transferable plasmid
modification enzyme	An enzyme that alters a base in the recognition site of a restriction enzyme. This protects the DNA from being cut
modified	When applied to DNA refers to chemically tagging by adding methyl groups to signal that the DNA is host cell DNA
monoclonal antibody	A pure antibody with a unique sequence that recognizes only a single antigen and which is made by a cell line derived from a single B-cell
montmorillonite clay	Type of clay which promotes polymer formation by vigorously sucking out water
multiallelic or hypervariable	A gene or site on the DNA with many variations or multiple alleles
multiple antibiotic resistance	Simultaneous resistance to several antibiotics. Often due to a single antibiotic resistance plasmid
multiple cloning site	Stretch of DNA with recognition sites for several restriction enzymes
multiregional model	Theory that multiple lines of humans, spread over the various continents, evolved simultaneously into modern man
mutagen	Agent that can cause mutations
mutant	Organism carrying a mutated gene
mutation	An alteration or defect in the genetic information
mutator gene	A gene that will cause an increased rate of mutation if it is defective
Myc protein	A transcription factor involved in regulation of cell division
Mycobacterium	The bacterial group whose members cause tuberculosis and leprosy
myeloma	Cancer originating from B-cells of the immune system
myoglobin	Oxygen carrier protein found in muscle tissue
N-terminus	The end of a polypeptide chain that is made first and has a free amino group

NAH plasmid	Bacterial plasmid that carries genes for the breakdown of the hydrocarbon naphthalene
natural classification	Classification based on genuine biological ancestry
natural selection	The evolutionary process in which the unfit fail to reproduce
necrosis	Unplanned death of cells as the result of injury
negative-strand RNA virus	RNA virus that contains a single minus, or non-coding, strand of RNA
neutrophil	Immune cell, mostly found in blood, which ingests and destroys invading microorganisms
nick	A break in the backbone of a DNA or RNA molecule
nitrite	A chemical that causes mutations by converting cytosine to uracil
Noah's Ark model	Theory that only a single ancestral line of humans, localized in a relatively restricted area, evolved into modern man
non-coding regions	DNA sequences that do not code for proteins or functional RNA molecules
non-template strand	The "other" strand of the DNA that is not read during transcription
nonsense mutation	When a sequence change in DNA results in the replacement of the codon for an amino acid with a stop codon, so producing a shortened protein
Northern blotting	Hybridization technique in which a DNA probe binds to an RNA target molecule
npt gene	Gene which confers resistance to the antibiotic neomycin
nuclear membrane	The membrane in eukaryotic cells which separates off the nucleus from the rest of the cell
nuclease	An enzyme that cuts nucleic acids into shorter pieces
nucleic acid	Polymeric molecule that carries genetic information as a sequence of bases
template strand	Strand of DNA read during transcription
nucleolus	The structural component of the nucleus where ribosomal RNA is made
nucleocapsid	Innermost protein shell of a virus plus the DNA or RNA inside it
nucleoside triphosphate (NTP)	Precursor used in synthesis of a nucleic acid, consisting of sugar + base + three phosphates
nucleosome	Subunit of a eukaryotic chromosome consisting of DNA coiled around histone proteins
nucleotide	Monomer or subunit of a nucleic acid, consisting of sugar + base + phosphate
nucleus	The nucleus of a cell is an internal compartment surrounded by the nuclear membrane and containing the chromosomes. Only the cells of higher organisms have nuclei.
null allele	A mutant version of a gene which completely lacks any activity
null mutation	A mutation that fully inactivates a gene
nullizygous	When both copies of a gene are fully inactivated
oil drop model	Model of protein structure in which the hydrophobic groups cluster together on the inside away from the water
Okazaki fragments	The short pieces of DNA that make up the lagging strand
oligo-dT	A short piece of DNA whose bases are all thymine
oligonucleotide array detector	Square array of short DNA probes fixed on a glass chip and used for carrying out multiple hybridizations at one time
oligonucleotide	Short stretch of DNA (or RNA) consisting of just a few nucleotides
oncogene	Mutant gene that promotes cancer
ONPG	*Ortho*-Nitrophenyl-galactoside, an artificial substrate that releases a yellow color when split by the enzyme β-galactosidase
open reading frame (ORF)	Sequence of mRNA or corresponding region of DNA, that can be translated to give a protein
operator	Site on DNA to which a repressor protein binds
operon	A cluster of genes transcribed together to give a single molecule of mRNA
opines	Special nutrient molecules that can only be used by bacteria possessing a Ti-plasmid
opportunistic infection	A disease not infecting healthy people, but which attacks patients with immune system defects

order	Group intermediate in rank between a class and a family
organic compounds	Chemical compounds characteristic of living creatures and containing carbon and hydrogen
organophosphates	Class of organic chemicals with phosphate groups and which are often neurotoxic
origin of replication	Site on a DNA molecule where replication begins
origin of transfer (*oriT*)	Site on a plasmid where the DNA is nicked, just before transfer begins. The origin of transfer enters the recipient cell first
orthomyxoviruses	Family of negative-stranded RNA viruses with an outer envelope surrounding the nucleocapsid which contains several pieces of ssRNA
p arm	The shorter of the two arms of a chromosome
p (peptide) site	Binding site on the ribosome for the tRNA that is holding the growing polypeptide chain
p21 protein	A protein that blocks cell division by binding to and inhibiting the cyclins
p53 gene	A notorious anti-oncogene often mutated in cancer cells
papillomavirus	Family of DNA containing viruses which sometimes cause tumors
Parkinson's disease	Degenerative disease of certain brain cells that make the neurotransmitter dopamine and which leads to muscle tremors and weakness
partial dominance	When a functional allele only partly masks a defective allele
particle bombardment	Use of micro-projectiles to insert genes into target animals or plants
particle gun (gene gun)	Device used to fire particles carrying DNA into cells
pathway engineering	Construction by genetic engineering of a complete biochemical pathway
paternity	Fatherhood
penicillin	An antibiotic that kills bacteria by destroying their cell walls and which is produced by certain kinds of mold
peptide bond	Type of chemical linkage holding amino acids together
permissive temperature	The temperature at which a temperature sensitive mutation has no effect or is relatively harmless
phenol	A corrosive chemical liquid that very enthusiastically dissolves proteins
phenotype	Characteristics due to the expression of our genes; usually refers to visible properties but may refer to characteristics revealed by laboratory tests
phoA gene	The gene that encodes alkaline phosphatase
phosphate group	Group of four oxygen atoms surrounding a central phosphorus atom found in the backbone of DNA and RNA
phosphoglucomutase	A blood enzyme that can easily be detected and is used in blood typing
photosynthesis	Process of trapping light energy to form food molecules
phylum (plural, phyla)	Major groups into which animals are divided, roughly equivalent in rank to the divisions of plants or bacteria
picogram	10^{-12} gram or a million-millionth of a gram
picornaviruses	Family of positive-stranded RNA viruses with a single protein shell surrounding the ssRNA
plasmid	Circular molecule of double stranded helical DNA which replicates independently of the host cell's chromosomes. Rare linear plasmids have been discovered
plus strand	The strand of DNA or RNA that carries the coding sequence for making protein
poly-A tail	String of adenine residues on the 3' end of messenger RNA from eukaryotes
polyacrylamide	Artificial polymer used to make gels for separating proteins by electrophoresis
polycistronic mRNA	mRNA carrying multiple cistrons and which may be translated to give several different protein molecules; only found in prokaryotic (bacterial) cells
polyhydroxyalkanoate (PHA)	Polymeric plastics made by certain types of bacteria by linking hydroxyacid subunits together
polyhydroxybutyrate (PHB)	Bioplastic polymer of hydroxybutyrate subunits

polylinker or multiple cloning site	Stretch of DNA with recognition sites for several restriction enzymes
polymer	Long macromolecule made of similar or identical subunits linked together
polymerase chain reaction (PCR)	Artificial amplification of a DNA sequence by repeated cycles of replication and strand separation
polypeptide chain	Polymeric chain of amino acids
polyprotein	Large protein made by stringing several genes together and which is cut up into individual proteins after synthesis
polysaccharide	Polymeric molecule made of sugar subunits
polysome	Group of ribosomes bound to and translating the same mRNA
positional cloning	Any cloning procedure based on knowing a gene's location rather than its function
positive-strand RNA virus	RNA virus that contains a single plus, or coding, strand of RNA
poxviruses	Family of viruses with dsDNA carrying up to 200 genes and an outer protein layer surrounding the nucleocapsid
preproinsulin	Insulin as first synthesized, with both a signal sequence and the connecting peptide
primary structure	The linear order in which the subunits of a polymer are arranged
primary transcript	The RNA molecule produced by transcription before it has been processed in any way
primase	Enzyme that starts a new chain of DNA by making an RNA primer
primer	Short segment of nucleic acid that binds to the template strand and allows synthesis of a new chain of DNA to get started. RNA primers are used by cells and DNA primers are used in PCR
primer walking	Sequencing a long stretch of DNA by using a series of primers
prion	Distorted disease-causing form of a normal brain protein which can transmit an infection
probe molecule	Molecule that is tagged in some way (usually radioactive or fluorescent) and is used to bind to and detect another molecule
procyclin	Protein that covers trypanosomes with a protective layer while they are inside an insect
proinsulin	Insulin without the signal sequence but still with the connecting peptide
prokaryotes	Lower organisms like bacteria with a primitive type of cell. Prokaryotes contain a single chromosome and have no nuclear membrane
prokaryotic cell	Primitive type of cell with a single chromosome and no nuclear membrane as in bacteria
promoters	Region of DNA in front of a gene that binds RNA polymerase and so promotes gene expression
prophase	Stage in mitosis which condensed chromosomes become visible, the centrioles divide and the nuclear membrane dissolves
protein	Polymers made from amino acids; they do most of the work in the cell
protein kinase	An enzyme that switches other enzymes on or off by attaching a phosphate group to them
proteinoids	Random proteins made by artificial chemical assembly of amino acids as opposed to genuine proteins made on ribosomes by cells
proto-oncogene	Original, healthy, form of gene which may give rise to an oncogene
provirus	Form of a virus in which the viral DNA is integrated into the host chromosome
pseudogene	Defective copy of a genuine gene
purine	Type of base with a double ring found in DNA and RNA
pyrimidine	Type of base with a single ring found in DNA and RNA
q arm	The longer of the two arms of a chromosome
quasi-species	Group of related RNA-based genomes that differ slightly in sequence, but which arose from the same parental RNA molecule
quaternary structure	Aggregation of more than one polymer chain in final structure
quorum sensing	Form of regulation where a gene is expressed in response to population density

R-body	Refractile body - a toxic protein found as a coiled body inside the kappa particles of killer *Paramecium*
R-group	Chemical group forming side chain of amino acid
R-loop analysis	When the DNA copy of a gene is base paired to the corresponding mRNA, the extra regions in the DNA, which have no partners in the mRNA, appear as loops
R-plasmids (or R-factors)	Plasmids that confer antibiotic resistance on their bacterial hosts
radical replacement	Replacement of an amino acid in a protein with another that is very different in its properties
radioactive	Emitting radiation due to unstable atoms which break down releasing α, β, or γ-rays
randomly amplified polymorphic DNA (RAPD)	Method for testing genetic relatedness using PCR to amplify arbitrarily chosen sequences
ras protein	A protein involved in cell proliferation which, when mutated, can cause cancer
reading frame	One of three possible ways to read off the bases of a gene in groups of three so as to give codons
recessive allele	The allele whose properties are not observed because they are masked by the dominant allele
recessive mutation	Defective copy of a gene whose properties are not observed because they are masked by a functional copy
recognition site	Specific base sequence where a restriction enzyme binds
recombinant	Genetically engineered; may refer to a whole organism or a single product
recombinant bovine somatotropin (rBST)	Genetically engineered version of cow growth hormone
recombinant human somatotropin (rHST)	Genetically engineered version of human growth hormone
recombinant tissue plasminogen activator (rTPA)	Genetically engineered version of tissue plasminogen activator, an enzyme that helps dissolve blood clots
recombination	Mixing of genetic information from two chromosomes as a result of crossing over
regulatory nucleotide	Signal molecule made using same chemical components as nucleic acids, *i.e.*, bases, phosphate groups and ribose or deoxyribose
regulatory protein	A protein that regulates the expression of a gene or the activity of another protein
regulatory region	DNA sequence in front of a gene, used for regulation rather than to encode a protein
release factor	Protein that supervises the release of a finished polypeptide chain from the ribosome
reoviruses	Family of viruses with two protein shells surrounding the double stranded RNA
repetitive sequences	DNA sequences that exist in many copies
replication	Duplication of DNA prior to cell division
replication fork	Region where the enzymes replicating a DNA molecule are bound to untwisted, single stranded DNA
replicative form (RF)	Double stranded circular DNA found as an intermediate form during the replication of viruses that contain single stranded DNA
replicon	Molecule of DNA or RNA that contains an origin of replication
reporter gene	A gene that is easy to detect and which is inserted for diagnostic purposes
repressor proteins	Proteins that switch off genes by binding to DNA and blocking the action of RNA polymerase
resolvase	Enzyme needed by complex transposons to cut into two separate DNA molecules the fused intermediate formed during the transposition process
respiration	Process of releasing energy from food molecules by reacting them with oxygen
restriction	Destruction of incoming foreign DNA by a bacterial cell
restriction enzyme	An enzyme that binds to DNA at a specific base sequence and then cuts the DNA
restriction fragment length polymorphism (RFLP)	Differences in lengths of fragments made by cutting the DNA of different individuals with restriction enzymes

restriction map	Diagram of DNA showing the cut sites for a series of restriction enzymes
restrictive temperature	The temperature at which a temperature sensitive mutation is lethal or detrimental
retinoblastoma (Rb) gene	A particular anti-oncogene responsible for retinal cancer if both copies are inactivated
retron	Genetic element found in bacteria whose mechanism of movement is still unknown but which has reverse transcriptase and uses it to make a bizarre RNA/DNA hybrid molecule
retrotransposon	Transposon resembling a retrovirus in having reverse transcriptase and which moves by making an intermediate RNA copy
retrovirus	Type of virus which has its genes as RNA in the virus particle but converts this to a DNA copy inside the host cell by using reverse transcriptase
reverse transcriptase	Enzyme that starts with RNA and makes a DNA copy of the genetic information
reversion	A second mutation that restores the original characteristics to a mutant organism
Rhodobacter	Type of aquatic bacterium that lives by photosynthesis
ribonuclease	An enzyme that cuts RNA into shorter pieces
ribonuclease P	An enzyme that processes the precursors to some transfer RNA molecules
ribosomal RNA	RNA molecules that make up part of the structure of a ribosome
ribosome binding site or (Shine-Dalgarno sequence)	Sequence on mRNA at the front of the message and which is recognized by the ribosome. It is only found in prokaryotic cells
ribosome	The cell's machinery for making proteins
ribozyme	RNA molecule that acts as an enzyme
RNA or ribonucleic acid	Nucleic acid that differs from DNA in having ribose in place of deoxyribose
RNA polymerase	Enzyme that synthesizes RNA using a DNA template
RNA world	Hypothetical stage of evolution when RNA played the role of both gene and enzyme
rolling circle replication	Mechanism of replicating double stranded circular DNA; starts by nicking and unrolling one strand and using the other, still circular strand, as a template for DNA synthesis
rotavirus	Member of reovirus family that causes infant diarrhea
Rous sarcoma virus (RSV)	A retrovirus which causes cancer in chickens
RT-PCR	Combination of reverse transcriptase with PCR which allows DNA copies to be manufactured in bulk from messenger RNA
S phase	Second stage in the cell cycle in which chromosomes are duplicated
scintillation counter	Machine that detects and counts pulses of light
scintillation counting	Detection and counting of individual microscopic pulses of light
scrapie	Brain degeneration disease of sheep caused by prion
second-site revertant	A revertant in which a second change in DNA base sequence cancels out the effects of the first
secondary structure	Initial folding up of a polymer due to hydrogen bonding
self-splicing intron	An intron that cuts itself out and throws itself away without any help from the spliceosome or any other proteins
selfish DNA	Any region of DNA which manages to replicate but which is of no use to the host cell it inhabits
semi-conservative replication	Replication of DNA in which each daughter molecule gets one of the two original strands and one new complementary strand
sequence	To find out the sequence of bases in a molecule of DNA
severe combined immunodeficiency (SCID)	Immune defect due to lack of T- and B-cells. About 20 percent of inherited SCID cases are due to adenosine deaminase deficiency
sex pilus	Long, thin, helical rod of protein used by the male bacterial cell to catch hold of female cell
sex-linked	A gene is sex-linked when it is carried on the sex chromosomes
Shigella	The bacterium that causes the bacterial form of dysentery
shuttle vector	A vector able to survive in more than one type of host cell

Shine-Dalgarno sequence	Sequence at the front of bacterial mRNA which is recognized by the ribosome
sigma subunit	Subunit of bacterial RNA polymerase that recognizes and binds to the promoter sequence
signal molecule	The molecule that activates an activator protein or deactivates a repressor protein
signal sequence	Sequence of about 20 amino acids found at front of proteins that are marked for export from the cell
silent mutation	A mutation that has no observable effect on cell growth or survival
Simian virus 40 (SV40)	A dsDNA-containing monkey tumor virus of the papovavirus family
SINES	The short, interspersed sequences that make up much of the highly-repetitive DNA of mammals
single strand binding protein (SSB)	A protein that keeps separated strands of DNA apart
sleeping sickness	One of several tropical diseases caused by trypanosomes; occurs in Africa
small nuclear RNA (snRNA)	Small molecules of RNA found only in the nucleus that oversees the splicing of mRNA
smallpox	A frequently fatal viral disease which only infects humans
snRNP	"Snurps" - small nuclear ribonucleo-proteins contain proteins plus snRNA and are responsible for RNA splicing
sodium dodecyl sulfate (SDS)	A detergent used to unfold proteins
somatic cells	Cells making up the body which are are not part of the germ cell line. They are usually diploid, *i.e.,* they have two sets of genes.
somatic hypermutation	Deliberately increased rate of mutation in non-reproductive cells, especially in the antibody encoding genes of antibody producing cells
somatic recombination	Gene rearrangement by recombination which occurs in somatic (non-reproductive) cells, in particular B-cells or T-cells of the immune system
Southern blotting	A method to detect single stranded DNA that has been transferred to nylon paper by using a probe that binds DNA
South-Western blotting	Detection technique in which a DNA probe binds to a protein target molecule
specialized transduction	Transduction in which only a few specially selected genes are transported
species	A group of closely related organisms with a relatively recent common ancestor
spiny and banded murex	Two related sea snails that make dyes of the indigo family
spliceosome	Assembly of proteins and small nuclear RNAs which cuts out introns during the processing of eukaryotic mRNA
splicing	Cutting out the unneeded parts of an RNA molecule and joining the useful segments together
spontaneous mutation	Mutations occuring without chemical damage or radiation due to errors during DNA replication
spruce budworm	Caterpillar that munches on spruce and related trees
Src oncogene	One particular oncogene which encodes an enzyme acting as a protein kinase
ssDNA	Single stranded DNA
ssRNA	Single stranded RNA
start codon	The special AUG codon that signals the start of a protein
stem and loop ("hairpin")	Structure generated by folding of an inverted repeat sequence
sticky ends	Ends of a DNA molecule with short single stranded overhangs
stop codon	Codon that signals the end of a protein
structural protein	A protein that forms part of a cellular structure
substrate	The molecule altered by the action of an enzyme
subtractive hybridization	Removal of unwanted genes by hybridization, so leaving behind the gene of interest
sugar	Sweet tasting molecules made from carbon, hydrogen and oxygen in the ratio 1:2:1
supercoiling	Higher level twisting of DNA that is already in a double helix

symbiotic theory	The idea that the organelles of eukaryotic cells are derived from bacteria that took up residence as symbionts inside the ancestral eukaryotic cell
T-cell	Type of immune system cell responsible for cell-mediated immunity and which makes T-cell receptors instead of antibodies
T-cell receptor	Protein found on the surface of T-cells that recognizes and binds to foreign protein fragments when they are displayed by the MHC proteins
T-DNA	The DNA segment from a Ti-plasmid that actually inserts into the chromosomes of a plant cell
T4 ligase	Form of DNA ligase coded by gene belonging to bacteriophage T4
tail specific protease	Enzyme which destroys mis-made proteins by eating them, tail first
Taq polymerase	Heat resistant DNA polymerase from *Thermus aquaticus*
target sequence	Short sequence of DNA recognized by the transposase and into which it inserts the transposon when it moves
taxonomy	The science of naming and classification
telomerase	Enzyme that adds DNA to the end, or telomere, of a chromosome
telomere	Specific sequence of DNA found at the end of linear eukaryotic chromosomes
telophase	A step in mitosis in which a new nuclear membrane is made to surround each set of newly divided chromosomes
temperature sensitive mutation	A mutation whose effects are harmless at one temperature but noticeable at another
template strand	Strand of DNA used as a guide for synthesizing a new strand by complementary base pairing
teratogen	Agent that can cause spectacular mutations or "monstrosities"
terminal deoxynucleotide transferase	An enzyme that adds a few extra bases when junctions are made between the gene segments for the V-, D-, and J-regions of antibody heavy chains
terminator sequence	DNA sequence at end of a gene which tells RNA polymerase to stop transcribing
tertiary structure	Final 3-D folding of a polymer chain
thalassemia	Hereditary disease causing hemoglobin deficiency
thermocycler	Machine used to rapidly shift samples between several temperatures in a pre-set order and so is used for PCR
Thermus aquaticus	A bacterium that only grows at very high temperatures and lives in hot springs
third base redundancy	Since many amino acids have several codons, the third codon base can often be changed without changing the amino acid for which it codes
third codon position	The base in the third position of a codon can often be changed without changing the amino acid encoded
thymine (T)	One of the pyrimidine bases found in DNA only and which pairs with adenine
thymus	Organ in the chest where T-cells mature after leaving the bone marrow
Ti-plasmid (tumor-inducing plasmid)	Plasmid which carries the genes necessary for causing tumors in plants
tissue graft	Tissue cut from one animal and inserted into another. It may be a whole organ or a portion of a tissue such as skin or bone marrow
tmRNA	a special RNA used to terminate protein synthesis when it finds a ribosome stalled by a bad mRNA
tobacco mosaic virus	A single stranded RNA virus that infects many plants, including tobacco; the most noticeable symptom is a mosaic of diseased blotches on the leaves
toxin	A toxic protein that acts against higher organisms
transcription factor	Protein that regulates gene expression by binding to DNA in the control region of the gene
transcription	Process by which information from DNA is converted into its RNA equivalent
transducing particle	Virus particle containing host cell DNA instead of the viruses' own genes

transduction	Transport of genes from one cell to another inside a virus particle
transfer (*tra*) system	Cluster of genes that code for the ability of a plasmid to transfer itself from one cell to another
transfer RNA	RNA molecules that carry amino acids to a ribosome
transferability	Ability of a plasmid to move itself from one cell to another
transformation	Conversion of a normal cell to a cancer cell; this word is also used to describe uptake of pure DNA by bacteria or even animal cells
transgene	A foreign gene that has been moved into a new host organism by genetic engineering
transgenic species	Individual into which genes from another species have been incorporated
translation	Making a protein using the information provided by messenger RNA
translocation	When a segment of DNA is removed and reinserted in a different place
transmissible antibiotic resistance	Resistance to an antibiotic carried on a transferable plasmid
transport protein	A protein that carries other molecules across membranes or around the body
transposable element or transposon	Segment of DNA that can move as a unit from one location to another, but which always remains part of another DNA molecule
transposase	The enzyme that carries out the transposition process
transposition	Process by which a transposon moves itself from one host molecule of DNA to another
transposon	Mobile segment of DNA that moves from place to place in other DNA molecules
trans-splicing	Making a messenger RNA by joining together segments from two original RNA molecules
true revertant	A revertant in which the original DNA base sequence is exactly restored
trypanosome	A single-celled eukaryotic microorganism that swims by means of a flagellum and alternates between insects and humans during its life cycle
tumor necrosis factor (TNF)	Short protein that kills cancer cells
tumor-infiltrating lymphocyte (TIL)	White blood cell that secretes TNF
TY-1	Transposon Yeast No. 1; a retrotransposon found in multiple copies in yeast cells
Type I restriction enzyme	A restriction enzyme that cuts the DNA far from its recognition sequence
Type II restriction enzyme	A restriction enzyme that cuts the DNA within its recognition sequence
tyrosine hydroxylase	Enzyme involved in the synthesis of the dopamine family of neurotransmitters, substances that carry signals between brain cells
U1 snRNA	The small nuclear RNA found in spliceosomes which is responsible for recognizing the front (5') splice site
ultraviolet light	Invisible radiation of higher energy than visible light
uncharged tRNA	tRNA without an amino acid attached
upstream element	DNA sequence found upstream of the TATA box in eukaryotic promoters and which is recognized by specific proteins
upstream region	Region of DNA in front of a gene; its bases are numbered negatively counting backwards from the start of transcription
urkaryote	The ancestral eukaryotic cell before it gained its mitochondria and chloroplasts
v-onc	Oncogene carried by a virus
vaccination	Artificial induction of the immune response by injecting foreign proteins or other antigens
vaccinia	Member of poxvirus family that causes cowpox
variable number tandem repeats (VNTRs)	Sequences that are repeated in the DNA at one site, and the number of repeats differs from one individual to another

variable region	Region of an antibody which is varied by gene shuffling in order to provide many alternative antigen binding sites
variant surface glycoprotein (VSG)	Protein which covers trypanosomes while inside humans and which is constantly altered to mislead the immune system
vector	Molecule of DNA which can replicate and is used to carry cloned genes or DNA fragments
vegetative replication	Replication which occurs in the absence of plasmid transfer or any sexual process
ventral surface	The belly or under surface
viral genome	The nucleic acid, either DNA or RNA, which carries the genetic information of a virus
virgin B-cell	B-cell that has an IgM antibody on its surface but which has not yet been stimulated to divide by encountering a matching antigen. Sometimes called a "naive" B-cell
virion	A virus particle
virulence gene	Gene involved in aiding infection or causing symptoms of disease
virulence plasmid	Plasmid that carries genes involved in aiding infection or causing symptoms of disease
virus	Subcellular parasite with genes of DNA or RNA and which replicates inside the host cell upon which it relies for energy and protein synthesis. In addition it has an extracellular form, in which the virus genes are contained inside a protective coat
Western blotting	Detection technique in which a probe, usually an antibody, binds to a protein target molecule
wild-type	The original or "natural" version of a gene or organism
wobble rules	Rules allowing less rigid base pairing but only for codon/anticodon pairing
X-gal	Substance split by β-galactosidase so yielding a blue dye
X-phos	Substance split by alkaline phosphatase so yielding a blue dye
X-ray crystallography or X-ray diffraction	Determination of 3-D crystal structure by using X-rays
yeast artificial chromosome or (YAC)	Linear DNA that mimics a yeast chromosome by having a yeast specific origin, centromere and telomere sequences
zinc finger	Finger-like bulge on a protein which binds and reads a short DNA sequence
zoo blotting	Comparative Southern blotting using DNA target molecules from several different animals to test whether the probe DNA is from a coding region
Zymomonas	A bacterium which ferments glucose to alcohol by a similar pathway to yeast

Index